MULTIDIMENSIONAL SIGNAL, IMAGE, AND VIDEO PROCESSING AND CODING

MULTIDIMENSIONAL SIGNAL, IMAGE, AND VIDEO PROCESSING AND CODING

JOHN W. WOODS

Rensselaer Polytechnic Institute
Troy, New York

ELSEVIER

AMSTERDAM • BOSTON • HEIDELBERG • LONDON
NEW YORK • OXFORD • PARIS • SAN DIEGO
SAN FRANCISCO • SINGAPORE • SYDNEY • TOKYO

Academic Press is an imprint of Elsevier

ACADEMIC PRESS

Cover image: The cover shows a rather short image sequence of five frames of the author's dog Heidi running in the back yard, as captured by a DV camcorder at 30 fps. The actual displayed "frame rate" here though is rather low and not recommended except to use as cover art.

Academic Press is an imprint of Elsevier
30 Corporate Drive, Suite 400, Burlington, MA 01803, USA
525 B Street, Suite 1900, San Diego, California 92101-4495, USA
84 Theobald's Road, London WC1X 8RR, UK

This book is printed on acid-free paper. ⊛

Library of Congress Cataloging-in-Publication Data
Application submitted.

British Library Cataloguing-in-Publication Data
A catalogue record for this book is available from the British Library.

ISBN 13: 978-0-12-088516-9
ISBN 10: 0-12-088516-6
ISBN 13: 978-0-12-372566-0 (CD-ROM)
ISBN 10: 0-12-372566-6 (CD-ROM)

For information on all Academic Press publications
visit our Web site at www.books.elsevier.com

Printed in the United States of America
06 07 08 09 10   9 8 7 6 5 4 3 2 1

CONTENTS

PREFACE

This is a textbook for a first- or second-year graduate course for electrical and computer engineering (ECE) students in the area of digital image and video processing and coding. The course might be called *Digital Image and Video Processing* (DIVP) or some such, and have its heritage in the signal processing and communications areas of ECE. The relevant image (and video) processing problems can be categorized as image-in/image-out, rather than image-in/analysis-out types of problems. The latter are usually studied in similarly titled courses such as picture processing, image analysis, or even computer vision, often given in a computer-science context. We do, however, borrow some concepts from image analysis and computer vision such as motion estimation, which plays a key role in advanced video signal processing, and to a lesser extent (at present), object classification.

The required background for the text is a graduate-level digital signal processing (DSP) course, a junior/senior-level course in probability, and a graduate course in discrete-time and continuous-time random processes. At Rensselaer, the course DIVP is offered in the Spring term as a graduate student's second course in DSP, coming just after a first graduate course in DSP and one on introduction to stochastic processes in the Fall term. A basic course in digital communications would also provide helpful background for the image- and video-coding chapters, however the presentation here is self-contained. Good students with deficiencies in one or more of these areas can however appreciate other aspects of the material, and have successfully completed our course, which usually involves a term project rather than final exam. It is hoped that the book is also suitable for self-study by graduate engineers in the areas of image and video processing and coding.

The DIVP course at Rensselaer has been offered for the last 12 years, having started as a course in multidimensional DSP and then migrated over to bring in an emphasis first on image and then on video processing and coding. The book, as well as the course, starts out with *two-dimensional signal processing* theory, comprising the first five chapters, including 2-D systems, partial difference equations, Fourier and Z-transforms, filter stability, discrete transforms such as DFT and DCT and their fast algorithms, ending up with 2-D or spatial filter design. We also introduce the subband/wavelet transform (SWT) here, along with coverage of the DFT and DCT. This material is contained in the first five chapters and constitutes the signal-processing or first part of the book. However, there is also a later chapter on 3-D and spatiotemporal signal processing, strategically positioned just ahead of the video processing chapters.

The second part, comprising the remaining six chapters, covers image and video processing and coding. We start out with a chapter introducing basic image processing, and include individual chapters on estimation/restoration and source coding of both images and video. Lastly we included a chapter on network transmission of video including consideration of packet loss and joint source-network issues.

This paragraph and the next provide detailed chapter information. Starting out the first part, Chapter 1 introduces 2-D systems and signals along with the stability concept, Fourier transform and spatial convolution. Chapter 2 covers sampling and considers both rectangular and general regular sampling patterns, e.g., diamond and hexagonal sample patterns. Chapter 3 introduces 2-D difference equations and the Z transform including recursive filter stability theorems. Chapter 4 treats the discrete Fourier and cosine transforms along with their fast algorithms and 2-D sectioned-convolution. Also we introduce the ideal subband/wavelet transform (SWT) here, postponing their design problem to the next chapter. Chapter 5 covers 2-D filter design, mainly through the separable and circular window method, but also introducing the problem of 2-D recursive filter design, along with some coverage of general or fully recursive filters.

The second part of the book, the part on image and video processing and coding starts out with Chapter 6, which presents basic concepts in image sensing, display, and human visual perception. Here, we also introduce the basic image processing operators: box, Prewitt, and Sobel filters. Chapter 7 covers image estimation and restoration, including adaptive or inhomogeneous approaches, and concludes with a section on image- and blur-model parameter identification via the EM algorithm. We also include material on compound Gauss-Markov models and their MAP estimation via simulated annealing. Chapter 8 covers image compression built up from the basic concepts of transform, scalar and vector quantization, and variable-length coding. We cover basic DCT coders and also include material on fully embedded coders such as EZW, SPIHT, and EZBC and introduce the main concepts of the JPEG 2000 standard. Then Chapter 9 on three-dimensional (3-D) and spatiotemporal or multidimensional signal processing (MDSP) extends the 2-D concepts of Chapters 1 to 5 to the 3-D case of video. Also included here are rational system models and spatiotemporal Markov models culminating in a spatiotemporal reduced-update Kalman filter. Next, Chapter 10 studies interframe estimation/restoration and introduces motion estimation and the technique of motion compensation. This technique is then applied to motion-compensated Kalman filtering, frame-rate change, and deinterlacing. The chapter ends with the Bayesian approach to joint motion estimation and segmentation. Chapter 11 covers video compression with both hybrid and spatiotemporal transform approaches, and includes coverage of video coding standards such as MPEG 2 and H.264/AVC. Also presented are highly scalable coders based on the motion-compensated temporal filter (MCTF). Finally, Chapter 12 is devoted to video on networks, first introducing network fundamentals and then presenting some robust methods for video transmission over networks. We include methods of error concealment and robust scalable approaches using MCTF and embedded source coding. Of course, this last chapter is not meant to replace an introductory course in computer networks, but rather to complement it. However, we have also tried to introduce the appropriate network terminology and concepts, so that the chapter will be accessible to signal processing and communication graduate students without a networking background.

This book also has an enclosed CD-ROM that contains many short MATLAB programs that complement examples and exercises on MDSP. There is also a .pdf document on the disk that contains high-quality versions of all the images in the book. There are numerous short video clips showing applications in video processing and coding. Enclosed is a copy of the *vidview* video player for playing .yuv video files on a Windows PC. Other video files can generally be decoded and played by the commonly available media decoder/players. Also included is an illustration of effect of packet loss on H.264/AVC coded bitstreams. (Your media decoder/player would need an H.264/AVC decoder component to play these, however, some .yuv files are included here in case it doesn't.)

This textbook can be utilized in several ways depending on the graduate course level and desired learning objectives. One path is to first cover Chapters 1 to 5 on MDSP, and then go on to Chapters 6, 7, and 8 to cover image processing and coding, followed by some material on video processing and coding from later chapters, and this is how we have most often used it at Rensselaer. Alternatively, after Chapters 1 to 5, one could go on to image and video processing in Chapters 6, 9, and 10. Or, and again after covering Chapters 1 to 5, go on to image and video compression in Chapter 7, part of Chapter 9, and 11. The material from Chapter 12 could also be included, time permitting. To cover the image and video processing and coding in Chapters 6 to 11 in a single semester, some significant sampling of the first five chapters would probably be needed. One approach may be to skip (or very lightly cover) Chapter 3 on 2-D systems and Z transforms and Chapter 5 on 2-D filter design, but cover Chapters 1, 2, and part of Chapter 4. Still another possibility is to cover Chapters 1 and 2, and then move on to Chapters 6 to 12, introducing topics from Chapters 3 to 5 only as needed. An on-line solutions manual is available to instructors at textbooks.elsevier.com with completion of registration in the Electronics and Electrical Engineering subject area.

John W. Woods
Rensselaer Polytechnic Institute
Spring 2006

ACKNOWLEDGMENTS

I started out on my book-writing project while teaching from the textbook *Two-Dimensional Digital Image Processing* by Jae Lim, and readers familiar with that book will certainly see a similarity to mine, especially in the first four chapters. Thanks Jae. Thanks to colleagues Janusz Konrad and Aggelos Katsaggelos with Eren Soyak, for providing some key examples in Chapters 10 and 12, respectively. Thanks also to Peisong Chen, Seung-Jong Choi, Allesandro Dutra, Soo-Chul Han, Shih-Ta Hsiang, Fure-Ching Jeng, Jaemin Kim, Ju-Hong Lee, Anil Murching, T. Naveen, Yufeng Shan, Yongjun Wu, and Jie Zou, who through their research, have helped me to provide many examples all through the book. Special thanks to former student Ivan Bajic for writing Section 12.2 on robust subband/wavelet video packetizing for wired networks. Additional thanks go out to colleagues Maya Bystrom and Shiv Kalayanaraman, who provided valuable feedback on Chapter 12. Thanks also to Tobias Lutz and Christoph Schmid for finding errors in a prior draft version of the text. If you find any further errors, please bring them to my attention via email woods@ecse.rpi.edu. Finally, thanks to my wife Harriet who has cheerfully put up with my book writing for many years now.

TWO-DIMENSIONAL SIGNALS AND SYSTEMS

1

This chapter sets forth the main concepts of two-dimensional (2-D) signals and systems as extensions of the linear systems concepts of 1-D signals and systems. We concentrate on the discrete-space case of digital data, including the corresponding 2-D Fourier transform. We also introduce the continuous-space Fourier transform to deal with angular rotations and prove the celebrated projection-slice theorem of computer tomography. Also, later we will find that the central role of motion compensation in image sequences and video, where motion is often not an exact number of pixels, makes the interplay of discrete and continuous parameters quite important. While this first chapter is all discrete space, Chapter 2 focuses on sampling of 2-D continuous-space functions.

1.1 TWO-DIMENSIONAL SIGNALS

A scalar 2-D signal $s(n_1, n_2)$ is mathematically a complex *bi-sequence*, or a mapping of the 2-D integers into the complex plane. Our convention is that the signal s is defined for all finite values of its integer arguments n_1, n_2 using zero padding as necessary. Occasionally we will deal with finite-extent signals, but will clearly say so. We will adopt the simplified term *sequence* over the more correct bi-sequence. A simple example of a 2-D signal is the *impulse* $\delta(n_1, n_2)$, defined as follows:

$$\delta(n_1, n_2) \triangleq \begin{cases} 1 & \text{for } (n_1, n_2) = (0, 0) \\ 0 & \text{for } (n_1, n_2) \neq (0, 0), \end{cases}$$

a portion of which is plotted in Figure 1.1.

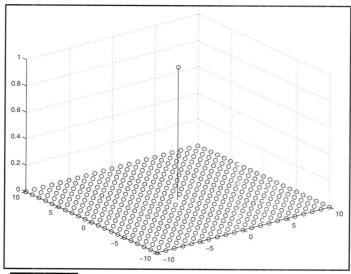

FIGURE 1.1   *MATLAB plot of 2-D spatial impulse bi-sequence $\delta(n_1, n_2)$*

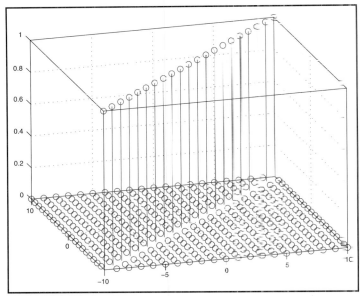

FIGURE 1.2   *Section of the unit impulse line $\delta(n_1 - n_2)$ at 45°*

A general 2-D signal can be written as an infinite sum over shifted impulses, which will be found useful later:

$$x(n_1, n_2) = \sum_{k_1, k_2} x(k_1, k_2)\delta(n_1 - k_1, n_2 - k_2), \qquad (1.1\text{-}1)$$

where the summation is taken over all integer pairs (k_1, k_2). By the definition of the 2-D impulse, for each point (n_1, n_2), there is exactly one term on the right side that is nonzero, the term $x(n_1, n_2) \cdot 1$, and hence (1.1-1) is correct. This equality is called the *shifting representation* of the signal x.

Another basic 2-D signal is the impulse line, e.g., a straight line at 45°,

$$\delta(n_1 - n_2) = \begin{cases} 1, & n_1 = n_2 \\ 0, & \text{else}, \end{cases}$$

as sketched in Figure 1.2. We can also consider line impulses at 0°, 90°, and −45° in discrete space, but other angles give "gaps in the line." Toward the end of this chapter, when we look at continuous-space signals, there the line impulse can be at any angle.

Apart from impulses, a basic 2-D signal class is step functions, perhaps the most common of which is the first-quadrant *unit step function* $u(n_1, n_2)$, defined

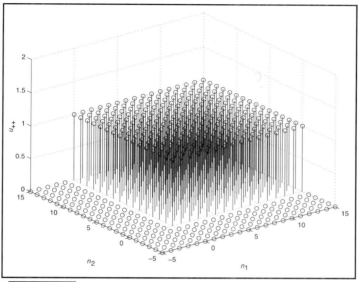

FIGURE 1.3   *Portion of unit step bi-sequence $u(n_1, n_2)$*

as follows:

$$u(n_1, n_2) \triangleq \begin{cases} 1, & n_1 \geqslant 0, \, n_2 \geqslant 0 \\ 0, & \text{elsewhere,} \end{cases}$$

a portion of which is plotted in Figure 1.3. Actually, in two dimensions, there are several step functions of interest. This one is called the first-quadrant unit step function, and is more generally denoted as $u_{++}(n_1, n_2)$.

We will find it convenient to use the word *support* to denote the set of all argument values for which the function is nonzero. In the case of the first-quadrant unit step $u_{++}(n_1, n_2)$, this becomes

$$\text{supp}(u_{++}) = \{n_1 \geqslant 0, \, n_2 \geqslant 0\}.$$

Three other unit step functions can be defined for the other three quadrants. They are denoted u_{-+}, u_{+-}, and u_{--}, with support on the second, fourth, and third quadrants, respectively. A plot of a portion of u_{-+} is shown in Figure 1.4.

A real example of a finite support 2-D sequence is the image *Eric* shown in three different ways in Figures 1.5–1.7. Figure 1.5 is a contour plot of *Eric*, a 100×76 pixel, 8-bit gray-level image. Figure 1.6 is a perspective or mesh plot. Figure 1.7 is an *image* or intensity plot, with largest value white and smallest value black.

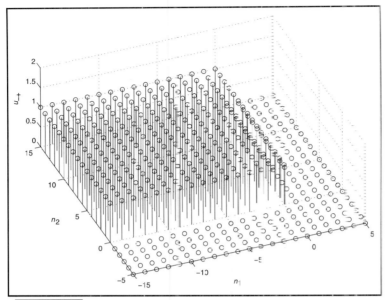

FIGURE 1.4   Portion of second-quadrant unit step bi-sequence $u_{--}(n_1, n_2)$

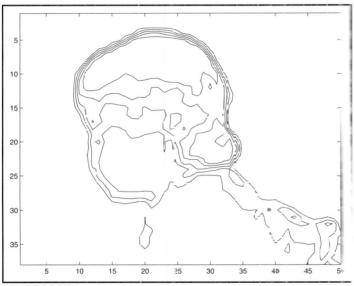

FIGURE 1.5   Contour plot of Eric

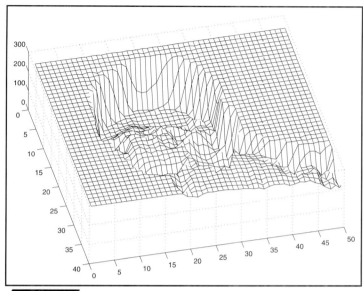

FIGURE 1.6 Mesh plot of Eric

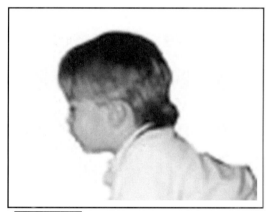

FIGURE 1.7 Image or intensity plot of Eric

1.1.1 SEPARABLE SIGNALS

A separable signal (sequence) satisfies the equation

$$x(n_1, n_2) = x_1(n_1)x_2(n_2) \quad \text{for all } n_1 \text{ and } n_2,$$

for some 1-D signals $x_1(n_1)$ and $x_2(n_2)$. If we think of the finite support case, where $x(n_1, n_2)$ can be represented by a matrix, then $x_1(n_1)$ and $x_2(n_2)$ can be represented as column and row vectors, respectively. So, we see that separability is the same as saying that the corresponding matrix x can be written as the outer product $x = x_1 x_2^T$, which is the same as saying that the matrix x has only one singular value in its *singular value decomposition* (SVD).[1] Clearly this is very special. Note that while an $N \times N$ matrix x has N^2 degrees of freedom (number of variables), the outer product $x_1 x_2^T$ has only $2N$ degrees of freedom. Nevertheless, separable signals play important roles in multidimensional signal processing (MDSP) as some representation bases (e.g., Fourier transform) and some filter impulse responses. The image *Eric* in Figure 1.7 regarded as a 100×76 matrix is clearly not separable and would have many terms in its SVD.

1.1.2 PERIODIC SIGNALS

A 2-D sequence $x(n_1, n_2)$ is periodic with period (N_1, N_2), also written $N_1 \times N_2$, if the following equalities hold for all integers n_1 and n_2:

$$x(n_1, n_2) = x(n_1 + N_1, n_2)$$
$$= x(n_1, n_2 + N_2),$$

where N_1 and N_2 are positive integers.

The period (N_1, N_2) defines a 2-D (either spatial or space–time) grid over which the signal repeats or is periodic. Since we are in discrete space, the period must be composed of *positive integers*. This type of periodicity occurs often for 2-D signals and is referred to as *rectangular periodicity*. We call the resulting period the *rectangular period*.

EXAMPLE 1.1-1 (*sine wave*)
An example is the signal $\sin(2\pi n_1/8 + 2\pi n_2/16)$, for which the rectangular period is easily seen to be $(8, 16)$. A separable signal with the same period is $\sin(2\pi n_1/8) \sin(2\pi n_2/16)$. As a counterexample to periodicity, note, however, removing the factor of 2π from the argument, that neither $\sin(n_1/8 + n_2/16)$ nor $\sin(n_1/8) \sin(n_2/16)$ is periodic at all, because we only admit integer values for the period, since we are operating in a discrete parameter space.

Given a periodic function, the period effectively defines a *basic cell* in the plane, which then repeats to form the function over all integers n_1 and n_2. As such, we often want the minimum size unit cell for efficiency of specification and storage. In the case of rectangular period, we thus want the smallest nonzero values that will suffice for N_1 and N_2.

---

1. The SVD is a representation of a matrix in terms of its eigenvalues and eigenvectors and is written for a real square matrix x as $x = \sum \lambda_i e_i e_i^T$, where the λ_i are the eigenvalues and the e_i are the eigenvectors of x.

EXAMPLE 1.1-2 (*horizontal wave*)
Consider the sine wave $x(n_1, n_2) = \sin(2\pi n_1/4)$. The horizontal period is $N_1 = 4$. In the vertical direction, the signal is constant though. So we can use any positive integer N_2, and we choose the smallest such $N_2 = 1$. Thus the rectangular period is $N_1 \times N_2 = 4 \times 1$, and the basic cell consists of the set of points $[(n_1, n_2) = \{(0, 0), (1, 0), (2, 0), (3, 0)\}]$ or any translate of same.

GENERAL PERIODICITY

There is a more general definition of periodicity that we will encounter from time to time in this course. It is a repetition of not necessarily rectangular blocks on a rectangular repeat grid. For this general case, we need to represent the periodicity with two integer vectors $\mathbf{N}_1$ and $\mathbf{N}_2$:

$$\mathbf{N}_1 = \begin{bmatrix} N_{11} \\ N_{21} \end{bmatrix} \quad \text{and} \quad \mathbf{N}_2 = \begin{bmatrix} N_{12} \\ N_{22} \end{bmatrix}.$$

Then we can say the 2-D signal $x(n_1, n_2)$ is *general periodic* with period $[\mathbf{N}_1, \mathbf{N}_2] \triangleq \mathbf{N}$ if the following equalities hold for all integers n_1 and n_2:

$$x(n_1, n_2) = x(n_1 + N_{11}, n_2 + N_{21})$$
$$= x(n_1 + N_{12}, n_2 + N_{22}).$$

To avoid degenerate cases, we restrict the not necessarily positive integers N_{ij} with the condition

$$\det(\mathbf{N}_1, \mathbf{N}_2) \neq 0.$$

The matrix $\mathbf{N}$ is called the *periodicity matrix* and in matrix notation we regard the corresponding periodic signal as one that satisfies

$$x(\mathbf{n}) = x(\mathbf{n} + \mathbf{N}\mathbf{r}),$$

for all integer vectors $\mathbf{r} = \binom{r_1}{r_2}$. In words, we can say that the signal repeats itself at all multiples of the shift vectors $\mathbf{N}_1$ and $\mathbf{N}_2$.

EXAMPLE 1.1-3 (*sine wave*)
An example is the signal $\sin(2\pi n_1/8 + 2\pi n_2/16)$, which is constant along the line $2n_1 + n_2 = 16$. We can compute shift vectors $\mathbf{N}_1 = \binom{4}{8}$ and $\mathbf{N}_2 = \binom{1}{-2}$.

We note that the special case of rectangular periodicity occurs when the periodicity matrix $\mathbf{N}$ is diagonal, for then

$$\mathbf{N} = \begin{bmatrix} N_1 & 0 \\ 0 & N_2 \end{bmatrix},$$

and the rectangular period is (N_1, N_2) as above. Also in this case, the two period vectors $\mathbf{N}_1 = \binom{N_1}{0}$ and $\mathbf{N}_2 = \binom{0}{N_2}$ lie along the horizontal and vertical axes, respectively.

EXAMPLE 1.1-4 (*cosine wave*)
Consider the signal $g(n_1, n_2) = \cos 2\pi (f_1 n_1 + f_2 n_2) = \cos 2\pi \mathbf{f}^T \mathbf{n}$. In continuous space, this signal is certainly periodic and the rectangular period would be $N_1 \times N_2$ with period vectors $\mathbf{N}_1 = \binom{f_1^{-1}}{0}$ and $\mathbf{N}_2 = \binom{0}{f_2^{-1}}$. However, this is not a correct answer in discrete space unless the resulting values f_1^{-1} and f_2^{-1} are integers. Generally, if f_1 and f_2 are rational numbers, we can get an integer period as follows: $N_i = p_i f_i^{-1}$, where $f_i = p_i / q_i, i = 1, 2$. If either of the f_i are not rational, there will be no exact rectangular period for this cosine wave. Regarding general periodicity, we are tempted to look for repeats in the direction of the vector $(f_1, f_2)^T$ since $\mathbf{f}^T \mathbf{n}$ is maximized if we increment the vector $\mathbf{n}$ in this direction. However, again we have the problem that this vector would typically not have integer components. We are left with the conclusion that common cos and sin waves are generally not periodic in discrete space, at least not exactly periodic. The analogous result is also true, although not widely appreciated, in the 1-D case.

1.1.3 2-D DISCRETE-SPACE SYSTEMS

As shown in Figure 1.8, a *2-D system* is defined as a general mathematical operator $\mathbf{T}$ that maps each input signal $x(n_1, n_2)$ into a unique output signal $y(n_1, n_2)$. In equations, we write

$$y(n_1, n_2) = \mathbf{T}\big[x(n_1, n_2)\big].$$

The signals are assumed to be defined over the entire 2-D discrete-space $(-\infty, +\infty) \times (-\infty, +\infty)$, unless otherwise indicated. There is only one restriction on the general system operator $\mathbf{T}$; it must provide a unique mapping, i.e., for each *input sequence x* there is one and only one *output sequence y*. Of course, two input sequences may agree only over some area of the plane, but differ elsewhere; then there can be different outputs corresponding to these two inputs, since these two inputs are not equal everywhere. In mathematics, an operator such as $\mathbf{T}$ is just the generalization of the concept of function where the input and output spaces are now sequences instead of numbers. The operator $\mathbf{T}$ may have an inverse or

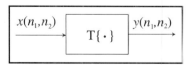

FIGURE 1.8   *A general 2-D discrete-space system*

not. We say that $\mathbf{T}$ is invertible if to each output sequence y there corresponds only one input sequence x, i.e., that the output determines the input. We denote the inverse operator, if it exists, by $\mathbf{T}^{-1}$.

EXAMPLE 1.1-5 (*systems*)
For example, consider the following 2-D systems:

1 $y(n_1, n_2) = \mathbf{T}[x(n_1, n_2)] = 2x(n_1, n_2).$
2 $y(n_1, n_2) = \mathbf{T}[x(n_1, n_2)] = x^2(n_1, n_2) + 3x(n_1, n_2) + 1.$
3 $y(n_1, n_2) = \mathbf{T}[x(n_1, n_2)] = 3x(n_1, n_2) + 5.$
4 $y(n_1, n_2) = \mathbf{T}[x(n_1, n_2)] = \frac{1}{2}[x(n_1, n_2) + x(n_1 - 1, n_2)].$

We note by inspection that the 2-D system operators given in items 1 and 3 are invertible, while that given in item 2 is not. The system given by item 4 has *memory* while those in items 1–3 do not. If we call (n_1, n_2) the *present*, then a *memoryless system* is one whose present output only depends on present inputs.

A special case of the general 2-D system is the linear system, defined as follows:

DEFINITION 1.1-1 (*linear system*)
A 2-D discrete system is *linear* if the following equation holds for all pairs of input–output pairs x_i–y_i and all complex scaling constants a_i:

$$L[a_1 x_1(n_1, n_2) + a_2 x_2(n_1, n_2)] = a_1 L[x_1(n_1, n_2)] + a_2 L[x_2(n_1, n_2)],$$

where we have denoted the linear operator by L.

As in 1-D signal processing, linear systems are very important to the theory of 2-D signal processing. The reasons are the same, i.e., that (1) we know the most about linear systems, (2) approximate linear systems arise a lot in practice, and (3) many adaptive nonlinear systems are composed from linear pieces, designed using linear system theory. In Example 1.1-5, systems 1 and 4 are linear by this definition. A simple necessary condition for linearity is that the output doubles when the input doubles, and this rules out the systems 2 and 3.

DEFINITION 1.1-2 (*shift-invariant system*)
A system $\mathbf{T}$ is *shift-invariant* if any shift of an arbitrary input x produces the identical shift in the corresponding output y, i.e., if $\mathbf{T}[x(n_1, n_2)] = y(n_1, n_2)$, then for all (integer) shifts (m_1, m_2) we have $\mathbf{T}[x(n_1 - m_1, n_2 - m_2)] = y(n_1 - m_1, n_2 - m_2)$.

Often we think in terms of a *shift vector*, denoted $\mathbf{m} = (m_1, m_2)^T$. In fact, we can just as well write the two preceding definitions in the more compact vector notation as follows:

Linearity: $\quad L[a_1 x_1(\mathbf{n}) + a_2 x_2(\mathbf{n})] = a_1 L[x_1(\mathbf{n})] + a_2 L[x_2(\mathbf{n})].$

Shift-invariance: $\quad T[x(\mathbf{n} - \mathbf{m})] = y(\mathbf{n} - \mathbf{m}) \quad$ for shift vector $\mathbf{m}$.

The vector notation is very useful for 3-D systems as occurring in video signal processing (cf. Chapter 9).

1.1.4 TWO-DIMENSIONAL CONVOLUTION

Shift-invariant linear systems can be represented by convolution. If a system is *linear shift-invariant* (LSI), then we can write, using the shifting representation (1.1-1),

$$y(n_1, n_2) = L\left[\sum_{\text{all } k_1, k_2} x(k_1, k_2) \delta(n_1 - k_1, n_2 - k_2) \right]$$

$$= \sum_{k_1, k_2} x(k_1, k_2) L[\delta(n_1 - k_1, n_2 - k_2)]^2$$

$$= \sum_{\text{all } k_1, k_2} x(k_1, k_2) h(n_1 - k_1, n_2 - k_2),$$

where the sequence h is called the *impulse response*, defined as $h(n_1, n_2) = L[\delta(n_1, n_2)]$. We define 2-D convolution as

$$(x * h)(n_1, n_2) = \sum_{k_1, k_2} x(k_1, k_2) h(n_1 - k_1, n_2 - k_2). \qquad (1.1\text{-}2)$$

It then follows that for a 2-D LSI system we have

$$y(n_1, n_2) = (x * h)(n_1, n_2)$$

$$= (h * x)(n_1, n_2),$$

where the latter equality follows easily by substitution in (1.1-2). Note that all of the summations over k_1, k_2 range from $-\infty$ to $-\infty$.

---

2. Bringing a linear operator inside a general *infinite sum* involves convergence issues. Here we simply assume that both series converge. A more theoretical treatment can be found in an advanced mathematics text [6].

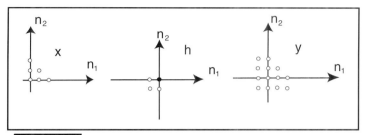

FIGURE 1.9   *Example of 2-D or spatial convolution*

PROPERTIES OF 2-D CONVOLUTION OR CONVOLUTION ALGEBRA

1 Commutativity: $x * y = y * x$.
2 Associativity: $(x * y) * z = x * (y * z)$.
3 Distributivity: $x * (y + z) = x * y + x * z$.
4 Identity element: $\delta(n_1, n_2)$ with property $\delta * x = x$.
5 Zero element: $0(n_1, n_2) = 0$ with property $0 * x = 0$.[3]

All of the following properties of convolution hold for any 2-D signals x, y, and z, for which convolution is defined, i.e., for which the infinite sums exist.

EXAMPLE 1.1-6 (*spatial convolution*)
In Figure 1.9, we see an example of 2-D or spatial convolution. The impulse response h has been reversed or reflected thorough the origin in two dimensions to yield $h(n_1 - k_1, n_2 - k_2)$, shown in the figure for $n_1 = n_2 = 0$. This sequence is then shifted around the $k_1 \times k_2$ plane via integer vector shifts (n_1, n_2) and then a sum of products is taken with input $x(k_1, k_2)$ to give output $y(n_1, n_2)$ as shown.

In general, for signals with rectangular support, we have the following result for the support of their convolution:

If

$$\text{supp}(x) = N_1 \times N_2 \quad \text{and} \quad \text{supp}(h) = M_1 \times M_2,$$

then

$$\text{supp}(y) = (N_1 + M_1 - 1) \times (N_2 + M_2 - 1).$$

For signals of nonrectangular support, this result can be used to rectangularly bound the support of the output.

---

3. Note that 0 here is the zero sequence defined as $0(n_1, n_2) = 0$ for all (n_1, n_2).

1.1.5 STABILITY OF 2-D SYSTEMS

Stable systems are those for which a small change in the input gives a small change in the output. As such, they are most often used for applications. We can mathematically define *bounded-input bounded-output* (BIBO) stability for 2-D systems analogously to that in 1-D system theory. A spatial or 2-D system will be *stable* if the response to every *uniformly* bounded input is itself uniformly bounded. It is generally very difficult to verify this condition, but for an LSI system the condition is equivalent to the impulse response being absolutely summable, i.e.,

$$\sum_{k_1, k_2} |h(k_1, k_2)| < \infty. \tag{1.1-3}$$

THEOREM 1.1-1 (*LSI STABILITY*)

A 2-D LSI system is BIBO stable if and only if its impulse response $h(n_1, n_2)$ is absolutely summable.

Proof: We start by assuming that Eq. (1.1-3) is true. Then by the convolution representation, we have

$$|y(n_1, n_2)| = \left| \sum_{k_1, k_2} h(k_1, k_2) x(n_1 - k_1, n_2 - k_2) \right|$$

$$\leqslant \sum_{k_1, k_2} |(k_1, k_2)| |x(n_1 - k_1, n_2 - k_2)|$$

$$\leqslant \left(\sum_{k_1, k_2} |h(k_1, k_2)| \right) \max |x(n_1, n_2)| < \infty,$$

for any uniformly bounded input x, thus establishing sufficiency of (1.1-3). To show necessity, following Oppenheim and Schafer [5] in the one-dimensional case, we can choose the uniformly bounded input signal $x(n_1, n_2) = \exp -j\theta(-n_1, -n_2)$, where $\theta(n_1, n_2)$ is the argument function of the complex function $h(n_1, n_2)$, then the system output at $(n_1, n_2) = (0, 0)$ will be given by (1.1-3), thus showing that absolute summability of the impulse response is necessary for a bounded output. ☐

In mathematics, the *function space* of absolutely summable 2-D sequences is often denoted l^1, so we can also write $h \in l^1$ as the condition for an LSI system with impulse response h to be BIBO stable. Thus, this well-known 1-D mathematical result [5] easily extends to the 2-D case.

1.2 2-D DISCRETE-SPACE FOURIER TRANSFORM

The Fourier transform is important in 1-D signal processing because it effectively explains the operation of linear time-invariant (LTI) systems via the concept of frequency response. This frequency response is just the Fourier transform of the system impulse response. While convolution provides a complicated description of the LTI system operation, where generally the input at all locations affects the output at all locations, the frequency response provides a simple interpretation as a scalar weighting in the Fourier domain, where the output at each frequency ω depends only on the input at that same frequency. A similar result holds for 2-D systems that are LSI, as we show in the following discussions.

DEFINITION 1.2-1 (*Fourier transform*)
We define the *2-D Fourier transform*,

$$X(\omega_1, \omega_2) \triangleq \sum_{\substack{n_1, n_2 \\ -\infty}}^{+\infty} x(n_1, n_2) \exp -j(\omega_1 n_1 + \omega_2 n_2). \tag{1.2-1}$$

The radian frequency variable ω_1 is called *horizontal frequency*, and the variable ω_2 is called *vertical frequency*. The domain of (ω_1, ω_2) is the entire plane $(-\infty, +\infty) \times (-\infty, +\infty) \triangleq (-\infty, +\infty)^2$.

One of the easy properties of the 2-D Fourier transform is that it is periodic with rectangular period $2\pi \times 2\pi$, a property originating from the integer argument values of n_1 and n_2. To see this, simply note

$$X(\omega_1 \pm 2\pi, \omega_2 \pm 2\pi) = \sum_{\substack{n_1, n_2 \\ -\infty}}^{+\infty} x(n_1, n_2) \exp -j\big[(\omega_1 \pm 2\pi)n_1 + (\omega_2 \pm 2\pi)n_2\big]$$

$$= \sum_{\substack{n_1, n_2 \\ -\infty}}^{+\infty} x(n_1, n_2) \exp -j\omega_1 n_1 \exp -j2\pi n_1$$

$$\times \exp -j\omega_2 n_2 \exp -j2\pi n_2$$

$$= \sum_{\substack{n_1, n_2 \\ -\infty}}^{+\infty} x(n_1, n_2)(\exp -j\omega_1 n_1) \cdot 1 \cdot (\exp -j\omega_2 n_2) \cdot 1$$

$$= X(\omega_1, \omega_2).$$

Thus the 2-D Fourier transform needs to be calculated for only one period, usually taken to be $[-\pi, +\pi] \times [-\pi, +\pi]$. It is analogous to the 1-D case, where the Fourier transform $X(\omega)$ has period 2π and is usually just calculated for $[-\pi, +\pi]$.

Upon close examination, we can see that the 2-D Fourier transform is closely related to the 1-D Fourier transform. In fact, we can rewrite (1.2-1) as

$$X(\omega_1, \omega_2) = \sum_{n_1} \sum_{n_2} x(n_1, n_2) \exp{-j(\omega_1 n_1 + \omega_2 n_2)}$$

$$= \sum_{n_2} \sum_{n_1} x(n_1, n_2) \exp{-j(\omega_1 n_1 + \omega_2 n_2)}$$

$$= \sum_{n_2} \left[\sum_{n_1} x(n_1, n_2) \exp{-j\omega_1 n_1} \right] \exp{-j\omega_2 n_2}$$

$$= \sum_{n_2} X(\omega_1; n_2) \exp{-j\omega_2 n_2}.$$

where $X(\omega_1; n_2) \triangleq \sum_{n_1} x(n_1, n_2) \exp{-j\omega_1 n_1}$, the Fourier transform of row n_2 of x. In words, we say that the 2-D Fourier transform can be decomposed as a set of 1-D Fourier transforms on all the columns of x, followed by another set of 1-D Fourier transforms on the rows $X(\omega_1; n_2)$ that result from the first set of transforms. We call the 2-D Fourier transform a *separable operator*, because it can be performed as the concatenation of 1-D operations on the rows followed by 1-D operations on the columns, or *vice versa*. Such 2-D operators are common in multidimensional signal processing and offer great simplification when they occur.

The Fourier transform has been studied for convergence of its infinite sum in several manners. First, if we have a stable signal, then the Fourier transform sum will converge in the sense of *uniform convergence*, and as a result will be a continuous function of ω_1 and ω_2. A weaker form of convergence of the Fourier transform sum is *mean-square convergence* [6]. This is the sense of convergence appropriate when the Fourier transform is a discontinuous function. A third type of convergence is as a *generalized function*, e.g., $\delta(\omega_1, \omega_2)$, the 2-D Dirac impulse. It is used for periodic signals such as $\exp{j(\alpha_1 n_1 + \alpha_2 n_2)}$, whose Fourier transform will be shown to be the scaled and shifted 2-D Dirac impulse $2\pi\delta(\omega_1 - \alpha_1, \omega_2 - \alpha_2)$ when $|\alpha_1| < \pi$ and $|\alpha_2| < \pi$, and when attention is limited to the unit cell $[-\pi, +\pi] \times [-\pi, +\pi]$. Outside this cell, the function must repeat.

In operator notation, we can write the Fourier transform relation as

$$X = FT[x],$$

indicating that the Fourier transform operator FT maps 2-D sequences into 2-D functions X that are continuous parameter and periodic functions with period $2\pi \times 2\pi$.

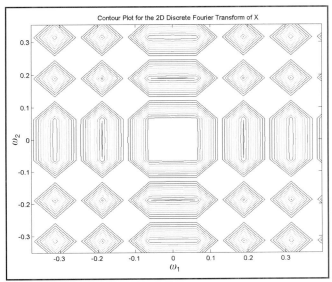

Contour Plot for the 2D Discrete Fourier Transform of X

FIGURE 1.10   *Zoomed-in contour plot of log magnitude of Fourier transform of rectangular pulse with $N_1 = N_2 = 50$*

EXAMPLE 1.2-1 (*Fourier transform of rectangular pulse function*)
Let

$$x(n_1, n_2) = \begin{cases} 1, & 0 \leqslant n_1 \leqslant N_1 - 1, \ 0 \leqslant n_2 \leqslant N_2 - 1 \\ 0, & \text{else,} \end{cases}$$

then its 2-D Fourier transform can be determined as

$$X(\omega_1, \omega_2) = \sum_{n_1=0}^{N_1-1} \sum_{n_2=0}^{N_2-1} 1 \exp{-j(\omega_1 n_1 + \omega_2 n_2)}$$

$$= \left(\sum_{n_1=0}^{N_1-1} \exp{-j\omega_1 n_1} \right) \left(\sum_{n_2=0}^{N_2-1} \exp{-j\omega_2 n_2} \right)$$

$$= \left(\frac{1 - e^{-j\omega_1 N_1}}{1 - e^{-j\omega_1}} \right) \left(\frac{1 - e^{-j\omega_2 N_2}}{1 - e^{-j\omega_2}} \right)$$

$$= e^{-j\omega_1(N_1-1)/2} \frac{\sin \omega_1 N_1/2}{\sin \omega_1/2} e^{-j\omega_2(N_2-1)/2} \frac{\sin \omega_2 N_2/2}{\sin \omega_2/2}$$

$$= e^{-j([\omega_1(N_1-1)/2 + \omega_2(N_2-1)/2])} \frac{\sin \omega_1 N_1/2}{\sin \omega_1/2} \frac{\sin \omega_2 N_2/2}{\sin \omega_2/2},$$

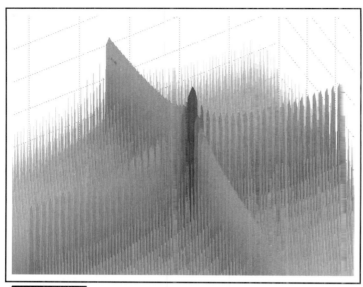

FIGURE 1.11   *3-D perspective plot of log magnitude of Fourier transform of rectangular pulse (origin at center) (see color insert)*

where the second-to-last line follows by factoring out the indicated phase term from each of the two factors in the line above, and then using Euler's equality $\sin\theta = (1/2j)(e^{i\theta} - e^{-i\theta})$ in the top and bottom terms of each factor. We see the result is just the product of two 1-D Fourier transforms for this separable and rectangular 2-D pulse function. If N_1 and N_2 are odd numbers, we can shift the pulse to be centered on the origin $(0, 0)$ and thereby remove the linear phase shift term out in front, corresponding to a shift (delay) of $(\frac{1}{2}(N_1 - 1), \frac{1}{2}(N_2 - 1))$. But the way we have written the signal x as starting at $(0, 0)$ there remains this linear phase shift. A contour plot of the log magnitude of this function is provided in Figure 1.10 and the corresponding 3-D perspective plot is shown in Figure 1.11. Both of these plots were produced with MATLAB.

EXAMPLE 1.2-2 (*Fourier transform of line impulse*)
Consider the line impulse signal

$$x(n_1, n_2) = \delta(n_1 - n_2) = \begin{cases} 1, & n_2 = n_1 \\ 0, & \text{else,} \end{cases}$$

which is a line of slope 1 in the $n_2 \times n_1$ plane or an angle of $45°$. We can take the Fourier transform as

$$X(\omega_1, \omega_2) = \sum_{n_1, n_2} \delta(n_1 - n_2) \exp{-j(\omega_1 n_1 + \omega_2 n_2)} \qquad (1.2\text{-}2)$$

$$= \sum_{n_1 = -\infty}^{+\infty} \exp{-j\omega_1 n_1} \exp{-j\omega_2 n_1}$$

$$= \sum_{n_1 = -\infty}^{+\infty} \exp{-j(\omega_1 + \omega_2)n_1}$$

$$= 2\pi \delta(\omega_1 + \omega_2) \quad \text{for } (\omega_1, \omega_2) \in [-\pi, +\pi]^2, \qquad (1.2\text{-}3)$$

which is a Dirac impulse in the 2-D frequency domain, along the line $\omega_2 = -\omega_1$. Thus the Fourier transform of an impulse line signal is a Dirac impulse line in the 2-D frequency domain. Note that the angle of this line is that of the spatial domain line plus/minus $90°$, i.e., the two lines are perpendicular to one another. Can you give a frequency interpretation of this result?

1.2.1 INVERSE 2-D FOURIER TRANSFORM

The inverse 2-D Fourier transform is given as

$$x(n_1, n_2) = \frac{1}{(2\pi)^2} \int_{[-\pi, +\pi] \times [-\pi, +\pi]} X(\omega_1, \omega_2) \exp{+j(\omega_1 n_1 + \omega_2 n_2)} \, d\omega_1 \, d\omega_2.$$
$$(1.2\text{-}4)$$

Note that we only integrate over a "unit" cell $[-\pi, +\pi] \times [-\pi, +\pi]$ in the $\omega_1 \times \omega_2$ frequency domain. To justify this formula, we can plug in (1.2-1) and interchange the order of summation and integration, just as was done in the 1-D case. Alternatively, we can note that the 2-D **FT** is a separable operator and just use the known results for transform and inverse in the 1-D case to arrive at this same answer for the inverse 2-D Fourier transform. In operator notation, we denote the inverse Fourier transform as **IFT** and write

$$x = \mathbf{IFT}[X].$$

A 2-D proof of this result follows closely the 1-D result [5]. First we insert (1.2-1) into (1.2-4) to obtain

$$x(n_1, n_2) = \frac{1}{(2\pi)^2} \int_{[-\pi,+\pi]\times[-\pi,+\pi]} \left[\sum_{l_1,l_2}^{-\infty} \sum_{-\infty} x(l_1, l_2) \exp{-j(\omega_1 l_1 + \omega_2 l_2)} \right]$$

$$\times \exp{+j(\omega_1 n_1 + \omega_2 n_2)}\, d\omega_1\, d\omega_2$$

$$= \sum_{l_1,l_2} x(l_1, l_2) \left[\frac{1}{(2\pi)^2} \int_{[-\pi,+\pi]\times[-\pi,+\pi]} \right.$$

$$\left. \times \exp{+j(\omega_1(n_1 - l_1) + \omega_2(n_2 - l_2))}\, d\omega_1\, d\omega_2 \right]$$

$$= \sum_{l_1,l_2} x(l_1, l_2) \left\{ \left[\frac{1}{2\pi} \int_{[-\pi,+\pi]} \exp{+j\omega_1(n_1 - l_1)}\, d\omega_1 \right] \right.$$

$$\left. \times \left[\frac{1}{2\pi} \int_{[-\pi,+\pi]} \exp{+j\omega_2(n_2 - l_2)}\, d\omega_2 \right] \right\}.$$

Now,

$$\left(\frac{1}{2\pi} \int_{[-\pi,+\pi]} \exp{+j\omega m}\, d\omega_1 \right) = \frac{\exp{-j\omega m}}{j2\pi m} \Big|_{-\pi}^{+\pi} = \frac{\sin \pi m}{\pi m} = \delta(m),$$

so substituting this result in the above equation yields

$$x(n_1, n_2) = \sum_{l_1,l_2} x(l_1, l_2)\delta(l_1)\delta(l_2)$$

$$= x(n_1, n_2),$$

as was to be shown.

A main application of 2-D Fourier transforms is to provide a simplified view of spatial convolution as used in linear filtering.

1.2.2 FOURIER TRANSFORM OF 2-D OR SPATIAL CONVOLUTION

THEOREM 1.2-1 (*Fourier convolution theorem*)
If $y(n_1, n_2) = h(n_1, n_2) * x(n_1, n_2)$, then $Y(\omega_1, \omega_2) = H(\omega_1, \omega_2)X(\omega_1, \omega_2)$.

Proof:

$$Y(\omega_1, \omega_2) = \sum_{\substack{n_1, n_2 \\ -\infty}}^{+\infty} y(n_1, n_2) \exp -j(\omega_1 n_1 + \omega_2 n_2)$$

$$= \sum_{n_1, n_2} \left[\sum_{k_1, k_2} h(k_1, k_2) x(n_1 - k_1, n_2 - k_2) \right] \exp -j(\omega_1 n_1 + \omega_2 n_2)$$

$$= \sum_{n_1, n_2} \sum_{k_1, k_2} h(k_1, k_2) x(n_1 - k_1, n_2 - k_2)$$

$$\times \exp -j\left\{ \omega_1 \left[k_1 + (n_1 - k_1) \right] + \omega_2 \left[k_2 + (n_2 - k_2) \right] \right\}$$

$$= \sum_{n_1, n_2} \sum_{k_1, k_2} h(k_1, k_2) \exp -j(\omega_1 k_1 + \omega_2 k_2) x(n_1 - k_1, n_2 - k_2)$$

$$\times \exp -j\left[\omega_1 (n_1 - k_1) + \omega_2 (n_2 - k_2) \right]$$

$$= \sum_{k_1, k_2} h(k_1, k_2) \exp -j(\omega_1 k_1 + \omega_2 k_2)$$

$$\times \left\{ \sum_{n_1, n_2} x(n_1 - k_1, n_2 - k_2) \exp -j\left[\omega_1 (n_1 - k_1) + \omega_2 (n_2 - k_2) \right] \right\}$$

$$= \sum_{k_1, k_2} h(k_1, k_2) \exp -j(\omega_1 k_1 + \omega_2 k_2)$$

$$\times \left[\sum_{n_1', n_2'} x(n_1', n_2') \exp -j(\omega_1 n_1' + \omega_2 n_2') \right]$$

with $n_1' \triangleq n_1 - k_1$ and $n_2' \triangleq n_2 - k_2$.

Now, since the limits of the sums are infinite, the shift by any finite value (k_1, k_2) makes no difference, and the limits on the inside sum over (n_1', n_2') remain at $(-\infty, +\infty) \times (-\infty, +\infty)$. Thus we can bring its double sum outside the sum over (k_1, k_2) and recognize it as $X(\omega_1, \omega_2)$, the Fourier transform of the input signal x. What is left inside is the *frequency response* $H(\omega_1, \omega_2)$, and so we have finally

$$= \left[\sum_{k_1, k_2} h(k_1, k_2) \exp -j(\omega_1 k_1 + \omega_2 k_2) \right]$$

$$\times \left[\sum_{n_1', n_2'} x(n_1', n_2') \exp -j(\omega_1 n_1' + \omega_2 n_2') \right]$$

$$= H(\omega_1, \omega_2) \cdot X(\omega_1, \omega_2).$$

as was to be shown.   □

Since we have seen that a 2-D or spatial LSI system is characterized by its impulse response $h(n_1, n_2)$, we now see that its frequency response $H(\omega_1, \omega_2)$ also suffices to characterize such a system. And the Fourier transform Y of the output equals the product of the frequency response H and the Fourier transform X of the input. When the frequency response H takes on only values 1 and 0, then we call the system an *ideal filter*, since such an LSI system will "filter" out some frequencies and pass others unmodified. More generally, the term filter has grown to include all such LSI systems, and has been extended to shift-variant and even nonlinear systems through the concept of the Voltera series of operators [7].

EXAMPLE 1.2-3 (*Fourier transform of complex plane wave*)
Let $x(n_1, n_2) = A \exp j(\omega_1^0 n_1 + \omega_2^0 n_2) \triangleq e(n_1, n_2)$, where $|\omega_i^0| < \pi$, then $X(\omega_1, \omega_2) = c\delta(\omega_1 - \omega_1^0, \omega_2 - \omega_2^0)$ in the basic square $[-\pi, +\pi]^2$. Finding the constant c by inverse Fourier transform, we conclude $c = (2\pi)^2 A$. Inputting this *plane wave* into an LSI system or filter with frequency response $H(\omega_1, \omega_2)$, we obtain the Fourier transform output signal $Y(\omega_1, \omega_2) = (2\pi)^2 A H(\omega_1^0, \omega_2^0)\delta(\omega_1 - \omega_1^0, \omega_2 - \omega_2^0) = H(\omega_1^0, \omega_2^0)E(\omega_1^0, \omega_2^0)$, or in the space domain, $y(n_1, n_2) = H(\omega_1^0, \omega_2^0) e(n_1, n_2)$, thus showing that complex exponentials (plane waves) are the *eigenfunctions* of spatial LSI systems, and the frequency response H evaluated at the plane wave frequency (ω_1^0, ω_2^0) becomes the corresponding *eigenvalue*.

SOME IMPORTANT PROPERTIES OF THE FT OPERATOR

1   Linearity: $ax + by \Leftrightarrow aX(\omega_1, \omega_2) + bY(\omega_1, \omega_2)$.

2   Convolution: $x * y \Leftrightarrow X(\omega_1, \omega_2)Y(\omega_1, \omega_2)$.

3   Multiplication: $xy \Leftrightarrow (X * Y)(\omega_1, \omega_2) \triangleq \frac{1}{(2\pi)^2} \int_{-\pi}^{+\pi}\int_{-\pi}^{+\pi} X(\nu_1, \nu_2)Y(\omega_1 - \nu_1, \omega_2 - \nu_2) \, d\nu_1 \, d\nu_2$.

4   Modulation: $x \exp j(\nu_1 n_1 - \nu_2 n_2) \Leftrightarrow X(\omega_1 - \nu_1, \omega_2 - \nu_2)$ for integers ν_1 and ν_2.

5   Shift (delay): $x(n_1 - m_1, n_2 - m_2) \Leftrightarrow X(\omega_1, \omega_2) \exp -j(\omega_1 m_1 + \omega_2 m_2)$ for integers m_1 and m_2.

6   Differentiation in frequency domain:

$$-jn_1 x(n_1, n_2) \quad \Leftrightarrow \quad \frac{\partial X}{\partial \omega_1}$$

$$-jn_2 x(n_1, n_2) \quad \Leftrightarrow \quad \frac{\partial X}{\partial \omega_2}$$

7   "Initial" value:

$$x(0, 0) = \frac{1}{(2\pi)^2} \int_{[-\pi, +\pi] \times [-\pi, -\pi]} X(\omega_1, \omega_2) \, d\omega_1 \, d\omega_2.$$

8   "DC" value:

$$X(0,0) = \sum_{\substack{n_1,n_2 \\ -\infty}}^{+\infty} x(n_1,n_2).$$

9   Parseval's theorem:

$$\sum_{n_1,n_2} x(n_1,n_2)y^*(n_1,n_2)$$

$$= \frac{1}{(2\pi)^2} \int_{[-\pi,+\pi]\times[-\pi,+\pi]} X(\omega_1,\omega_2)Y^*(\omega_1,\omega_2)\,d\omega_1\,d\omega_2,$$

with "power" version (special case)

$$\sum_{n_1,n_2} |x(n_1,n_2)|^2 = \frac{1}{(2\pi)^2} \int_{[-\pi,+\pi]\times[-\pi,+\pi]} |X(\omega_1,\omega_2)|^2\,d\omega_1\,d\omega_2.$$

10   Separable signal:

$$x_1(n_1)x_2(n_2) \quad \Leftrightarrow \quad X_1(\omega_1)X_2(\omega_2).$$

SOME USEFUL FOURIER TRANSFORM PAIRS

  1   Constant c:

$$\mathbf{FT}\{c\} = (2\pi)^2 c\delta(\omega_1,\omega_2)$$

in the unit cell $[-\pi,+\pi]^2$.

  2   Complex exponential—for spatial frequency $(v_1,v_2) \in [-\pi,+\pi]^2$:

$$\mathbf{FT}\{\exp j(v_1 n_1 + v_2 n_2)\} = (2\pi)^2 \delta(\omega_1 - v_1, \omega_2 - v_2)$$

in the unit cell $[-\pi,+\pi]^2$.

  3   Constant in n_2 dimension:

$$\mathbf{FT}\{x_1(n_1)\} = 2\pi X_1(\omega_1)\delta(\omega_2),$$

where $X_1(\omega_1)$ is a 1-D Fourier transform and $\delta(\omega_2)$ is a 1-D Dirac impulse (and ω_1, ω_2).

  4   Ideal lowpass filter (square passband):

$$H_s(\omega_1,\omega_2) = I_{\omega_c}(\omega_1)I_{\omega_c}(\omega_2),$$

in $[-\pi,+\pi]^2$, where

$$I_{\omega_c} \triangleq \begin{cases} 1, & |\omega| \leqslant \omega_c \\ 0, & \text{else,} \end{cases}$$

with ω_c the *cut-off frequency* of the filter with $|\omega_c| \leq \pi$. The function I_{ω_c} is sometimes called an *indicator function*, since it indicates the passband. Taking the inverse 2-D Fourier transform of this separable function, we proceed as follows to obtain the ideal impulse response,

$$h_s(n_1, n_2) = \frac{1}{(2\pi)^2} \int_{[-\pi,+\pi] \times [-\pi,+\pi]} H_s(\omega_1, \omega_2) \exp + j(\omega_1 n_1 + \omega_2 n_2) \, d\omega_1 \, d\omega_2$$

$$= \left[\frac{1}{2\pi} \int_{[-\pi,+\pi]} I_{\omega_c}(\omega_1) \exp + j\omega_1 n_1 \, d\omega_1 \right]$$

$$\times \left[\frac{1}{2\pi} \int_{[-\pi,+\pi]} I_{\omega_c}(\omega_2) \exp + j\omega_2 n_2 \, d\omega_2 \right]$$

$$= \frac{\sin \omega_c n_1}{\pi n_1} \cdot \frac{\sin \omega_c n_2}{\pi n_2}, \quad -\infty < n_1, n_2 < +\infty.$$

5 Ideal lowpass filter (circular passband):

$$H_c(\omega_1, \omega_2) = \begin{cases} 1, & \sqrt{\omega_1^2 + \omega_2^2} \leq \omega_c \\ 0, & \text{else} \end{cases} \quad \text{for } (\omega_1, \omega_2) \in [-\pi, +\pi] \times [-\pi, +\pi].$$

The inverse Fourier transform of this circular symmetric frequency response can be represented as the integral

$$h_c(n_1, n_2) = \frac{1}{(2\pi)^2} \iint_{\sqrt{\omega_1^2 + \omega_2^2} \leq \omega_c} 1 \exp + j(\omega_1 n_1 - \omega_2 n_2) \, d\omega_1 \, d\omega_2$$

$$= \frac{1}{(2\pi)^2} \int_0^{\omega_c} \int_{-\pi}^{+\pi} \exp + j u(n_1 \cos \theta + n_2 \sin \theta) u \, du \, d\theta$$

$$= \frac{1}{(2\pi)^2} \int_0^{\omega_c} u \left\{ \int_{-\pi}^{-\pi} \exp + j[ur\cos(\theta - \phi)] \, d\theta \right\} du$$

$$= \frac{1}{(2\pi)^2} \int_0^{\omega_c} u \left[\int_{-\pi}^{+\pi} \exp + j(ur\cos \theta) \, d\theta \right] du,$$

where we have used, first, polar coordinates in frequency $\omega_1 = u\cos\theta$ and $\omega_2 = u\sin\theta$, and then in the next line, polar coordinates in space, $n_1 = r\cos\phi$ and $n_2 = r\sin\phi$. Finally, the last line follows because the integral over θ does not depend on ϕ, since it is an integral over 2π. The inner integral over θ can now be recognized in terms of the zeroth-order Bessel function of the first kind $J_0(x)$, with integral representation [1,3]

$$J_0(x) = \frac{1}{2\pi} \int_{-\pi}^{+\pi} \cos(x \cos \theta) \, d\theta = \frac{1}{2\pi} \int_{-\pi}^{+\pi} \cos(x \sin \theta) \, d\theta.$$

To see this, we note that the integral

$$\int_{-\pi}^{+\pi} \exp +j(ur\cos\theta)\, d\theta = \int_{-\pi}^{+\pi} \cos(ur\cos\theta)\, d\theta,$$

since the imaginary part, via Euler's formula, is an odd function integrated over even limits, and hence is zero. So, continuing, we can write

$$h_c(n_1, n_2) = \frac{1}{(2\pi)^2} \int_0^{\omega_c} u\left(\int_{-\pi}^{+\pi} \exp +j(ur\cos\theta)\, d\theta\right) du$$

$$= \frac{1}{2\pi} \int_0^{\omega_c} u J_0(ur)\, du$$

$$= \frac{\omega_c}{2\pi r} J_1(\omega_c r)$$

$$= \frac{\omega_c}{2\pi \sqrt{n_1^2 + n_2^2}} J_1\left(\omega_c \sqrt{n_1^2 + n_2^2}\right),$$

where J_1 is the first-order Bessel function of the first kind $J_1(x)$, satisfying the known relation [1, p. 484]

$$x J_1(x) = \int_0^x u J_0(u)\, du.$$

Comparing these two ideal lowpass filters, one with square passband and the other circular with diameter equal to a side of the square, we get the two impulse responses along the n_1 axis:

$$h_s(n_1, 0) = \frac{\omega_c}{\pi} \frac{\sin \omega_c n_1}{\pi n_1},$$

and

$$h_c(n_1, 0) = \frac{\omega_c}{2\pi n_1} J_1(\omega_c n_1).$$

These 1-D sequences are plotted via MATLAB in Figure 1.12. We note their similarity.

EXAMPLE 1.2-4 (*Fourier transform of separable signal*)
Let $x(n_1, n_2) = x_1(n_1) x_2(n_2)$, then when we compute the Fourier transform,

the following simplification arises:

$$X(\omega_1, \omega_2) = \sum_{n_1} \sum_{n_2} x(n_1, n_2) \exp -j(\omega_1 n_1 + \omega_2 n_2)$$

$$= \sum_{n_1} \sum_{n_2} x_1(n_1) x_2(n_2) \exp -j(\omega_1 n_1 + \omega_2 n_2)$$

$$= \sum_{n_1} x_1(n_1) \left[\sum_{n_2} x_2(n_2) \exp -j\omega_2 n_2 \right] \exp -j\omega_1 n_1$$

$$= \left[\sum_{n_1} x_1(n_1) \exp -j\omega_1 n_1 \right] \left[\sum_{n_2} x_2(n_2) \exp -j\omega_2 n_2 \right]$$

$$= X_1(\omega_1) X_2(\omega_2).$$

A consequence of this result is that in 2-D signal processing, multiplication in the spatial domain does not always lead to convolution in the frequency domain! Can you reconcile this fact with the basic 2-D Fourier transform property 3? (See problem 7 at the end of this chapter.)

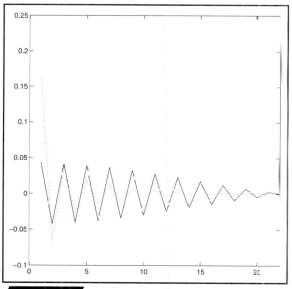

FIGURE 1.12 *Two impulse responses of 2-D ideal lowpass filters. Solid curve is "square" and dotted curve is "circular"*

1.2.3 SYMMETRY PROPERTIES OF FOURIER TRANSFORM

We consider the symmetry properties of the Fourier transform of complex signal $x(n_1, n_2)$. We start with the original FT pair:

$$x(n_1, n_2) \quad \Leftrightarrow \quad X(\omega_1, \omega_2).$$

First, *reflection through the origin* (a generalization of time reversal in the 1-D case): We ask what is the FT of the signal $x(-n_1, -n_2)$, where each of the axes are reversed. For an image, this corresponds to flipping it right-to-left, and then flipping it top-to-bottom. We seek

$$\sum_{n_1} \sum_{n_2} x(-n_1, -n_2) \exp -j(\omega_1 n_1 + \omega_2 n_2)$$

$$= \sum_{n_1'} \sum_{n_2'} x(n_1', n_2') \exp -j[\omega_1(-n_1') + \omega_2(-n_2')]$$

upon setting $n_1' = -n_1$ and $n_2' = -n_2$, so moving the minus signs to the ω_i terms, we have

$$\sum_{n_1'} \sum_{n_2'} x(n_1', n_2') \exp -j[(-\omega_1)n_1' + (-\omega_2)n_2'],$$

which can be seen as $X(-\omega_1, -\omega_2)$. Thus we have the new transform pair

$$x(-n_1, -n_2) \quad \Leftrightarrow \quad X(-\omega_1, -\omega_2).$$

If the original spatial signal is instead conjugated, we can find

$$x^*(n_1, n_2) \quad \Leftrightarrow \quad X^*(-\omega_1, -\omega_2), \tag{1.2-5}$$

which generalizes the 1-D FT pair $x^*(n) \Leftrightarrow X^*(-\omega)$. Combining these two 2-D transform pairs, we get the transform of $x^*(-n_1, -n_2)$,

$$x^*(-n_1, -n_2) \quad \Leftrightarrow \quad X^*(\omega_1, \omega_2),$$

resulting in the conjugate of the original $X(\omega_1, \omega_2)$. We can organize these four basic transform pairs using the concept of *conjugate symmetric* and *conjugate antisymmetric* parts, as in one dimension [5].

DEFINITION 1.2-2 (*signal symmetries*)
Conjugate symmetric part of x: $x_e(n_1, n_2) \triangleq \frac{1}{2}[x(n_1, n_2) + x^*(-n_1, -n_2)]$; *conjugate antisymmetric part of* x: $x_o(n_1, n_2) \triangleq \frac{1}{2}[x(n_1, n_2) - x^*(-n_1, -n_2)]$.

From this we get the following two transform pairs:

$$x_e(n_1, n_2) \quad \Leftrightarrow \quad \operatorname{Re} X(\omega_1, \omega_2),$$

$$x_o(n_1, n_2) \quad \Leftrightarrow \quad j \operatorname{Im} X(\omega_1, \omega_2).$$

Similarly, the conjugate symmetric and antisymmetric parts in the 2-D frequency domain are defined.

DEFINITION 1.2-3 (Fourier symmetries)

Conjugate symmetric part of X: $X_e(\omega_1, \omega_2) \triangleq \frac{1}{2}[X(\omega_1, \omega_2) + X^*(-\omega_1, -\omega_2)]$;
conjugate antisymmetric part of X: $X_o(\omega_1, \omega_2) \triangleq \frac{1}{2}[X(\omega_1, \omega_2) - X^*(-\omega_1, -\omega_2)]$.

Using these symmetry properties, we get the following two general transform pairs:

$$\operatorname{Re} x(n_1, n_2) \quad \Leftrightarrow \quad X_e(\omega_1, \omega_2),$$

$$j \operatorname{Im} x(n_1, n_2) \quad \Leftrightarrow \quad X_o(\omega_1, \omega_2).$$

SYMMETRY PROPERTIES OF REAL VALUED SIGNALS

There is a special case of great interest for *real valued* signals, i.e., $x(n_1, n_2) = \operatorname{Re} x(n_1, n_2)$, which implies that $X_o(\omega_1, \omega_2) = 0$, so when the signal x is real valued, we must have

$$X(\omega_1, \omega_2) = X^*(-\omega_1, -\omega_2). \qquad (1.2\text{-}6)$$

Directly from this equation we get the following transform pairs, or symmetry properties, by taking the real, imaginary, magnitude, and phase of both sides of (1.2-6) and setting the results equal:

$$\operatorname{Re} X(\omega_1, \omega_2) = \operatorname{Re} X(-\omega_1, -\omega_2) \quad \text{(read "real part is even");}$$

$$\operatorname{Im} X(\omega_1, \omega_2) = -\operatorname{Im} X(-\omega_1, -\omega_2) \quad \text{(read "imaginary part is odd");}$$

$$\left| X(\omega_1, \omega_2) \right| = \left| X(-\omega_1, -\omega_2) \right| \quad \text{(read "magnitude is even");}$$

$$\arg X(\omega_1, \omega_2) = -\arg X(-\omega_1, -\omega_2)$$

(read "argument (phase) *may be taken as* odd").

Note that "even" really means *symmetric through the origin*, i.e., $x(n_1, n_2) = x(-n_1, -n_2)$. Also note that this type of symmetry is the necessary and sufficient condition for the Fourier transform of a 2-D real valued signal to be real valued, since x_0 is zero in this case.

1.2.4 CONTINUOUS-SPACE FOURIER TRANSFORM

While this text is one on digital image and video, we need to be aware of the generalization of continuous Fourier transforms to two and higher dimensions. Here, we look at the two-dimensional continuous parameter Fourier transform, with application to continuous-space images, e.g., film. We will denote the two continuous parameters as t_1 and t_2, but time is not the target here, although one of the two parameters could be time, e.g., acoustic array processing in geophysics [2]. We could equally have used the notation x_1 and x_2 to denote these free parameters. We have chosen t_1 and t_2 so that we can use x for signal.

2-D FOURIER CONTINUOUS TRANSFORM

We start with a continuous parameter function of two dimensions t_1 and t_2, often called a bi-function, and herein denoted $f_c(t_1, t_2)$ defined over $-\infty < t_1 < +\infty$, $-\infty < t_2 < +\infty$. We write the corresponding 2-D Fourier transform as the double integral

$$F_c(\Omega_1, \Omega_2) \triangleq \int_{-\infty}^{+\infty} \int_{-\infty}^{+\infty} f_c(t_1, t_2) \exp -j(\Omega_1 t_1 + \Omega_2 t_2) \, dt_1 \, dt_2.$$

We recognize that this **FT** operator is separable and consists of the 1-D **FT** repeatedly applied in the t_1 domain for each fixed t_2, followed by the 1-D **FT** repeatedly applied in the t_2 domain for each fixed Ω_1, finally culminating in the value $F_c(\Omega_1, \Omega_2)$. Twice applying the inverse 1-D Fourier operator, once for Ω_1 and once for Ω_2, we arrive at the inverse continuous parameter Fourier transform,

$$f_c(t_1, t_2) = \frac{1}{(2\pi)^2} \int_{-\infty}^{+\infty} \int_{-\infty}^{+\infty} F_c(\Omega_1, \Omega_2) \exp +j(\Omega_1 t_1 + \Omega_2 t_2) \, d\Omega_1 \, d\Omega_2,$$

assuming only that the various integrals converge, i.e., are well defined in some sense. One very useful property of the **FT** operator on such bi-functions is the so-called rotation theorem, which states that the **FT** of any rotated version of the original function $f_c(t_1, t_2)$ is just the rotated **FT**, i.e., the corresponding rotation operation applied to $F_c(\Omega_1, \Omega_2)$. We call such an operation *rotationally invariant*. Notationally we can economically denote such rotations via the matrix equation

$$\mathbf{t}' \triangleq \mathbf{Rt}, \tag{1.2-7}$$

where the vectors $\mathbf{t} \triangleq (t_1, t_2)^T$, $\mathbf{t}' \triangleq (t_1', t_2')^T$, and the rotation matrix $\mathbf{R}$ is given in terms of the rotation angle θ as

$$\mathbf{R} \triangleq \begin{bmatrix} \cos\theta & \sin\theta \\ -\sin\theta & \cos\theta \end{bmatrix},$$

with reference to Figure 1.13.

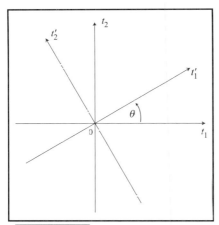

FIGURE 1.13   *An illustration of rotation of coordinates from t to t' by angle +θ*

THEOREM 1.2-2 (*ROTATIONAL INVARIANCE OF FT OPERATOR*)
The continuous parameter Fourier transform operator **FT** is rotationally invariant, i.e., if $g_c(\mathbf{t}) \triangleq f_c(\mathbf{Rt})$, with **R** a rotation matrix, then $G_c(\mathbf{\Omega}) = F_c(\mathbf{R\Omega})$.

Proof: We start at

$$g_c(\mathbf{t}) \triangleq f_c(\mathbf{Rt})$$
$$= f_c(t_1 \cos\theta + t_2 \sin\theta \quad -t_1 \sin\theta + t_2 \cos\theta),$$

which corresponds to a rotation of axes by $+\theta$ degrees. Then its Fourier transform is given by

$$G_c(\Omega_1, \Omega_2) = \int_{-\infty}^{+\infty}\int_{-\infty}^{+\infty} g_c(t_1, t_2) \exp{-j(\Omega_1 t_1 + \Omega_2 t_2)}\, dt_1\, dt_2$$

$$= \int_{-\infty}^{+\infty}\int_{-\infty}^{+\infty} g_c(\mathbf{t}) \exp{-j(\mathbf{\Omega}^T \mathbf{t})}\, d\mathbf{t}$$

$$= \int_{-\infty}^{+\infty}\int_{-\infty}^{+\infty} f_c(\mathbf{Rt}) \exp{-j(\mathbf{\Omega}^T \mathbf{t})}\, d\mathbf{t},$$

where $\mathbf{\Omega} \triangleq (\Omega_1, \Omega_2)^T$. Then we use (1.2-7) to rotate coordinates to get

$$G_c(\Omega_1, \Omega_2) = \int_{-\infty}^{+\infty}\int_{-\infty}^{+\infty} f_c(\mathbf{t}') \exp{-j(\mathbf{\Omega}^T \mathbf{R}^{-1} \mathbf{t}')}|\mathbf{R}^{-1}|\, d\mathbf{t}'$$

where $|\mathbf{R}^{-1}|$ denotes the determinant of $\mathbf{R}^{-1}$

$$= \int_{-\infty}^{+\infty} \int_{-\infty}^{+\infty} f_{c}(\mathbf{t}') \exp -j\left(\boldsymbol{\Omega}^{T} \mathbf{R}^{T} \mathbf{t}'\right)\left|\mathbf{R}^{-1}\right| d\mathbf{t}',$$

since the inverse of $\mathbf{R}$ is just its transpose $\mathbf{R}^{T}$,

$$= \int_{-\infty}^{+\infty} \int_{-\infty}^{+\infty} f_{c}(\mathbf{t}') \exp -j\left\{(\mathbf{R}\boldsymbol{\Omega})^{T} \mathbf{t}'\right\} d\mathbf{t}',$$

since the determinant of $\mathbf{R}$ is one,

$$= F_{c}(\mathbf{R}\boldsymbol{\Omega})$$

$$= F_{c}(\Omega_{1}\cos\theta + \Omega_{2}\sin\theta, -\Omega_{1}\sin\theta + \Omega_{2}\cos\theta)$$

$$= F_{c}(\boldsymbol{\Omega}'),$$

upon setting $\boldsymbol{\Omega}' \triangleq \mathbf{R}\boldsymbol{\Omega}$. This completes the proof. $\square$

PROJECTION-SLICE THEOREM

In the medical/industrial 3-D imaging technique computer tomography (CT), two-dimensional slices are constructed from a series of X-ray images taken on an arc. A key element in this approach is the projection-slice theorem presented here for the 2-D case. First we define the Radon transform.

DEFINITION 1.2-4 (*Radon transform and projection*)
We define the *Radon transform* for angles: $-\pi \leqslant \theta \leqslant +\pi$, as

$$p_{\theta}(t) \triangleq \int_{-\infty}^{+\infty} f_{c}(t\cos\theta + u\sin\theta, -t\sin\theta + u\cos\theta)\, du,$$

where $f_{c}(t, u)$ is an integrable and continuous function defined over all of R^{2}.

Using the rotation matrix $\mathbf{R}$ and writing it as a function of θ, we have $\mathbf{R} = R(\theta)$, and can express the Radon transform as

$$p_{\theta}(t) = \int_{-\infty}^{+\infty} f_{c}\left(\mathbf{R}(\theta)\begin{bmatrix} t \\ u \end{bmatrix}\right) du.$$

We can then state the projection-slice theorem as follows:

THEOREM 1.2-3 (*projection-slice*)
Let continuous function $f_{c}(t, u)$ have Fourier transform $F_{c}(\Omega_{1}, \Omega_{2})$. Let the

Radon transform of f_c be defined and given as $p_\theta(t)$ with 1-D Fourier transform $P_\theta(\Omega)$ for each angle θ. Then we have the equality:

$$P_\theta(\Omega) = F_c(\Omega \cos\theta, \Omega \sin\theta). \tag{1.2-8}$$

Proof: By the rotational invariance of the continuous FT operator, we only have to show (1.2-8) for $\theta = 0$. So

$$P_0(\Omega) = \text{FT}\big[p_0(t)\big]$$

$$= \int_{-\infty}^{+\infty}\left[\int_{-\infty}^{+\infty} f_c(t, u)\, du\right] e^{-j\Omega t}\, dt$$

$$= F_c(\Omega, 0),$$

which agrees with (1.2-8), as was to be shown.   □

In words, we can sweep out the full 2-D spatial transform F_c in terms of the 1-D frequency components P_θ of the Radon transform at all angles $\theta \in [-\pi, +\pi]$.

1.3 CONCLUSIONS

This chapter has introduced 2-D or spatial signals and systems. We looked at how the familiar step and impulse functions of 1-D digital signal processing (DSP) generalize to the 2-D case. We then extended the Fourier transform and studied some of its important properties. We looked at some ideal spatial filters and other important discrete-space Fourier transform pairs. We then turned to the continuous-space Fourier transform and treated an application in computer tomography, which is understood using the rotational invariant feature of this transform. The projection-slice theorem is only approximated in the discrete-space Fourier transform when the sampling rate is high. We discuss sampling in the next chapter. A good alternate reference for these first few chapters is Lim [4].

1.4 PROBLEMS

1   Plot the following signals:
   (a) $s(n_1, n_2) = u(n_1, n_2)u(5 - n_1, n_2)$.
   (b) $s(n_1, n_2) = \delta(n_1 - n_2)$.
   (c) $s(n_1, n_2) = \delta(n_1, n_2 - 1)$.
   (d) $s(n_1, n_2) = \delta(n_1 - 3n_2)u(-n_1, 2 - n_2)$.

2   Let a given periodic signal satisfy

$$s(n_1, n_2) = s(n_1 + 5, n_2 + 3)$$

$$= s(n_1 - 1, n_2 - 3) \quad \text{for all } n_1, n_2.$$

(a) What is the corresponding periodicity matrix $\mathbf{N}$?

(b) Is s periodic also with vectors

$$\mathbf{n}_1 = \begin{bmatrix} 4 \\ 0 \end{bmatrix} \quad \text{and} \quad \mathbf{n}_2 = \begin{bmatrix} 3 \\ -3 \end{bmatrix}?$$

3 This problem concerns the period of cosine waves in two dimensions, both in continuous space and in discrete space:

(a) What is the period in continuous space of the cosine wave

$$\cos 2\pi \left(\frac{4t_1 + t_2}{16} \right)?$$

(b) What is the period of the cosine wave

$$\cos 2\pi \left(\frac{4n_1 + n_2}{16} \right)?$$

You may find several answers. Which one is best? Why?

4 Convolve the spatial impulse response

$$h(n_1, n_2) = \begin{cases} \frac{1}{4}, & (n_1, n_2) = (0, 0),\ (1, 0),\ (0, 1),\ \text{or}\ (1, 1) \\ 0, & \text{otherwise}, \end{cases}$$

and the input signal $u_{++}(n_1, n_2)$, the first-quadrant unit step function.

5 Which of the following 2-D systems is linear shift-invariant?

(a) $y(n_1, n_2) = 3x(n_1, n_2) - x(n_1 - 1, n_2)$?

(b) $y(n_1, n_2) = 3x(n_1, n_2) - y(n_1 - 1, n_2)$? Any additional information needed?

(c) $y(n_1, n_2) = \displaystyle\sum_{(k_1, k_2) \in \mathcal{W}(n_1, n_2)} x(k_1, k_2)$?

Here $\mathcal{W}(n_1, n_2) \triangleq \{(n_1, n_2),\ (n_1 - 1, n_2),\ (n_1, n_2 - 1),\ (n_1 - 1, n_2 - 1)\}$.

(d) For the same region $\mathcal{W}(n_1, n_2)$, but now

$$y(n_1, n_2) = \sum_{(k_1, k_2) \in \mathcal{W}(n_1, n_2)} x(n_1 - k_1, n_2 - k_2)?$$

What can you say about the stability of each of these systems?

6 Consider the 2-D signal

$$x(n_1, n_2) = 4 + 2\cos\left[\frac{2\pi}{8}(n_1 + n_2) \right] + 2\cos\left[\frac{2\pi}{8}(n_1 - n_2) \right],$$

for all $-\infty < n_1, n_2 < +\infty$. Find the 2-D Fourier transform of x and give a labeled sketch of it.

7   In general, the Fourier transform of a product of 2-D sequences is given by the periodic convolution of their transforms, i.e.,

$$x(n_1, n_2)y(n_1, n_2) \quad \Leftrightarrow \quad X(\omega_1, \omega_2) \circledast Y(\omega_1, \omega_2).$$

However, if the product is a separable product, then the product of 1-D sequences corresponds through Fourier transformation to the product of their Fourier transforms, i.e.,

$$x(n_1)y(n_2) \quad \Leftrightarrow \quad X(\omega_1)Y(\omega_2).$$

Please reconcile these two facts by writing $x(n_1)$ and $y(n_2)$ as 2-D sequences, taking their 2-D Fourier transforms, and showing that the resulting periodic convolution in the 2-D frequency domain reduces to the product of two 1-D Fourier transforms.

8   Take the inverse Fourier transform of (1.2-3) to check that we obtain the $45°$ impulse line $\delta(n_1 - n_2)$.

9   For a general complex signal $x(n_1, n_2)$, show that (1.2-5) is correct.

10   Let the signal $x(n_1, n_2) = 1\delta(n_1, n_2) - 0.5\delta(n_1 - 1, n_2) + 0.5\delta(n_1, n_2 - 1)$ have Fourier transform $X(\omega_1, \omega_2)$. What is the inverse Fourier transform of $|X(\omega_1, \omega_2)|^2$?

11   A certain *ideal* lowpass filter with cutoff frequency $\omega_c = \pi/2$ has impulse response:

$$h(n_1, n_2) = \frac{1}{\sqrt{n_1^2 + n_2^2}} J_1\left(\frac{\pi}{2}\sqrt{n_1^2 + n_2^2}\right).$$

What is the passband gain of this filter? Hint:

$$\lim_{x \to 0} \frac{J_1(x)}{x} = \frac{1}{2}.$$

12   Consider the ideal filter with elliptically shaped frequency response

$$H_e(\omega_1, \omega_2) = \begin{cases} 1, & (\omega_1/\omega_{c1})^2 + (\omega_2/\omega_{c2})^2 \leqslant 1 \\ 0, & \text{else} \end{cases} \quad \text{in } [-\pi, +\pi] \times [-\pi, +\pi]$$

and find the corresponding impulse response $h_e(n_1, n_2)$.

REFERENCES

[1] M. Abramowitz and I. A. Stegun, In *Handbook of Mathematical Functions*, Dover, New York, NY, 1965, pp. 360 and 484.

[2] D. E. Dudgeon and R. M. Mersereau, *Multidimensional Digital Signal Processing*, Prentice-Hall, Englewood Cliffs, NJ, 1983.

[3] F. B. Hildebrand, In *Advanced Calculus for Applications*, Prentice-Hall, Englewood Cliffs, NJ, 1962. pp. 254–255.

[4] J. Lim, *Two-Dimensional Signal and Image Processing*, Prentice-Hall, Englewood Cliffs, NJ, 1990.

[5] A. Oppenheim, R. W. Schafer, and J. R. Buck, *Discrete-Time Signal Processing*, 2nd edn., Prentice-Hall, Englewood Cliffs, NJ, 1999.

[6] W. Rudin, *Real and Complex Analysis*, McGraw-Hill, New York, NY, 1966.

[7] S. Thurnhofer and S. K. Mitra, "A General Framework for Quadratic Volterra Filters for Edge Enhancement," *IEEE Trans. Image Process*, 5, 950–963, June 1996.

SAMPLING IN TWO DIMENSIONS **2**

In two- and higher-dimensional signal processing, sampling and the underlying continuous space play a bigger role than in 1-D signal processing. This is due to the fact that some digital images, and many digital videos, are undersampled. When working with motion in video, half-pixel accuracy is often the minimal accuracy needed. Also, there is a variety of sampling possible in space that does not exist in the case of the 1-D time axis. We start out with the so-called rectangular sampling theorem, and then move on to more general but still regular sampling patterns and their corresponding sampling theorems. An example of a regular nonrectangular grid is the hexagonal array of phosphor dots on the shadow mask cathode ray tube (CRT). In spatiotemporal processing, diamond sampling is used in interlaced video formats, where one dimension is the vertical image axis and the second dimension is time. Two-dimensional continuous-space Fourier transform theory is often applied in the study of lens systems in optical devices, e.g., in cameras and projectors, wherein the optical intensity field at a distance of one focal length from the lens is approximated well by the Fourier transform. This study is called *Fourier optics*, a basic theory used in the design of lenses.

2.1 SAMPLING THEOREM—RECTANGULAR CASE

We assume that the continuous space function $x_c(t_1, t_2)$ is given. It could correspond to a film image or some other continuous spatial data, e.g., a focused image through a lens system. We can think of the axes t_1 and t_2 as being orthogonal, but that is not necessary. We proceed to produce samples via the rectangular sampling pattern:

$$t_1 = n_1 T_1 \quad \text{and} \quad t_2 = n_2 T_2,$$

thereby producing the discrete-space data

$$x(n_1, n_2) \triangleq x_c(t_1, t_2)|_{t_1=n_1 T_1, \, t_2=n_2 T_2}.$$

We call this process *rectangular sampling*.

THEOREM 2.1-1 (*rectangular sampling theorem*)
Let $x(n_1, n_2) \triangleq x_c(t_1, t_2)|_{t_1=n_1 T_1, \, t_2=n_2 T_2}$, with $T_1, T_2 > 0$, a regular sampling on the continuous-space function x_c on the space (space–time axes) t_1 and t_2. Then the Fourier transform of the discrete-space sequence $x(n_1, n_2)$ can be given as

$$X(\omega_1, \omega_2) = \frac{1}{T_1 T_2} \sum_{\text{all } k_1, k_2} X_c\left(\frac{\omega_1 - 2\pi k_1}{T_1}, \, \frac{\omega_2 - 2\pi k_2}{T_2}\right).$$

Proof: We start by writing $x(n_1, n_2)$ in terms of the samples of the inverse Fourier transform of $X_c(\Omega_1, \Omega_2)$:

$$x(n_1, n_2) = \frac{1}{(2\pi)^2} \int_{-\infty}^{+\infty} \int_{-\infty}^{+\infty} X(\Omega_1, \Omega_2) \exp -j(\Omega_1 n_1 T_1 + \Omega_2 n_2 T_2)\, d\Omega_1\, d\Omega_2.$$

Next we let $\omega_1 \triangleq \Omega_1 T_1$ and $\omega_2 \triangleq \Omega_2 T_2$ in the integral to get

$$x(n_1, n_2) = \frac{1}{(2\pi)^2} \int_{-\infty}^{+\infty} \int_{-\infty}^{+\infty} \frac{1}{T_1 T_2} X\left(\frac{\omega_1}{T_1}, \frac{\omega_2}{T_2}\right) \exp +j(\omega_1 n_1 + \omega_2 n_2)\, d\omega_1\, d\omega_2$$

$$= \frac{1}{(2\pi)^2} \sum_{\text{all } k_1, k_2} \int_{SQ(k_1, k_2)} \frac{1}{T_1 T_2} X\left(\frac{\omega_1}{T_1}, \frac{\omega_2}{T_2}\right)$$

$$\times \exp +j(\omega_1 n_1 + \omega_2 n_2)\, d\omega_1\, d\omega_2, \tag{2.1-1}$$

where $SQ(k_1, k_2)$ is a $2\pi \times 2\pi$ square centered at position $2\pi k_1, 2\pi k_2)$, i.e.,

$$SQ(k_1, k_2) \triangleq [-\pi + 2\pi k_1, +\pi + 2\pi k_1] \times [-\pi + 2\pi k_2, +\pi + 2\pi k_2].$$

Then, making the change of variables $\omega_1' \triangleq \omega_1 - 2\pi k_1$ and $\omega_2' \triangleq \omega_2 - 2\pi k_2$ separately and *inside* each of the above integrals, we get

$$x(n_1, n_2) = \frac{1}{(2\pi)^2} \sum_{\text{all } k_1, k_2} \int_{-\pi}^{+\pi} \int_{-\pi}^{+\pi} \frac{1}{T_1 T_2} X\left(\frac{\omega_1' + 2\pi k_1}{T_1}, \frac{\omega_2' + 2\pi k_2}{T_2}\right)$$

$$\times \exp +j(\omega_1' n_1 + \omega_2' n_2)\, d\omega_1'\, d\omega_2'$$

$$= \frac{1}{(2\pi)^2} \int_{-\pi}^{+\pi} \int_{-\pi}^{+\pi} \left[\sum_{\text{all } k_1, k_2} \frac{1}{T_1 T_2} X\left(\frac{\omega_1' + 2\pi k_1}{T_1}, \frac{\omega_2' + 2\pi k_2}{T_2}\right) \right]$$

$$\times \exp +j(\omega_1' n_1 + \omega_2' n_2)\, d\omega_1'\, d\omega_2'$$

$$= \text{IFT}\left[\sum_{\text{all } k_1, k_2} \frac{1}{T_1 T_2} X\left(\frac{\omega_1 + 2\pi k_1}{T_1}, \frac{\omega_2 + 2\pi k_2}{T_2}\right) \right],$$

as was to be shown. $\qquad\square$

Equivalently, we have established the important and basic Fourier sampling relation

$$X(\omega_1, \omega_2) = \frac{1}{T_1 T_2} \sum_{\text{all } k_1, k_2} X_c\left(\frac{\omega_1 - 2\pi k_1}{T_1}, \frac{\omega_2 - 2\pi k_2}{T_2}\right), \tag{2.1-2}$$

which can also be written in terms of analog frequency as

$$X(T_1\Omega_1, T_2\Omega_2) = \frac{1}{T_1 T_2} \sum_{k_1,k_2} X_c\left(\Omega_1 - \frac{2\pi k_1}{T_1}, \Omega_2 - \frac{2\pi k_2}{T_2}\right), \qquad (2.1\text{-}3)$$

showing more clearly where the aliased components in $X(\omega_1, \omega_2)$ come from in the analog frequency domain. The aliased components are each centered on analog frequency locations $(\frac{2\pi k_1}{T_1}, \frac{2\pi k_2}{T_2})$ for all integer grid locations (k_1, k_2). Of course, for X_c lowpass, and for the case where the sampling density is not too low, we would expect that the main contributions to aliasing would come from the eight nearest neighbor bands corresponding to $(k_1, k_2) = (\pm 1, 0), (0, \pm 1),$ and $(\pm 1, \pm 1)$ in (2.1-2) and (2.1-3), which are sketched in Figure 2.1. This rectangular sampling produces a 2-D or spatial aliasing of the continuous-space Fourier transform. As such, we would expect to avoid most of the aliasing, for a nonideal lowpass signal, by choosing the sampling periods T_1, T_2 small enough. We notice that a variety of aliasing can occur. It can be horizontal, vertical, and/or diagonal aliasing.

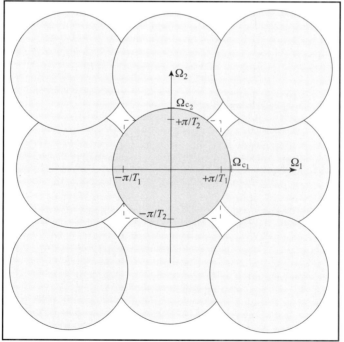

FIGURE 2.1   *Illustrating effect of nearest neighbor aliases for circular symmetric case*

We notice that if the input signal is rectangularly bandlimited in the following sense,

$$X_c(\Omega_1, \Omega_2) = 0 \quad \text{for } |\Omega_1| \geqslant \frac{\pi}{T_1} \text{ or } |\Omega_2| \geqslant \frac{\pi}{T_2},$$

or, equivalently, that $\text{supp}\{X_c(\Omega_1, \Omega_2)\} = (-\pi/T_1, +\pi/T_1) \times (-\pi/T_2, +\pi/T_2)$, then we have no aliasing in the baseband, and

$$X(\omega_1, \omega_2) = \frac{1}{T_1 T_2} X_c\left(\frac{\omega_1}{T_1}, \frac{\omega_2}{T_2}\right) \quad \text{for } |\omega_1| \leqslant \pi \text{ and } |\omega_2| \leqslant \pi.$$

or, equivalently, in terms of analog frequencies.

$$X(T_1 \Omega_1, T_2 \Omega_2) = \frac{1}{T_1 T_2} X_c(\Omega_1, \Omega_2) \quad \text{for } |T_1 \Omega_1| \leqslant \pi \text{ and } |T_2 \Omega_2| \leqslant \pi,$$

so that X_c can be recovered exactly, where it is nonzero, from the FT of its sampled version, by

$$X_c(\Omega_1, \Omega_2) = T_1 T_2 X(T_1 \Omega_1, T_2 \Omega_2) \quad \text{for } |\Omega_1| \leqslant \pi/T_1 \text{ and } |\Omega_2| \leqslant \pi/T_2.$$

More generally, these exact reconstruction results will be true for any signal rectangularly bandlimited to $[-\Omega_{c_1}, +\Omega_{c_1}] \times [-\Omega_{c_2}, +\Omega_{c_2}]$, whenever

$$\Omega_{c_1} \leqslant \pi/T_1 \quad \text{and} \quad \Omega_{c_2} \leqslant \pi/T_2.$$

This is illustrated in Figure 2.2 for a circularly bandlimited signal.

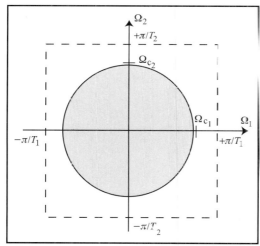

FIGURE 2.2 *Continuous Fourier transform that will not alias when sampled at multiples of* (T_1, T_2)

2.1.1 RECONSTRUCTION FORMULA

At this point we have found the effect of rectangular sampling in the frequency domain. We have seen that no information is lost if the horizontal and vertical sampling rate is high enough for rectangular bandlimited signals. Next we investigate how to reconstruct the original signal from these samples. To obtain x_c we start out by writing the inverse continuous-space IFT:

$$x_c(t_1, t_2) = \frac{1}{(2\pi)^2} \int_{-\infty}^{+\infty} \int_{-\infty}^{+\infty} X_c(\Omega_1, \Omega_2) \exp +j(\Omega_1 t_1 + \Omega_2 t_2) \, d\Omega_1 \, d\Omega_2$$

$$= \frac{1}{(2\pi)^2} \int_{-\infty}^{+\infty} \int_{-\infty}^{+\infty} T_1 T_2 X(T_1 \Omega_1, T_2 \Omega_2) I_{\Omega_{c_1}, \Omega_{c_2}}(\Omega_1, \Omega_2)$$
$$\times \exp +j(\Omega_1 t_1 + \Omega_2 t_2) \, d\Omega_1 \, d\Omega_2,$$

making use of the indicator function

$$I_{\Omega_{c_1}, \Omega_{c_2}}(\Omega_1, \Omega_2) \triangleq \begin{cases} 1, & |\Omega_1| < \Omega_{c_1} \text{ and } |\Omega_2| < \Omega_{c_2} \\ 0, & \text{elsewhere.} \end{cases}$$

Continuing,

$$x_c(t_1, t_2) = \frac{T_1 T_2}{(2\pi)^2} \int_{-\Omega_{c_1}}^{+\Omega_{c_1}} \int_{-\Omega_{c_2}}^{+\Omega_{c_2}} X(T_1 \Omega_1, T_2 \Omega_2) \exp +j(\Omega_1 t_1 + \Omega_2 t_2) \, d\Omega_1 \, d\Omega_2$$

$$= \frac{T_1 T_2}{(2\pi)^2} \int_{-\Omega_{c_1}}^{+\Omega_{c_1}} \int_{-\Omega_{c_2}}^{+\Omega_{c_2}} X(T_1 \Omega_1, T_2 \Omega_2) \exp +j(\Omega_1 t_1 + \Omega_2 t_2) \, d\Omega_1 \, d\Omega_2.$$

Next we substitute X with its FT expression in terms of the samples $x(n_1, n_2)$,

$$X(T_1 \Omega_1, T_2 \Omega_2) = \sum_{n_1=-\infty}^{+\infty} \sum_{n_2=-\infty}^{+\infty} x(n_1, n_2) \exp -j(T_1 \Omega_1 n_1 + T_2 \Omega_2 n_2),$$

to obtain

$$x_c(t_1, t_2) = \frac{T_1 T_2}{(2\pi)^2} \int_{-\Omega_{c_1}}^{+\Omega_{c_1}} \int_{-\Omega_{c_2}}^{+\Omega_{c_2}} \left[\sum_{n_1=-\infty}^{+\infty} \sum_{n_2=-\infty}^{+\infty} x(n_1, n_2) \exp -j(T_1 \Omega_1 n_1 + T_2 \Omega_2 n_2) \right]$$
$$\times \exp +j(\Omega_1 t_1 + \Omega_2 t_2) \, d\Omega_1 \, d\Omega_2,$$

and then interchange the sums and integrals to obtain

$$x_c(t_1, t_2) = T_1 T_2 \sum_{n_1=-\infty}^{+\infty} \sum_{n_2=-\infty}^{+\infty} x(n_1, n_2) \left\{ \frac{1}{(2\pi)^2} \int_{-\Omega_{c_1}}^{+\Omega_{c_1}} \int_{-\Omega_{c_2}}^{+\Omega_{c_2}} \right.$$

$$\left. \times \exp +j\left[\Omega_1(t_1 - n_1 T_1) + \Omega_2(t_2 - n_2 T_2)\right] d\Omega_1 \, d\Omega_2 \right\}$$

$$= T_1 T_2 \sum_{n_1=-\infty}^{+\infty} \sum_{n_2=-\infty}^{-\infty} x(n_1, n_2) h(t_1 - n_1 T_1, t_2 - n_2 T_2), \quad (2.1\text{-}4)$$

where

$$h(t_1, t_2) = \frac{\sin \Omega_{c_1} t_1}{\pi t_1} \frac{\sin \Omega_{c_2} t_2}{\pi t_2}, \quad -\infty < t_1, t_2 < +\infty.$$

The interpolation function h is the continuous-space IFT

$$\frac{1}{(2\pi)^2} \int_{-\Omega_{c_1}}^{+\Omega_{c_1}} \int_{-\Omega_{c_2}}^{+\Omega_{c_2}} \exp +j(\Omega_1 t_1 - \Omega_2 t_2) \, d\Omega_1 \, d\Omega_2,$$

which is the impulse response of the ideal rectangular lowpass filter

$$I_{\Omega_{c_1}, \Omega_{c_2}}(\Omega_1, \Omega_2) \triangleq \begin{cases} 1, & |\Omega_1| \leq \Omega_{c_1} \text{ and } |\Omega_2| \leq \Omega_{c_2} \\ 0, & \text{elsewhere.} \end{cases}$$

Equation (2.1-4) is known as the *reconstruction formula* for the rectangular sampling theorem, and is valid whenever the sampling rate satisfies

$$T_1^{-1} \geq \frac{\Omega_{c_1}}{\pi} \quad \text{and} \quad T_2^{-1} \geq \frac{\Omega_{c_2}}{\pi}.$$

We note that the reconstruction consists of an infinite weighted sum of delayed ideal lowpass filter impulse responses multiplied by a gain term of $T_1 T_2$ centered at each sample location and weighted by the sample value $x(n_1, n_2)$. As in the 1-D case [4], we can write this in terms of filtering as follows. First we define the continuous-space impulse train, sometimes called the *modulated impulse train*,

$$x_\delta(t_1, t_2) \triangleq \sum_{n_1=-\infty}^{+\infty} \sum_{n_2=-\infty}^{+\infty} x(n_1, n_2) \delta(t_1 - n_1 T_1, t_2 - n_2 T_2),$$

and then we put this impulse train function through the filter with impulse response $T_1 T_2 h(t_1, t_2)$.

When the sample rate is minimal to avoid aliasing, then we have the *critical sampling* case, and the interpolation function

$$T_1 T_2 h(t_1, t_2) = T_1 T_2 \frac{\sin \Omega_{c_1} t_1}{\pi t_1} \frac{\sin \Omega_{c_2} t_2}{\pi t_2}$$

becomes

$$= T_1 T_2 \frac{\sin(\pi/T_1)t_1}{\pi t_1} \frac{\sin(\pi/T_2)t_2}{\pi t_2}$$

$$= \frac{\sin(\pi/T_1)t_1}{\frac{\pi}{T_1} t_1} \frac{\sin(\pi/T_2)t_2}{\frac{\pi}{T_2} t_2}$$

$$= \frac{\sin \Omega_{c_1} t_1}{\Omega_{c_1} t_1} \frac{\sin \Omega_{c_2} t_2}{\Omega_{c_2} t_2}.$$

For this critical sampling case the reconstruction formula becomes

$$x_c(t_1, t_2) = \sum_{n_1=-\infty}^{+\infty} \sum_{n_2=-\infty}^{+\infty} x(n_1, n_2) \frac{\sin \frac{\pi}{T_1}(t_1 - n_1 T_1)}{\frac{\pi}{T_1}(t_1 - n_1 T_1)} \frac{\sin \frac{\pi}{T_2}(t_2 - n_2 T_2)}{\frac{\pi}{T_2}(t_2 - n_2 T_2)} \quad (2.1\text{-}5)$$

$$= \sum_{n_1=-\infty}^{+\infty} \sum_{n_2=-\infty}^{+\infty} x(n_1, n_2) \frac{\sin \Omega_{c_1}(t_1 - n_1 T_1)}{\Omega_{c_1}(t_1 - n_1 T_1)} \frac{\sin \Omega_{c_2}(t_2 - n_2 T_2)}{\Omega_{c_2}(t_2 - n_2 T_2)},$$

$$(2.1\text{-}6)$$

since $\pi/T_i = \Omega_{c_i}$ in this case.

For this critical sampling case, the interpolating functions

$$\frac{\sin \Omega_{c_1}(t_1 - n_1 T_1)}{\Omega_{c_1}(t_1 - n_1 T_1)} \frac{\sin \Omega_{c_2}(t_2 - n_2 T_2)}{\Omega_{c_2}(t_2 - n_2 T_2)}$$

have the property that each one is equal to 1 at its sample location $(n_1 T_1, n_2 T_2)$ and equal to 0 at all other sample locations. Thus at a given sample location, only one of the terms in the double infinite sum is nonzero.

If this critical sampling rate is not achieved, then aliasing can occur as shown in Figure 2.1, showing alias contributions from the horizontally and vertically nearest neighbors in (2.1-3).

Note that even if the analog signal is sampled at high enough rates to avoid any aliasing, there is a type of alias error that can occur on reconstruction due to an inadequate or nonideal reconstruction filter. Sometimes in the one-dimensional case, this distortion, which is not due to undersampling, is called *imaging distortion* to distinguish it from true aliasing error [2]. This term may not be the best to use in an image and video processing text, so if we refer to this spectral imaging error in the sequel, we will put "image" in quote marks.

2.1.2 IDEAL RECTANGULAR SAMPLING

In the case where the continuous-space Fourier transform is not bandlimited, one simple expedient is to provide an ideal continuous-space lowpass filter (Figure 2.3) prior to the spatial sampler. This sampler would be *ideal* in the sense that the maximum possible bandwidth of the signal is preserved alias free. The lowpass filter (LPF) would pass the maximum band that can be represented with rectangular sampling with period $T_1 \times T_2$, which is the analog frequency band $[-\pi/T_1, +\pi/T_1] \times [-\pi/T_2, +\pi/T_2]$ with passband gain $= 1$.

EXAMPLE 2.1-1 (*nonisotropic signal spectra*)

Consider a nonisotropic signal with continuous Fourier transform support, as shown in Figure 2.4, that could arise from rotation of a texture that is relatively lowpass in one direction and broadband in the perpendicular direction. If we apply the rectangular sampling theorem to this signal, we get the minimal or Nyquist sampling rates

$$T_1^{-1} = \Omega_{c_1}/\pi \quad \text{and} \quad T_2^{-1} = \Omega_{c_2}/\pi,$$

resulting in the discrete-space Fourier transform support shown in Figure 2.5.

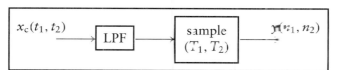

FIGURE 2.3   *System diagram for ideal rectangular sampling to avoid all aliasing error*

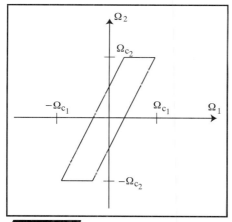

FIGURE 2.4   *Continuous-space Fourier transform with "diagonal" support*

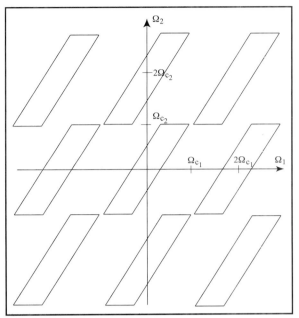

FIGURE 2.5   *Effect of rectangular sampling at rectangular Nyquist rate*

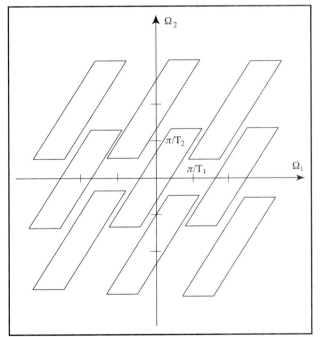

FIGURE 2.6   *After lowering vertical sampling rate below the rectangular Nyquist rate*

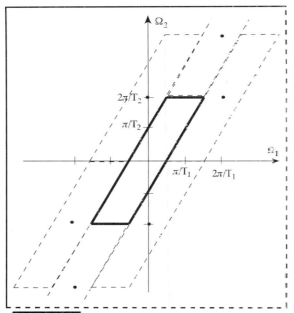

FIGURE 2.7   *Basic cell, indicated by heavy lines, for diagonal analog Fourier transform support*

Clearly there is a lot of wasted spectrum here. If we lower the sampling rates judiciously, we can move to a more efficiently sampled discrete-space Fourier transform with support as shown in Figure 2.6. This figure shows aliased replicas that do not overlap, but yet cannot be reconstructed properly with the ideal rectangular reconstruction formula, (2.1-5). If we reconstruct with an appropriate ideal diagonal support filter though, we can see that it is still possible to reconstruct this analog signal exactly from this lower sample-rate data. Effectively, we are changing the basic cell in the analog frequency domain from $[-\pi/T_1, +\pi/T_1] \times [-\pi/T_2, +\pi/T_2]$ to the diagonal basic cell shown in Figure 2.7. The dashed-line aliased repeats of the diagonal basic cell show the spectral aliasing resulting from the rectangular sampling, illustrating the resulting periodicity in the sampled Fourier transform.

From this example we see that, more so than in one dimension, there is a wider variety of the support or shape of the incoming analog Fourier-domain data, and it is this analog frequency domain support that, together with the sampling rates, determines whether the resulting discrete-space data are aliased or not. In Example 2.1-1, a rectangular Fourier support would lead to aliased discrete-space data for such low spatial sampling rates, but for the indicated diagonal analog Fourier support shown in Figure 2.6, we see that aliasing will not occur if we use this new basic cell. Example 2.1-2 shows one way that such diagonal analog frequency domain support arises naturally.

EXAMPLE 2.1-2 (*propagating plane waves*)
Consider the geophysical spatiotemporal data given as

$$s_c(t, x) = g(t - x/v), \qquad (2.1\text{-}7)$$

where v is a given velocity, $v \neq 0$. We can interpret this as a plane wave propagating in the $+x$ direction when v is positive, with wave crests given by the equation $t - x/v = $ constant. Here g is a given function indicating the wave shape. At position $x = 0$, the signal value is just $g(t) = s_c(t, 0)$. At a general position x, we see this same function delayed by the *propagation time x/v*. Taking the continuous-parameter Fourier transform, we have

$$\mathbf{FT}[s_c] = \iint s_c(t, x) \exp{-j(\Omega t + Kx)}\, dt\, dx,$$

where Ω as usual denotes continuous-time radian frequency, and the continuous variable K denotes continuous-space radian frequency, and is referred to as *wavenumber*. Plugging in the plane-wave equation (2.1-7), we obtain

$$
\begin{aligned}
S_c(\Omega, K) &= \iint g(t - x/v) \exp{-j(\Omega t + Kx)}\, dt\, dx \\
&= \int \left[\int g(t - x/v) \exp{-j\Omega t}\, dt \right] \exp{-jKx}\, dx \\
&= \int G(\Omega) \exp(-j\Omega x/v) \exp{-jKx}\, dx \\
&= G(\Omega) \int \exp{-j(\Omega x/v + Kx)}\, dx \\
&= 2\pi G(\Omega)\delta(K + \Omega/v),
\end{aligned}
$$

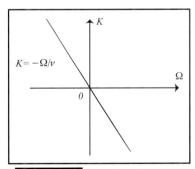

FIGURE 2.8 *FT of ideal plane wave at velocity $v > 0$*

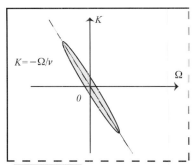

FIGURE 2.9   *Fourier transform illustration of approximate plane wave at velocity ν*

FIGURE 2.10   *2000 × 1000 pixel image with aliasing*

a Dirac delta function in the frequency domain, concentrated along the line $K + \Omega/v = 0$, plotted in Figure 2.8.

If we relax the assumption of an exact plane wave in this example we get some spread out from this ideal impulse line, and hence find a diagonal Fourier transform support as illustrated in Figure 2.9. More on multidimensional geophysical processing is contained in [1].

EXAMPLE 2.1-3 *(alias error in images)*
In image processing, aliasing energy looks like ringing that is perpendicular to high frequency or sharp edges. This can be seen in Figure 2.10, where there was excessive high-frequency information around the palm fronds.

FIGURE 2.11   *Zoomed-in section of aliased image*

We can see the aliasing in the zoomed-in image shown in Figure 2.11, where we see ringing parallel to the fronds, caused by some prior filtering, and aliased energy approximately perpendicular to the edges, appearing as a ringing approximately perpendicular to the fronds. A small amount of aliasing is not really a problem in images.

To appreciate how the aliasing (or "imaging" error) energy can appear nearly perpendicular to a local diagonal component, please refer to Figure 2.12, where we see two alias components coming from above and below in quadrants opposite to those of the main signal energy, giving the alias error signal a distinct high-frequency directional component.

2.2 SAMPLING THEOREM—GENERAL REGULAR CASE

Now consider more general, but still regular, nonorthogonal sampling on the regular grid of locations or *lattice*,

$$\begin{bmatrix} t_1 \\ t_2 \end{bmatrix} = \begin{bmatrix} \mathbf{v}_1 & \mathbf{v}_2 \end{bmatrix} \begin{bmatrix} n_1 \\ n_2 \end{bmatrix},$$

for sampling vectors $\mathbf{v}_1$ and $\mathbf{v}_2$, or what is the same $\mathbf{t} = n_1 \mathbf{v}_1 + n_2 \mathbf{v}_2$ for all integers n_1 and n_2. Thus we have the sampled data

$$x(\mathbf{n}) \triangleq x_c(\mathbf{Vn}), \tag{2.2-1}$$

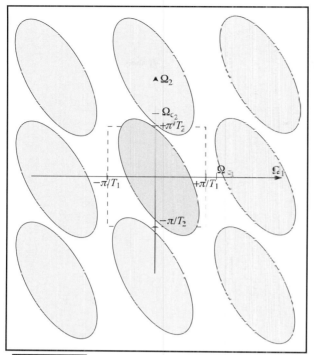

FIGURE 2.12   *Illustration of how alias (or "imaging") error can arise in more directional case*

with *sampling matrix* $\mathbf{V} \triangleq [\mathbf{v}_1 \ \mathbf{v}_2]$. The sampling matrix is assumed always invertible, since otherwise, the sampling locations would not cover the plane, i.e., would not be a lattice. The rectangular or Cartesian sampling we encountered in the preceding section is the special case $\mathbf{v}_1 = (T_1, 0)^T$ and $\mathbf{v}_2 = (0, T_2)^T$.

EXAMPLE 2.2-1 (*hexagonal sampling lattice*)
One sampling matrix of particular importance,

$$\mathbf{V} = \begin{bmatrix} 1 & 1 \\ 1/\sqrt{3} & -1/\sqrt{3} \end{bmatrix},$$

results in a hexagonal sampling pattern. The resulting hexagonal sampling pattern is sketched in Figure 2.13, where we note that the axes n_1 and n_2 are no longer orthogonal to one another. The hexagonal sampling grid shape comes from the fact that the two sampling vectors have angle $\pm30°$ with the horizontal axis, and that they are of equal length. It is easily seen by example, though, that these sample vectors are not unique to produce this grid. One could as well use this $\mathbf{v}_1 = (1, 1/\sqrt{3})^T$ together with $\mathbf{v}_2 = (0, 2/\sqrt{3})^T$ to generate this same lattice. We will see later that the hexagonal sampling grid can

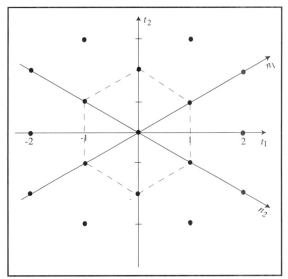

FIGURE 2.13   *Hexagonal sampling grid in space*

be more efficient than the Cartesian grid in many common image processing situations.

To develop a theory for the general sampling case of (2.2-1), we start by writing the continuous-space inverse Fourier transform

$$x_c(\mathbf{t}) = \frac{1}{(2\pi)^2} \int_{-\infty}^{+\infty}\int_{-\infty}^{+\infty} X_c(\boldsymbol{\Omega}) \exp +j(\boldsymbol{\Omega}^T\mathbf{t})\, d\boldsymbol{\Omega},$$

with analog vector frequency $\boldsymbol{\Omega} \triangleq (\Omega_1, \Omega_2)^T$. Then, by definition of the sample locations, we have the discrete data as

$$x(\mathbf{n}) = \frac{1}{(2\pi)^2} \int_{-\infty}^{+\infty}\int_{-\infty}^{+\infty} X_c(\boldsymbol{\Omega}) \exp +j(\boldsymbol{\Omega}^T\mathbf{V}\mathbf{n})\, d\boldsymbol{\Omega},$$

which, upon writing $\boldsymbol{\omega} \triangleq \mathbf{V}^T\boldsymbol{\Omega}$, becomes

$$x(\mathbf{n}) = \frac{1}{(2\pi)^2} \int_{-\infty}^{+\infty}\int_{-\infty}^{+\infty} X_c(\mathbf{V}^{-T}\boldsymbol{\omega}) \exp +j(\boldsymbol{\omega}^T\mathbf{n})\frac{d\boldsymbol{\omega}}{|\det \mathbf{V}|}. \qquad (2.2\text{-}2)$$

Here $\mathbf{V}^{-T}$ denotes the inverse of the transpose sampling matrix $\mathbf{V}^T$, where the notational simplification is permitted because the order of transpose and inverse commute for invertible matrices. Next we break up the integration region in this equation into squares of support $[-\pi, +\pi]^2 \triangleq [-\pi, +\pi] \times [-\pi, +\pi]$ and write

$$x(\mathbf{n}) = \frac{1}{(2\pi)^2} \int_{-\pi}^{+\pi} \int_{-\pi}^{+\pi} \frac{1}{|\det \mathbf{V}|} \left\{ \sum_{\text{all } k} X_c [\mathbf{V}^{-T} (\omega - 2\pi k)] \right\} \cdot \exp +j(\omega^T \mathbf{n}) \, d\omega,$$

just as in (2.1-1) of the previous section, and valid for the same reason. We can now invoke the uniqueness of the inverse Fourier transform for 2-D discrete-space to conclude that the discrete-space Fourier transform $X(\omega)$ must be given as

$$X(\omega) = \frac{1}{|\det \mathbf{V}|} \left\{ \sum_{\text{all } k} X_c [\mathbf{V}^{-T} (\omega - 2\pi k)] \right\},$$

where the discrete-space Fourier transform $X(\omega) = \sum_{\mathbf{n}} x(\mathbf{n}) \exp(-j\omega^T \mathbf{n})$ just as usual.

We now introduce the *periodicity matrix* $\mathbf{U} \triangleq 2\pi \mathbf{V}^{-T}$ and note the fundamental equation

$$\mathbf{U}^T \mathbf{V} = 2\pi \mathbf{I} \tag{2.2-3}$$

where $\mathbf{I}$ is the identity matrix. This equation relates any regular sampling matrix to its corresponding periodicity matrix in the analog frequency domain.

In terms of periodicity matrix $\mathbf{U}$ we can write

$$X(\omega) = \frac{1}{|\det \mathbf{V}|} \left\{ \sum_{\text{all } k} X_c \left[\frac{1}{2\pi} \mathbf{U}(\omega - 2\pi k) \right] \right\}, \tag{2.2-4}$$

which can be written also in terms of the analog frequency variable $\boldsymbol{\Omega}$ as

$$X(\mathbf{V}^T \boldsymbol{\Omega}) = \frac{1}{|\det \mathbf{V}|} \left\{ \sum_{\text{all } k} X_c (\boldsymbol{\Omega} - \mathbf{U}k) \right\}$$

which shows that the alias location points of the discrete-space Fourier transform have the periodicity matrix $\mathbf{U}$ when written in the analog frequency variable $\boldsymbol{\Omega}$.

For the rectangular sampling case with sampling matrix

$$\mathbf{V} = \begin{bmatrix} T_1 & 0 \\ 0 & T_2 \end{bmatrix}$$

the periodicity matrix is

$$\mathbf{U} = \begin{bmatrix} 2\pi/T_1 & 0 \\ 0 & 2\pi/T_2 \end{bmatrix},$$

showing consistency with the results of the previous section, i.e., (2.2-4) simplifies to (2.1-2).

EXAMPLE 2.2-2 (*general hexagonal case*)

Consider the case where the sampling matrix is hexagonal,

$$\mathbf{V} = \begin{bmatrix} T & T \\ T/\sqrt{3} & -T/\sqrt{3} \end{bmatrix},$$

with T being an arbitrary scale coefficient for the sampling vectors. The corresponding periodicity matrix is easily seen to be

$$\mathbf{U} = \begin{bmatrix} \pi/T & \pi/T \\ \pi\sqrt{3}/T & -\pi\sqrt{3}/T \end{bmatrix} = [\mathbf{u}_1 \ \mathbf{u}_2],$$

and $|\det \mathbf{V}| = 2T^2/\sqrt{3}$. The resulting repetition or alias anchor points in analog frequency space are shown in Figure 2.14. Note that the hexagon appears rotated (by 30°) from the hexagonal sampling grid in the spatial dimension in Figure 2.13. At each alias anchor point, a copy of the analog Fourier transform would be seated. Aliasing will not occur if the support of the analog Fourier transform was circular and less than $\pi\sqrt{3}/T$ in radius. We next turn to the reconstruction formula for the case of hexagonal sampling.

2.2.1 HEXAGONAL RECONSTRUCTION FORMULA

First we have to scale down the hexagonal grid with scaling parameter T till there is no aliasing (assuming finite analog frequency domain support). Then we have

$$X(\mathbf{V}^T \boldsymbol{\Omega}) = \frac{1}{|\det \mathbf{V}|} X_c(\boldsymbol{\Omega})$$

$$= \frac{\sqrt{3}}{2T^2} X_c(\boldsymbol{\Omega})$$

and so

$$x_c(t) = \frac{2T^2}{\sqrt{3}} \left(\frac{1}{2\pi} \right)^2 \sum_{\mathbf{n}} x(\mathbf{n}) \int_{\mathcal{B}} \exp\left[j\boldsymbol{\Omega}^T(t - \mathbf{V}\mathbf{n}) \right] d\boldsymbol{\Omega},$$

where $\mathcal{B}$ is the hexagonal frequency domain basic cell shown in Figure 2.15. This basic cell is determined by bounding planes placed to bisect lines joining spectral alias anchor points and the origin. The whole $\boldsymbol{\Omega}$ plane can be covered without gaps or overlaps by repeating this basic cell centered at each anchor point (*tessellated*). It is worthwhile noting that an analog Fourier transform with circular support would fit nicely in such a hexagonal basic cell.

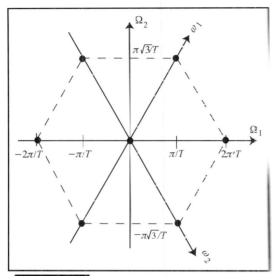

FIGURE 2.14 Hexagonal alias repeat grid in analog frequency domain

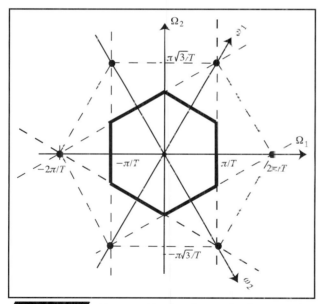

FIGURE 2.15 Hexagonal basic cell in analog frequency domain (heavy line)

Upon defining the spatial impulse response

$$h(\mathbf{t}) \triangleq \frac{2T^2}{\sqrt{3}} \left(\frac{1}{2\pi}\right)^2 \int_{\mathcal{B}} \exp(i\boldsymbol{\Omega}^T \mathbf{t}) \, d\boldsymbol{\Omega},$$

we can write the reconstruction formula for hexagonal sampling as

$$x_c(\mathbf{t}) = \sum_{\mathbf{n}} x(\mathbf{n}) h(\mathbf{t} - \mathbf{Vn}).$$

Of course, we could go on to subsequently evaluate $\mathbf{t}$ on a different sampling lattice, and thereby have a way to convert sampling lattices.

EXAMPLE 2.2-3 (*sampling efficiency for spherical baseband*)
We can generalize the hexagonal lattice to three dimensions, where it is called 3-D rhombic dodecahedron, and continue on to four and higher dimensions. We would notice then that a spherical baseband fits snugly into a generalized hexagon, much better than into a generalized square or cube. Dudgeon and Mersereau [1] obtained the following results showing the sampling efficiency of these generalized hexagonal lattices with respect to the Cartesian lattice:

Dimension M	1	2	3	4
Efficiency	1.000	0.866	0.705	0.505

We see a 30% advantage for the 3-D case (e.g., spatiotemporal or 3-D space) and a 50% advantage in 4-D (e.g., the three spatial dimensions + time, or so-called 3-D video).

In spite of its increased sampling efficiency, the 3-D generalization of the hexagonal lattice has not found much use in applications that sample 3-D data. However, a variation of the diamond sampling lattice has found application in the interlaced sensors, commonly found at the time of this writing in television cameras, both standard definition (SD) and high definition (HD).

EXAMPLE 2.2-4 (*diamond-shaped sampling lattice*)
A special sampling lattice is given by sampling matrix

$$\mathbf{V} = \begin{bmatrix} T & T \\ T & -T \end{bmatrix}, \quad \text{and periodicity matrix} \quad \mathbf{U} = \begin{bmatrix} \pi/T & \pi/T \\ \pi/T & -\pi/T \end{bmatrix},$$

resulting in the so-called diamond-shaped lattice, illustrated in Figure 2.16 for the normalized case $T = 1$. If we look to the periodicity matrix in analog frequency space we obtain Figure 2.17.

If we restrict our 2-D sampling to the vertical and temporal axes of video data, we get the following example of diamond sampling in conventional video.

EXAMPLE 2.2-5 (*interlaced video*)
Consider a video signal $s_c(x, y, t)$, sometimes sampled on a rectangular or Cartesian grid to get the so-called *progressive* digital video signal

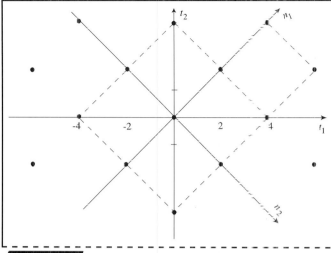

FIGURE 2.16   *Diamond sampling grid with* $T = 1$

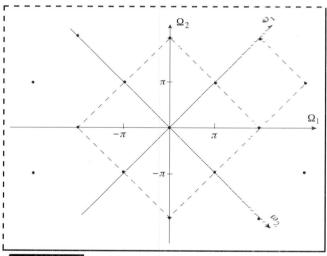

FIGURE 2.17   *Alias anchor points for diamond sampling with* $T = 1$

$s(n_1, n_2, n) \triangleq s_c(n_1 \Delta_1, n_2 \Delta_2, n\Delta)$, where $(\Delta_1, \Delta_2)^T$ is the sample vector in space and Δ is the sampling interval in time. But in 3D and HD video, it is currently more common for it to be sampled on an *interlaced* lattice consisting of two *fields* making up each *frame*. One field just contains the even lines, while the second field gets the odd scan lines. In equations, we have

$$s(n_1, 2n_2, 2n) \triangleq s_c(n_1 \Delta_1, 2n_2 \Delta_2, 2n\Delta)$$

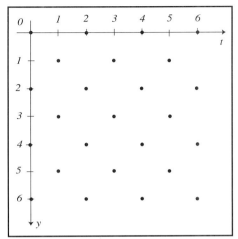

FIGURE 2.18   *Illustration of interlaced sampling in 2-D vertical–temporal domain*

and

$$s(n_1, 2n_2 + 1, 2n + 1) \triangleq s_c\big(n_1 \Delta_1, (2n_2 + 1)\Delta_2, (2n + 1)\Delta\big),$$

whose sampling period is illustrated in Figure 2.18, where we show only the vertical and time axes. We can fix the horizontal variable at, say $n_1 = n_1^0$, and then regard the data as 2-D (vertical–time) data. In the two mentioned cases, this becomes *progressive*,

$$g_c(y, t) \triangleq s_c(x^0, y, t)$$

and

$$g(n_2, n) = g_c(n_2 \Delta_2, n\Delta),$$

or *interlaced*,

$$g(2n_2, 2n) \triangleq g_c(2n_2 \Delta_2, 2n\Delta)$$

and

$$g(2n_2 + 1, 2n + 1) \triangleq g_c\big((2n_2 + 1)\Delta_2, (2n + 1)\Delta\big).$$

The sampling matrix for this 2-D interlaced data is easily seen to be $\mathbf{V} = \begin{bmatrix} \Delta_2 & \Delta_2 \\ \Delta & -\Delta \end{bmatrix}$, corresponding to diamond sampling in the vertical–time domain. Applying the diamond-shaped sample lattice theory, we get a diamond or rotated square baseband to keep alias free. So, if sampled at a high enough spatiotemporal rate, we can reconstruct the continuous space–time signal from the diamond sampled or interlaced digital video data. We would do this, conceptually at least, via 2-D processing in the vertical–temporal domain, for each horizontal value n_1.

In performing the processing in Example 2.2-5, we consider the 3-D spatiotemporal data as a set of 2-D $y \times t$ data planes at each horizontal location $x = x^0$. Thus the 2-D processing would be done on each such $y \times t$ plane separately, and then all the results would be combined. In practice, since there is no interaction between the $y \times t$ planes, this processing is actually performed interleaved in time, so that the data are processed in time locality, i.e., those frames near frame n are processed together.

Interlace has gotten a bad name in video despite its increased spectral efficiency over progressive sampling. This is because of three reasons: The proper space–time filtering to reconstruct the interlaced video is hardly ever done, and in its place is inserted a lowpass vertical filter, sacrificing typically about 30% of the vertical resolution. The second reason is that almost all computer monitors and all the current flat-panel display devices are progressive, thus interlaced images have to be transformed to progressive by the display hardware, often with less than optimal results, in addition to the vertical resolution loss mentioned previously. Lastly, today arguably the best video source is moving picture film (35 mm and higher) and this video is not interlaced. So people have come to expect that the best images are noninterlaced or progressive.

2.3 CHANGE OF SAMPLE RATE

Returning to the more common case of rectangular sampling, it is often necessary in image and video signal processing, to change the sample rate either up or down. This is commonly the case in the resizing of images and videos for the purpose of display on a digital image or video monitor.

2.3.1 DOWNSAMPLING BY INTEGERS $M_1 \times M_2$

We define decimation or downsampling by integers M_1 and M_2 as follows:

$$x_c(n_1, n_2) \triangleq x(M_1 n_1, M_2 n_2),$$

denoted via the system element shown in Figure 2.19. The correspondence to rectangular downsampling in the frequency domain is

$$X_d(\omega_1, \omega_2) = \frac{1}{M_1 M_2} \sum_{i_1=0}^{M_1-1} \sum_{i_2=0}^{M_2-1} X\left(\frac{\omega_1 - 2\pi i_1}{M_1}, \frac{\omega_2 - 2\pi i_2}{M_2}\right). \qquad (2.3\text{-}1)$$

One way of deriving this equation is to first imagine an underlying continuous-space function $x_c(t_1, t_2)$, of which the values $x(n_1, n_2)$ are its samples on the grid $T_1 \times T_2$. Then we can regard the downsampled signal x_d as the result of sampling

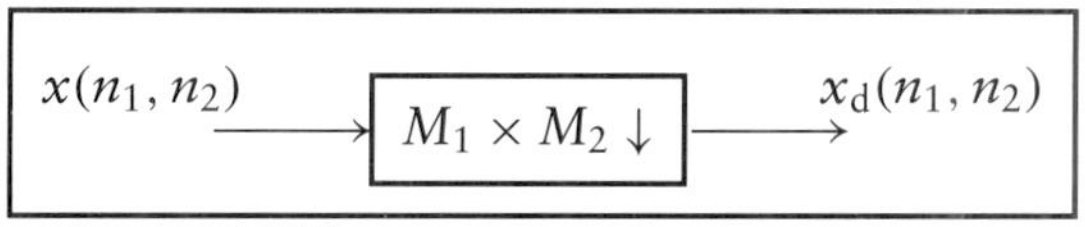

$$x(n_1, n_2) \longrightarrow \boxed{M_1 \times M_2 \downarrow} \longrightarrow x_d(n_1, n_2)$$

FIGURE 2.19 *Downsample system element*

x_c on the less dense sample grid $M_1 T_1 \times M_2 T_2$. From the rectangular sampling theorem, we have

$$X_d(\omega_1, \omega_2) = \frac{1}{M_1 T_1 M_2 T_2} \sum_{k_1,k_2=-\infty}^{+\infty} X_c \left(\frac{\omega_1 - 2\pi k_1}{M_1 T_1}, \frac{\omega_2 - 2\pi k_2}{M_2 T_2} \right). \quad (2.3\text{-}2)$$

Comparing this equation with (2.1-2), we can see that this equation includes the alias components of (2.1-2) when $k_1 = M_1 l_1$ and $k_2 = M_2 l_2$ with l_1 and l_2 integers, plus many others in between. So we can substitute into (2.3-2):

$$k_1 = i_1 + M_1 l_1, \quad \text{where } i_1: \quad 0, \dots, M_1 - 1,$$
$$k_2 = i_2 + M_2 l_2, \quad \text{where } i_2: \quad 0, \dots, M_2 - 1,$$

and then rewrite the equation as

$$X_d(\omega_1, \omega_2)$$

$$= \frac{1}{M_1 T_1 M_2 T_2} \sum_{l_1,l_2=-\infty}^{+\infty} \sum_{i_1=0}^{M_1-1} \sum_{i_2=0}^{M_2-1} X_c \left(\frac{\omega_1 - 2\pi i_1}{M_1 T_1} - \frac{2\pi l_1}{T_1}, \frac{\omega_2 - 2\pi i_2}{M_2 T_2} - \frac{2\pi l_2}{T_2} \right)$$

$$= \frac{1}{M_1 M_2} \sum_{i_1=0}^{M_1-1} \sum_{i_2=0}^{M_2-1} \frac{1}{T_1 T_2} \sum_{l_1,l_2=-\infty}^{+\infty} X_c \left(\frac{\omega_1 - 2\pi i_1}{M_1 T_1} - \frac{2\pi l_1}{T_1}, \frac{\omega_2 - 2\pi i_2}{M_2 T_2} - \frac{2\pi l_2}{T_2} \right)$$

$$= \frac{1}{M_1 M_2} \sum_{i_1=0}^{M_1-1} \sum_{i_2=0}^{M_2-1} X \left(\frac{\omega_1 - 2\pi i_1}{M_1}, \frac{\omega_2 - 2\pi i_2}{M_2} \right), \quad (2.3\text{-}3)$$

in terms of the discrete-space Fourier transform X, by evaluation of (2.1-2) at the locations $(\frac{\omega_1 - 2\pi i_1}{M_1}, \frac{\omega_2 - 2\pi i_2}{M_2})$.

2.3.2 IDEAL DECIMATION

In order to avoid alias error and to preserve as much of the frequency content as possible, we can choose to insert an ideal rectangular lowpass filter ahead of the downsampler, as shown in Figure 2.20. Here the passband of the filter should

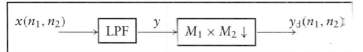

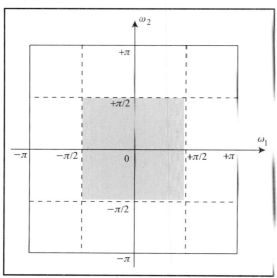

FIGURE 2.20 *System diagram for ideal decimation that avoids aliasing error in the decimated signal*

FIGURE 2.21 *Gray area is the LL subband preserved under 2 × 2 ideal decimation*

have gain 1 and the passband is chosen as $[-\pi/M_1, +\pi/M_1] \times [-\pi/M_1, +\pi/M_1]$ to avoid aliasing in the decimated signal. This process is called *ideal decimation*.

EXAMPLE 2.3-1 *(ideal 2 × 2 decimation)*
Here $M_1 = M_2 = 2$, so we have $x_d(n_1, n_2) \triangleq x(2n_1, 2n_2)$ and in the frequency domain

$$X_d(\omega_1, \omega_2) = \frac{1}{4} \sum_{i_1=0}^{1} \sum_{i_2=0}^{1} X\left(\frac{\omega_1 - 2\pi i_1}{2}, \frac{\omega_2 - 2\pi i_2}{2}\right)$$

$$= \frac{1}{4}\left[X\left(\frac{\omega_1}{2}, \frac{\omega_2}{2}\right) + X\left(\frac{\omega_1}{2}, \frac{\omega_2}{2} - \pi\right) + X\left(\frac{\omega_1}{2} - \pi, \frac{\omega_2}{2}\right)\right.$$

$$\left. + X\left(\frac{\omega_1}{2} - \pi, \frac{\omega_2}{2} - \pi\right)\right]. \tag{2.3-4}$$

We can see that the baseband that can be preserved alias free in the decimated signal x_d is $[-\pi/2, +\pi/2] \times [-\pi/2, +\pi/2]$, which is one-fourth of the full

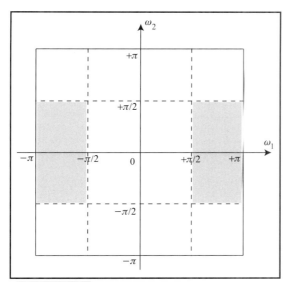

FIGURE 2.22 *Frequency domain support of the HL subband*

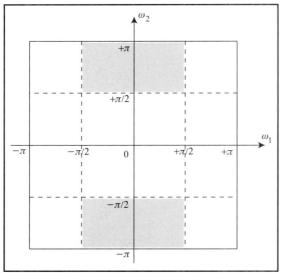

FIGURE 2.23 *Frequency domain support of the LH subband*

bandwidth. This region is shown as the gray area in Figure 2.21, a *subband* of the original full-band signal. Since this subband is lowpass in each frequency variable, it is called the LL *subband*. To actually separate out this subband, we prefilter with the ideal lowpass filter with frequency response support and

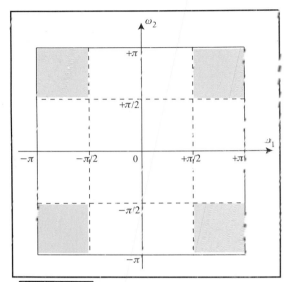

FIGURE 2.24   *Frequency domain support of the HH subband*

gain 1, whose frequency domain support is shown in Figure 2.21 (gray area).

$$H(\omega_1, \omega_2) = \begin{cases} 1, & (\omega_1, \omega_2) \in [-\pi/2, +\pi/2] \times [-\pi/2, +\pi/2] \\ 0, & \text{else on } [-\pi, +\pi]^2. \end{cases}$$

The other parts or subbands of the original signal can be obtained as follows. Consider the gray area shown in Figure 2.22. We can separate out this part, called the HL *subband*, by prefiltering before the decimation with the ideal bandpass filter with passband shown as the gray region in Figure 2.22. Similarly the LH and HH subbands can be separated out, using filters with supports as shown in Figures 2.23 and 2.24, respectively. Together, these four subbands contain all the information from the original full-band signal. We will see in the sequel that decomposition of a full-band signal into its subbands can offer advantages for both image processing and compression. In fact, the international image compression standard JPEG 2000, is based on subband analysis.

2.3.3 UPSAMPLING BY INTEGERS $L_1 \times L_2$

We define the upsampled signal by integers $L_1 \times L_2$ as

$$x_{\mathrm{u}}(n_1, n_2) \triangleq \begin{cases} x\left(\dfrac{n_1}{L_1}, \dfrac{n_2}{L_2}\right), & \text{when } L_1 \text{ divides } n_1 \text{ and } L_2 \text{ divides } n_2 \\ 0, & \text{else.} \end{cases}$$

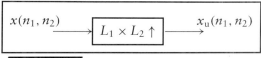

FIGURE 2.25   *Upsample system element*

So, just as in one dimension, we define 2-D upsampling as inserting zeros for the missing samples. The system diagram for this is shown in Figure 2.25. The expression for upsampling in the Fourier domain can be found as

$$X_u(\omega_1, \omega_2) = \sum_{\text{all } n_1, n_2} x_u(n_1, n_2) \exp{-j(\omega_1 n_1 + \omega_2 n_2)}$$

$$= \sum_{n_1 = L_1 k_1, \ n_2 = L_2 k_2} x_u(L_1 k_1, L_2 k_2) \exp{-j(\omega_1 L_1 k_1 + \omega_2 L_2 k_2)}$$

$$= \sum_{\text{all } k_1, k_2} x(k_1, k_2) \exp{-j(\omega_1 L_1 k_1 + \omega_2 L_2 k_2)}$$

$$= X(L_1 \omega_1, L_2 \omega_2).$$

We note that this expression is what we would expect from the 1-D case and that no aliasing occurs due to upsampling, although there are spectral "images" appearing. We see that upsampling effectively shrinks the frequency scale in both dimensions by the corresponding upsampling factors. Since the Fourier transform is $2\pi \times 2\pi$ periodic, this brings in many spectral repeats. Note that there is no overlap of components though, so that these repeats can be removed by a lowpass filter with bandwidth $[-\pi/L_1, +\pi/L_1] \times [-\pi/L_2, +\pi/L_2]$, which leads to ideal interpolation.

2.3.4 IDEAL INTERPOLATION

By ideal interpolation, we mean that corresponding to simply sampling a corresponding continuous-space function at the higher sample rate. Using the rectangular sampling reconstruction formula (2.1-5), we can evaluate at $(t_1, t_2) = (n_1 T_1 / L_1, n_2 T_2 / L_2)$ to obtain

$$x_c\left(\frac{n_1 T_1}{L_1}, \frac{n_2 T_2}{L_2}\right) = \sum_{k_1 = -\infty}^{+\infty} \sum_{k_2 = -\infty}^{+\infty} x(k_1, k_2) \frac{\sin \frac{\pi}{T_1}(\frac{n_1 T_1}{L_1} - k_1 T_1)}{\frac{\pi}{T_1}(\frac{n_1 T_1}{L_1} - k_1 T_1)} \frac{\sin \frac{\pi}{T_2}(\frac{n_2 T_2}{L_2} - k_2 T_2)}{\frac{\pi}{T_2}(\frac{n_2 T_2}{L_2} - k_2 T_2)}$$

$$= \sum_{k_1 = -\infty}^{+\infty} \sum_{k_2 = -\infty}^{+\infty} x(k_1, k_2) \frac{\sin \pi(\frac{n_1}{L_1} - k_1)}{\pi(\frac{n_1}{L_1} - k_1)} \frac{\sin \pi(\frac{n_2}{L_2} - k_2)}{\pi(\frac{n_2}{L_2} - k_2)},$$

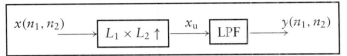

FIGURE 2.26   *System diagram for ideal interpolation by integer factor $L_1 \times L_2$*

which begins to look similar to a convolution. To achieve a convolution exactly, we can introduce the upsampled function x_u and write the ideal interpolation as

$$
y \triangleq x_c \left(\frac{n_1 T_1}{L_1}, \frac{n_2 T_2}{L_2} \right)
$$

$$
= \sum_{k_1=-\infty}^{+\infty} \sum_{k_2=-\infty}^{+\infty} x(k_1, k_2) \frac{\sin \pi (\frac{n_1}{L_1} - k_1)}{\pi (\frac{n_1}{L_1} - k_1)} \frac{\sin \pi (\frac{n_2}{L_2} - k_2)}{\pi (\frac{n_2}{L_2} - k_2)}
$$

$$
= \sum_{k_1', k_2' = \text{multiples of } L_1, L_2}^{+\infty} x \left(\frac{k_1'}{L_1}, \frac{k_2'}{L_2} \right) \frac{\sin \pi (\frac{n_1 - k_1'}{L_1})}{\pi (\frac{n_1 - k_1'}{L_1})} \frac{\sin \pi (\frac{n_2 - k_2'}{L_2})}{\pi (\frac{n_2 - k_2'}{L_2})}
$$

$$
= \sum_{\text{all } k_1', k_2'}^{+\infty} x_u(k_1', k_2') \frac{\sin \pi (\frac{n_1 - k_1'}{L_1})}{\pi (\frac{n_1 - k_1'}{L_1})} \frac{\sin \pi (\frac{n_2 - k_2'}{L_2})}{\pi (\frac{n_2 - k_2'}{L_2})},
$$

which is just the 2-D convolution of the upsampled signal x_u and an ideal impulse response h:

$$
h(n_1, n_2) = \frac{\sin \pi (\frac{n_1}{L_1})}{\pi (\frac{n_1}{L_1})} \frac{\sin \pi (\frac{n_2}{L_2})}{\pi (\frac{n_2}{L_2})},
$$

which corresponds to an ideal rectangular lowpass filter with bandwidth $[-\pi/L_1, +\pi/L_1] \times [-\pi/L_2, +\pi/L_2]$ and passband gain $= L_1 L_2$, with system diagram shown in Figure 2.26.

EXAMPLE 2.3-2 (*oversampling camera*)

Image and video cameras have only the optical system in front of a charge-coupled device (CCD) or complementary metal oxide semiconductor (CMOS) sensor to do the anti-aliasing filtering. As such, there is often alias energy evident in the image frame, especially in the lower sample-density video case. So-called *oversampling* camera chips have been introduced, which first capture the digital image at a high resolution and then do digital filtering combined with downsampling to produce the lower resolution output image. Since aliasing is generally confined to the highest spatial frequencies, the oversampled image sensor can result in a significant reduction in aliasing error.

2.4 SAMPLE-RATE CHANGE—GENERAL CASE

Just as 2-D sampling is not restricted to rectangular or Cartesian schemes, so also decimation and interpolation can be accomplished more generally in two dimensions. The 2-D grid is called a lattice. When we subsample a lattice we create a sublattice, i.e., a lattice contained in the original one. If the original data came from sampling a continuous-space curve, the overall effect is then the same as sampling the continuous-space function with the sublattice. General upsampling can be viewed similarly, but we wish to go to a superlattice, i.e., one for which the given lattice is a sublattice. Here we assume that we already have data, however obtained, and wish to either subsample it or to interpolate it. Applications occur in various areas of image and video processing. One is the conversion of sampling lattices. Some images are acquired from sensors on diamond sampling grids and must be upsampled to a Cartesian lattice for display on common image displays. Another application is to the problem of conversion between interlaced and progressive video by 2-D filtering in the vertical–time domain.

2.4.1 GENERAL DOWNSAMPLING

Let the 2×2 matrix $\mathbf{M}$ be nonsingular and contain only integer values. Then, given a discrete-space signal $x(\mathbf{n})$, where we have indicated position with the column vector $\mathbf{n}$, i.e., $(n_1, n_2)^T \triangleq \mathbf{n}$, a decimated signal $x_\mathrm{d}(\mathbf{n})$, can be obtained as follows:

$$x_\mathrm{d}(\mathbf{n}) \triangleq x(\mathbf{M}\mathbf{n}),$$

where we call $\mathbf{M}$ the *decimation matrix*.

If we decompose $\mathbf{M}$ into its two column vectors as

$$\mathbf{M} = [\mathbf{m}_1 \quad \mathbf{m}_2],$$

then we see that we are subsampling at the locations

$$n_1 \mathbf{m}_1 + n_2 \mathbf{m}_2,$$

and thus we say that these two vectors $\mathbf{m}_1$ and $\mathbf{m}_2$ are the basis vectors of the sublattice *generated* by $\mathbf{M}$.

EXAMPLE 2.4-1 (*diamond sublattice*)
If we choose to subsample with the decimation matrix

$$\mathbf{M} = \begin{bmatrix} 1 & 1 \\ -1 & 1 \end{bmatrix}, \tag{2.4-1}$$

then we get the sublattice consisting of all the points

$$n_1 \begin{bmatrix} 1 \\ -1 \end{bmatrix} + n_2 \begin{bmatrix} 1 \\ 1 \end{bmatrix},$$

a portion of which is shown in Figure 2.27. We see the generated sublattice, i.e., the retained sample locations denoted by large filled-in circles. The left-out points are denoted by small filled-in circles. Thus the original lattice consists of all the filled-in circles. Note that we also indicate the various multiples n_1 and n_2 via the plotted axes, which then become the coordinate axes of the subsampled signal.

The corresponding result in frequency, derived in [5], is

$$X_{\mathrm{d}}(\boldsymbol{\omega}) = \frac{1}{|\det \mathbf{M}|} \left\{ \sum_{\text{certain } \mathbf{k}} X[\mathbf{M}^{-T}(\boldsymbol{\omega} - 2\pi \mathbf{k})] \right\}, \qquad (2.4\text{-}2)$$

with $\boldsymbol{\omega} = (\omega_1, \omega_2)^T$, which is seen to be very close to (2.2-4) for the general sampling case, with the decimation matrix $\mathbf{M}$ substituted for the sampling matrix $\mathbf{V}$ of Section 2.2. Also, we use X in place of the continuous-space FT X_{c} used there. While in (2.2-4) the sum is over all $\mathbf{k}$, here the aliasing sum is only over the additional alias points introduced by the decimation matrix $\mathbf{M}$. For example, looking at Figure 2.27, we see that the small filled-in circles constitute another lattice that has been left behind by the chosen subsampling. There is a shift of $(1, 0)^T$ in this example. The derivation of (2.4-2) is considered in problem 11 at the end of this chapter.

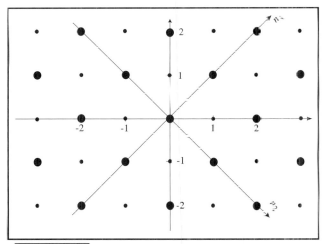

FIGURE 2.27   *Illustration of a portion of sublattice generated by diamond subsampling. Large filled-in circles are sublattice; large and small filled-in circles are original lattice*

EXAMPLE 2.4-2 (*effect of diamond subsampling on frequency*)
Since we have the generator matrix, (2.4-1), from the previous example, we calculate its determinant as $\det \mathbf{M} = 2$ and transposed inverse as

$$\mathbf{M}^{-T} = \frac{1}{2} \begin{bmatrix} 1 & 1 \\ -1 & 1 \end{bmatrix}.$$

Then substituting into (2.4-2), we sum over $\mathbf{k} = (0,0)^T$ and $\mathbf{k} = (1,0)^T$ to obtain

$$
\begin{aligned}
Y(\boldsymbol{\omega}) &= \frac{1}{2} X\left(\frac{\omega_1}{2} \begin{bmatrix} 1 \\ -1 \end{bmatrix} + \frac{\omega_2}{2} \begin{bmatrix} 1 \\ 1 \end{bmatrix} \right) + \frac{1}{2} X\left(\frac{(\omega_1 - 2\pi)}{2} \begin{bmatrix} 1 \\ -1 \end{bmatrix} + \frac{\omega_2}{2} \begin{bmatrix} 1 \\ 1 \end{bmatrix} \right) \\
&= \frac{1}{2} X\left(\frac{\omega_1 + \omega_2}{2}, \frac{-\omega_1 + \omega_2}{2} \right) \\
&\quad + \frac{1}{2} X\left(\frac{(\omega_1 - 2\pi)}{2} + \frac{\omega_2}{2}, \frac{-(\omega_1 - 2\pi)}{2} + \frac{\omega_2}{2} \right), \\
&= \frac{1}{2} X\left(\frac{\omega_1 + \omega_2}{2}, \frac{-\omega_1 + \omega_2}{2} \right) + \frac{1}{2} X\left(\frac{\omega_1 + \omega_2}{2} - \pi, \frac{-\omega_1 + \omega_2}{2} + \pi \right).
\end{aligned}
$$

We can interpret the first term as a diamond-shaped baseband and the second term as the single high-frequency aliased component. This alias signal comes from the fullband signal outside the diamond, with corners at $(0, \pi)$, $(\pi, 0)$, $(-\pi, 0)$, and $(0, -\pi)$.

2.5 CONCLUSIONS

This chapter has focused on how sampling theory extends to two dimensions. We first treated rectangular sampling and looked into various analog frequency supports that permit perfect reconstruction, i.e., an alias-free discrete-space representation. Then we turned to general but regular sampling patterns and focused on the commonly used hexagonal and diamond-shape patterns. Finally we looked at sample-rate change for the common rectangular sampled case, and also briefly looked at the general regular subsampling problem in two dimensions.

2.6 PROBLEMS

1   A certain continuous-space signal s_c is given as

$$s_c(x_1, x_2) = 100 + 20 \cos 6\pi x_1 + 40 \sin(10\pi x_1 + 6\pi x_2),$$

for $-\infty < x_1, x_2 < +\infty$.

(a)   What is the rectangular bandwidth of this signal? Express your answer in radians using cutoff variables $\Omega_{c_1}, \Omega_{c_2}$.

(b)   Sample the function s_c at sample spacing $(\Delta_1, \Delta_2) = (0.05, 0.10)$ to obtain discrete-space signal $s(n_1, n_2) \triangleq s_c(n_1 \Delta_1, n_2 \Delta_2)$. Write an expression for s. Has overlap aliasing error occurred? Why or why not?

(c)   Find and plot the Fourier transform $S \triangleq \mathbf{FT}\{s\}$.

2 A rectangularly bandlimited, continuous-space signal $s_c(x, y)$ is sampled at sample spacing T_1 and T_2, respectively, which is sufficient to avoid spatial aliasing. The resulting discrete-space signal (image) is $s(n_1, n_2)$. It is desired to process this signal to obtain the signal that would have been obtained had the sample spacing been halved, i.e., using $T_1/2$ and $T_2/2$, respectively. How should the signal s be upsampled? Then what should the ideal filter impulse response $h(n_1, n_2)$ be? Please call the low rate signal s_L and the high sample-rate signal s_H.

3 In the reconstruction formula (2.1-4), show directly that the interpolation functions provide the correct result by evaluating (2.1-4) at the sample locations, i.e., at $t_1 = n_1 T_1, t_2 = n_2 T_2$.

4 It is known that the continuous-time 1-D signal $x_c(t) = \exp -\alpha t^2$ has Fourier transform $X_c(\Omega) = \sqrt{\frac{\alpha}{\pi}} \exp -\frac{1}{4\alpha}\Omega^2$, where $\alpha > 0$. Next we sample $x(t)$ at times $t = n$ to obtain a discrete-time signal $x(n) = x_c(n) = \exp -\alpha n^2$.

   (a) Write the Fourier transform $X(\omega)$ of $x(n)$ in terms of the parameter α. This should be an aliased sum involving X_c. Use sum index k.

   (b) Find an upper bound on α so that the $k = \pm 1$ aliased terms in $X(\omega)$ are no larger than $10^{-3}\sqrt{\frac{\alpha}{\pi}}$?

   (c) Consider the 2-D signal $x(n_1, n_2) = \exp -\alpha(n_1^2 + n_2^2)$, and repeat part a, finding now $X(\omega_1, \omega_2)$. Use aliased sum index (k_1, k_2).

   (d) Find an upper bound on α so that the $(k_1, k_2) = (\pm 1, 0)$ and $(0, \pm 1)$ aliased terms in $X(\omega_1, \omega_2)$ are no larger than $10^{-3}\frac{\alpha}{\pi}$.

5 Consider the 2-D signal with continuous-space Fourier transform support, as shown in the gray area of the following diagram:

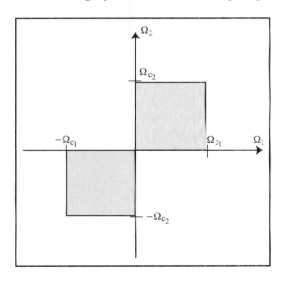

(a) What is the minimum permissible rectangular sampling pattern, based on an assumption of $[-\Omega_{c_1}, \Omega_{c_1}] \times [-\Omega_{c_2}, \Omega_{c_2}]$ Fourier transform support?

(b) What is the minimum permissible sampling pattern for the Fourier transform support shown in the diagram?

In each case, answer by specifying the sample matrix $\mathbf{V}$.

6   Use Figure 2.16 to find the diamond-shaped unit cell in the frequency domain for sampling matrix $\mathbf{V} = \begin{bmatrix} 1 & 1 \\ 1 & -1 \end{bmatrix}$. Hint: Draw straight lines connecting the alias anchor points. What does the midpoint of each line tell us about the boundary of the unit cell?

7   Find the impulse response corresponding to a continuous-space ideal diamond-shaped lowpass filter, with passband equal to the frequency-domain unit cell found in problem 6. This is the ideal anti-alias filter that should be used prior to such sampling. Assume $T_1 = T_2 = 1$.

8   Show that $X_d(\omega_1, \omega_2)$ given in (2.3-4) is rectangularly periodic with period $2\pi \times 2\pi$, as it must be to be correct.

9   Consider the 2-D signal from problem 6 of Chapter 1,

$$x(n_1, n_2) = 4 + 2\cos\left[\frac{2\pi}{8}(n_1 + n_2)\right] + 2\cos\left[\frac{2\pi}{8}(n_1 - n_2)\right],$$

for all $-\infty < n_1, n_2 < +\infty$.

(a) Let x now be input to the 2×2 decimator,

$$x(n_1, n_2) \longrightarrow \boxed{2 \times 2 \downarrow} \longrightarrow x_d(n_1, n_2),$$

and give a simple expression for the output $x_d(n_1, n_2)$ and its Fourier transform $X_d(\omega_1, \omega_2)$. Hint: Consider each of the terms in x separately, starting with the constant term 4.

(b) Check and verify that

$$X_d(\omega_1, \omega_2) = \frac{1}{4} \sum_{l_1=0}^{1} \sum_{l_2=0}^{1} X\left(\frac{\omega_1 - 2\pi l_1}{2}, \frac{\omega_2 - 2\pi l_2}{2}\right).$$

10   Consider a full-band signal with circular symmetric frequency domain contour plot as sketched in Figure 2.28. Use a filter with frequency domain support as shown in Figure 2.22 to generate the so-called HL subband. Usually this signal is then decimated 2×2 for efficiency, generating the subband signal $x_{HL}(n_1, n_2)$. Sketch the frequency domain contour plot of this signal. (Note that there has been an inversion of higher and lower horizontal frequencies.)

11   This problem relates to the formula for frequency domain effect of general
     subsampling (2.4-2) and its derivation. Things are made much simpler by
     using the alternative definition of the discrete-space Fourier transform [3]

$$X'(\omega) \triangleq \sum_n x(\mathbf{n}) \exp(-j\omega^T \mathbf{Vn})$$

$$= X(\mathbf{V}^T\omega)$$

With this definition, the warping evident in (2.2-4) does not occur, and we
have the simpler expression for aliasing due to general sampling:

$$X'(\omega) = \frac{1}{|\det \mathbf{V}|}\left\{ \sum_{\text{all } k} X_c(\omega - \mathbf{U}k)\right\}.$$

Note that in the rectangular case, there is no distinction between these two,
except for the familiar scaling of the analog frequency axes.

(a)   Using the alternative FT, find the frequency domain equation corre-
      sponding to direct sampling with sampling matrix $\mathbf{MV}$.

(b)   Note that this same equation must correspond to decimating by $\mathbf{M}$
      after first sampling with $\mathbf{V}$.

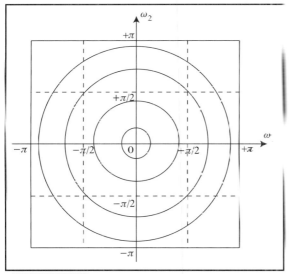

FIGURE 2.28   *Contour plot sketch of a full-band signal*

(c) Use your result in part b to justify (2.4-2) as expressed in terms of the alternative FT

$$X'_{\mathrm{d}}(\boldsymbol{\omega}) = \frac{1}{|\det \mathbf{M}|} \left\{ \sum_{\mathrm{certain\ k}} X(\boldsymbol{\omega} - 2\pi \mathbf{M}^{-T} \mathbf{k}) \right\}.$$

(d) Now, can you justify our using $\mathbf{k} = (0,0)^T$ and $\mathbf{k} = (1,0)^T$ in Example 2.4-2?

References

[1] D. E. Dudgeon and R. M. Mersereau, *Multidimensional Digital Signal Processing*, Prentice-Hall, Englewood Cliffs, NJ, 1983.

[2] A. Gersho and R. M. Gray, *Vector Quantization and Signal Compression*, Kluwer Academic Publ., Norwell, MA, 1992.

[3] R. M. Mersereau and T. C. Speake, "The Processing of Periodically Sampled Multidimensional Signals," *IEEE Trans. Accoust., Speech, Signal Process.*, **ASSP-31**, 188–194, Feb. 1983.

[4] A. Oppenheim, R. W. Schafer, and J. R. Buck, *Discrete-Time Signal Processing*, 2nd edn., Prentice-Hall, Englewood Cliffs, NJ, 1999.

[5] P. P. Vaidyanathan, *Multirate Systems and Filter Banks*, Prentice-Hall, Englewood Cliffs, NJ, 1993 (see Section 12.4).

Two-Dimensional Systems and Z-Transforms

3

In this chapter we look at the 2-D Z-transform. It is a generalization of the 1-D Z-transform used in the analysis and synthesis of systems based on 1-D linear constant-coefficient difference equations. In two and higher dimensions, the corresponding linear systems are partial difference equations. The analogous continuous-parameter systems are partial differential equations. In fact, one big application of partial difference equations is in the numerical or computer solution of the partial differential equations of physics. We also look at LSI system stability in terms of its Z-transform system function and present several stability conditions in terms of the zero-root locations of the system function.

3.1 LINEAR SPATIAL OR 2-D SYSTEMS

The spatial or 2-D systems we will mainly be concerned with are governed by difference equations in the two variables n_1 and n_2. These equations can be realized by logical interconnection of multipliers, adders, and shift or delay elements via either software or hardware. Mostly the coefficients of such equations will be constant, hence the name *linear constant-coefficient difference equations* (LCCDEs). The study of 2-D or partial difference equations is much more involved than that of the corresponding 1-D LCCDEs, and much less is known about the general case. Nevertheless, many practical results have emerged, the most basic of which will be presented here. We start with the general input/output equation

$$\sum_{(k_1,k_2)\in\mathcal{R}_a} a_{k_1,k_2} y(n_1 - k_1, n_2 - k_2) = \sum_{(k_1,k_2)\in\mathcal{R}_b} b_{k_1,k_2} x(n_1 - k_1, n_2 - k_2), \quad (3.1\text{-}1)$$

where x is the known input and y is the output to be determined. We consider the coefficients a_{k_1,k_2} and b_{k_1,k_2} to be arrays of real numbers, and call b_{k_1,k_2} the *feedforward* coefficients and a_{k_1,k_2} the *feedback* coefficients. We wish to solve (3.1-1) by finding output value y for every point in a prescribed region $\mathcal{R}_y$ given needed input values x plus output values y on the boundary of $\mathcal{R}_y$. We denote this boundary region somewhat imprecisely as $\mathcal{R}_{bc}$. The highest values of k_1 and k_2 on the left-hand side of (3.1-1) determine the order of the difference equation. In general, such equations have to be solved via matrix or iterative methods, but our main interest is 2-D *filters* for which the output y can be calculated in a recursive manner from the input x by scanning through the data points (n_1, n_2).

Keeping only the output value $y(n_1, n_2)$ on the left-hand side of (3.1-1), assuming $a_{0,0} \neq 0$, and that $(0,0) \in \mathcal{R}_a$, and moving the other y terms to the right-hand side, we can write

$$y(n_1, n_2) = -\sum_{(k_1,k_2)\in\mathcal{R}_a} a'_{k_1,k_2} y(n_1-k_1, n_2-k_2) + \sum_{(k_1,k_2)\in\mathcal{R}_b} b'_{k_1,k_2} x(n_1-k_1, n_2-k_2),$$

$$(3.1\text{-}2)$$

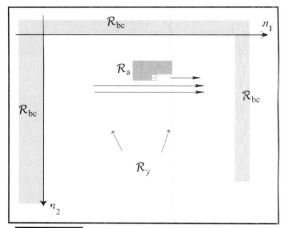

FIGURE 3.1 *Example of a spatial difference equation solution region using a nonsymmetric half-plane $\mathcal{R}_a$ coefficient support*

where the a'_{k_1,k_2} and b'_{k_1,k_2} are the normalized coefficients, i.e., those divided by $a_{0,0}$. Then, depending on the shape of the region $\mathcal{R}_a$, called the output mask, we may be able to calculate the solution recursively. For example, we would say that the *direction of recursion* of (3.1-2) is "downward and to the right" in Figure 3.1, which shows a scan proceeding left-to-right and top-to-bottom. Note that the special shape of the output mask $\mathcal{R}_a$ in Figure 3.1 permits such a recursion because of its property of not including any outputs that have not already been scanned and processed in the *past*, i.e., "above and to the left."

Example 3.1-1 shows how such a recursion proceeds in the case of a simple first-order 2-D difference equation.

EXAMPLE 3.1-1 (*simple difference equation*)
We now consider the simple LCCDE

$$y(n_1, n_2) = x(n_1, n_2) + \tfrac{1}{2}\big[y(n_1 - 1, n_2) + y(n_1, n_2 - 1)\big]. \qquad (3.1-3)$$

to be solved over the first quadrant, i.e., $\mathcal{R}_y = \{n_1 \geqslant 0 \ n_2 \geqslant 0\}$. In this example we assume that the input x is everywhere zero, but that the boundary conditions given on $\mathcal{R}_{bc} = \{n_1 = -1\} \cup \{n_2 = -1\}$ are nonzero and specified by

$$y(-1, 1) = y(-1, 2) = y(-1, 3) = 1$$

$$y(-1, \text{else}) = 0,$$

$$y(\text{else}, -1) = 0.$$

---

1. The vertical axis is directed downward, as is common in image processing, where typically the processing proceeds from top to bottom of the image

To calculate the solution recursively, we first determine a *scanning order*; in this case it is the so-called *raster scan* used in video monitors. First we process the row $n_2 = 0$, starting at $n_1 = 0$ and incrementing by one each time; then we increment n_2 by one, and process the next row. With this scanning order, the difference equation (3.1-3), is seen to use only previous values of y at the "present time," and so is recursively calculable. Proceeding to work out the solution, we obtain

$(n_1 = 0)$

		0	0	1	1	1	0	0	...
$(n_2 = 0)$		0	0	$\frac{1}{2}$	$\frac{3}{4}$	$\frac{7}{8}$	$\frac{7}{16}$	$\frac{7}{32}$	...
$n_2\downarrow$		0	0	$\frac{1}{4}$	$\frac{1}{2}$	$\frac{11}{16}$	$\frac{18}{32}$	$\frac{25}{64}$	... $\to n_1$.
		0	0	$\frac{1}{8}$	$\frac{5}{16}$	$\frac{16}{32}$	$\frac{34}{64}$	...	...
		0	0	$\frac{1}{16}$	...				

In this example we computed the solution to a spatial difference equation by recursively calculating out the values in a suitable scanning order, for a nonzero set of boundary "initial" conditions, but with zero input sequence. In Example 3.1-2 we consider the same 2-D difference equation to be solved over the same output region, but with zero initial boundary conditions and nonzero input. By linearity of the partial difference equation, the general case of nonzero boundaries and nonzero input follows by superposition of these two *zero-input* and *zero-state* solutions.

EXAMPLE 3.1-2 (*simple difference equation cont'd*)
We consider the simple LCCDE

$$y(n_1, n_2) = x(n_1, n_2) + \tfrac{1}{2}\big[y(n_1 - 1, n_2) + y(n_1, n_2 - 1)\big], \qquad (3.1\text{-}4)$$

to be solved over output solution region $\mathcal{R}_y = \{n_1 \geqslant 0, n_2 \geqslant 0\}$. The boundary conditions given on $\mathcal{R}_{bc} = \{n_1 = -1\} \cup \{n_2 = -1\}$ are taken as all zero. The input sequence is $x(n_1, n_2) = \delta(n_1, n_2)$. Starting at $(n_1, n_2) = (0, 0)$ we begin to generate the impulse response of the difference equation. Continuing the recursive calculation for the next few columns and rows we obtain

		0	0	0	0	0	0	0	...
		0	1	$\frac{1}{2}$	$\frac{1}{4}$	$\frac{1}{8}$	$\frac{1}{16}$	...	...
$n_2\downarrow$		0	$\frac{1}{2}$	$\frac{1}{2}$	$\frac{3}{8}$	$\frac{4}{16}$	$\frac{5}{32}$	...	... $\to n_1$.
		0	$\frac{1}{4}$	$\frac{3}{8}$	$\frac{6}{16}$	$\frac{10}{32}$	...	...	...
		0	$\frac{1}{8}$	$\frac{4}{16}$	$\frac{10}{32}$	...	...	...	...

It turns out that this spatial impulse response has a closed-form analytic solution [2,8],

$$y(n_1, n_2) = h(n_1, n_2) = \binom{n_1 + n_2}{n_1} 2^{-(n_1+n_2)} u_{++}(n_1, n_2).$$

where $\binom{n_1+n_2}{n_1}$ is the combinatorial symbol for "$n_1 + n_2$ things taken n_1 at a time,"

$$\binom{n_1 + n_2}{n_1} = \frac{(n_1 + n_2)!}{n_1! n_2!}, \quad \text{for } n_1 \geqslant 0, n_2 \geqslant 0$$

with 0! taken as 1, and where $u_{++}(n_1, n_2) = u(n_1, n_2)$ is the first-quadrant unit step function.

Though it is usually the case that 2-D difference equations do not have a closed-form impulse response, the first-order difference equation of this example is one of the few exceptions. From Examples 3.1-1 and 3.1-2, we can see it is possible to write the general solution to a spatial linear difference equation as a sum of a zero-input solution, given rise by the boundary values, plus a zero-state solution driven by the input sequence:

$$y(n_1, n_2) = y_{ZI}(n_1, n_2) + y_{ZS}(n_1, n_2).$$

This generalizes the familiar 1-D systems theory result. To see this, consider a third example with both nonzero input and nonzero boundary conditions. Then note that the sum of the two solutions from Examples 3.1-1 and 3.1-2 will solve this new problem.

In general, and depending on the output coefficient mask $\mathcal{R}_a$, there can be different recursive directions for (3.1-1), which we can obtain by bringing other terms to the left-hand side and recursing in other directions. For example, we can take (3.1-4) from the last example and bring $y(n_1, n_2 - 1)$ to the left-hand side to yield

$$y(n_1, n_2 - 1) = +2y(n_1, n_2) - 2x(n_1, n_2) - y(n_1 - 1, n_2),$$

or, equivalently,

$$y(n_1, n_2) = +2y(n_1, n_2 + 1) - y(n_1 - 1, n_2 + 1) - 2x(n_1, n_2 + 1),$$

with direction of recursion upward, to the right or left. So the direction in which a 2-D difference equation can be solved recursively, or recursed, depends on the support of the output or feedback coefficients, i.e., $\mathcal{R}_a$. For a given direction of recursion, we can calculate the output points in particular orders that are constrained by the shape of the coefficient output mask $\mathcal{R}_a$, resulting in an *order of computation*. In fact there are usually several such orders of computation that are consistent with a given direction of recursion. Further, usually several output points can be calculated in parallel to speed the recursion.

Such recursive solutions are appropriate when the boundary conditions are only imposed "in the past" of the recursion, i.e., not on any points that must be calculated. In particular, with reference to Figure 3.1, we see no boundary conditions on the bottom of the solution region. In the more general case where there are both "initial" and "final" conditions, we can fall back on the general matrix solution for a finite region.

To solve LCCDE (3.1-1) in a finite solution region, we can use linear algebra and form a vector of the solution $\mathbf{y}$ scanned across the region in any prespecified manner. Doing the same for the input $\mathbf{x}$ and the boundary conditions $\mathbf{y}_{bc}$, we can write all the equations with one LARGE vector equation

$$\mathbf{x} = \mathbf{Ay} + \mathbf{By}_{bc},$$

for appropriately defined coefficient matrices $\mathbf{A}$ and $\mathbf{B}$. For a 1000×1000 image, the dimension of $\mathbf{y}$ would be 1,000,000. Here $\mathbf{Ay}$ provides the terms of the equations, where $\mathbf{y}$ is inside the region and $\mathbf{By}_{bc}$ provides the terms when $\mathbf{y}$ is on the boundary. A problem at the end of the chapter asks you to prove this fact.

If the solution region of the LCCDE is infinite, then, as in the 1-D case, it is often useful to express the solution in terms of a Z-transform, which is our next topic.

3.2 Z-TRANSFORMS

DEFINITION 3.2-1 (*Z-transform*)
The 2-D Z-transform of a two-sided sequence $x(n_1, n_2)$ is defined as follows:

$$X(z_1, z_2) \triangleq \sum_{n_1=-\infty}^{+\infty} \sum_{n_2=-\infty}^{+\infty} x(n_1, n_2) z_1^{-n_1} z_2^{-n_2}, \qquad (3.2\text{-}1)$$

where $(z_1, z_2) \in \mathcal{C}^2$, the "two-dimensional" (really four-dimensional) complex Cartesian product space. In general, there will be only some values of $(z_1, z_2)^T \triangleq \mathbf{z}$, for which this double sum will converge. Only *absolute convergence*,

$$\sum_{n_1=-\infty}^{+\infty} \sum_{n_2=-\infty}^{+\infty} \left| x(n_1, n_2) z_1^{-n_1} z_2^{-n_2} \right|$$

$$= \sum_{n_1=-\infty}^{+\infty} \sum_{n_2=-\infty}^{+\infty} \left| x(n_1, n_2) \right| |z_1|^{-n_1} |z_2|^{-n_2} < \infty,$$

is considered in the theory of complex variables [6,7], so we look for joint values of $|z_1|$ and $|z_2|$ that will yield absolute convergence. The set of z for which this occurs is called the *region of convergence*, denoted $\mathcal{R}_x$. In summary, a 2-D Z-transform is specified by its functional form $X(z_1, z_2)$ and its convergence region $\mathcal{R}_x$.

Similar to the 1-D case, the Z-transform is simply related to the Fourier transform, when both exist:

$$X(z_1, z_2)\big|_{\substack{z_1=e^{j\omega_1} \\ z_2=e^{j\omega_2}}} = X(\omega_1, \omega_2),$$

with the customary abuse of notation.[2] A key difference from the 1-D case is that the 2-D complex variable z exists in a four-dimensional space, and is hard to visualize. The familiar unit circle becomes something a bit more abstract, the *unit bi-circle* in $\mathcal{C}^2$ [6]. The unit disk then translates over to the *unit bi-disk*, $\{|z_1|^2 + |z_2|^2 \leqslant 1\} \in \mathcal{C}^2$. Another key difference for two and higher dimensions is that the zeros of the Z-transform are no longer isolated. Two different *zero loci* can intersect.

EXAMPLE 3.2-1 (*zero loci*)
Consider the following signal $x(n_1, n_2)$,

$n_2 \backslash n_1$	0	1
0	1	2
1	2	1

with assumed support $\{0, 1\} \times \{0, 1\}$. This simple 4-point signal could serve, after normalization, as the impulse response of a simple directional spatial averager, giving an emphasis to structures at $45°$. Proceeding to take the Z-transform, we obtain

$$X(z_1, z_2) = 1 + 2z_1^{-1} + 2z_2^{-1} + z_1^{-1}z_2^{-1}.$$

This Z-transform X is seen to exist for all $\mathcal{C}^2$ except for $z_1 = 0$ or $z_2 = 0$. Factoring X, we obtain

$$X(z_1, z_2) = 1 + 2z_2^{-1} + z_1^{-1}(2 + z_2^{-1}),$$

---

2. To avoid confusion, when the same symbol X is being used for two different functions, we note that the Fourier transform $X(\omega_1, \omega_2)$ is a function of real variables, while the Z-transform $X(z_1, z_2)$ is a function of complex variables. A pitfall, for example $X(1, 0)$, can be avoided by simply writing either $X(\omega_1, \omega_2)\big|_{\substack{\omega_1=1 \\ \omega_2=0}}$; or $X(z_1, z_2)\big|_{\substack{z_1=1 \\ z_2=0}}$, whichever is appropriate.

which, upon equating to zero, gives the zero (z_1, z_2) locus,

$$z_1 = -\frac{2z_2 + 1}{z_2 + 2} \quad \text{for } z_2 \neq -2.$$

We notice that for each value of z_2 there is a corresponding value of z_1 for which the Z-transform X takes on the value of zero. Notice also that, with the possible exception of $z_2 = -2$, the zero locus value $z_1 = f(z_2)$ is a *continuous function* of the complex variable z_2. This first-order 2-D system thus has one zero locus.

We next look at a more complicated second-order case where there are two root loci and they intersect, but without being identical, so we cannot just cancel the factors out. In the 1-D case that we are familiar with, the only way there can be a pole and zero at the same z location is when the numerator and denominator have a common factor. Example 3.2-2 shows that this is not true in general for higher dimensions.

EXAMPLE 3.2-2 (*intersecting zero loci*)
Consider the Z-transform

$$X(z_1, z_2) = (1 + z_1)/(1 + z_1 z_2),$$

for which the zero locus is easily seen to be $(z_1, z_2) = (-1, *)$ and the pole locus is $(z_1, z_2) = (\alpha, -1/\alpha)$, where $*$ represents an arbitrary complex number and α is any nonzero complex number. These two distinct zero sets are seen to intersect at $(z_1, z_2) = (-1, 1)$. One way to visualize these root loci is *root mapping*, which we will introduce later on when we study stability of 2-D filters (see Section 3.5).

Next we turn to the topic of convergence for the 2-D Z-transform. As in the 1-D case, we expect that knowledge of the region in z space where the series converges will be essential to the uniqueness of the transform, and hence to its inversion.

3.3 REGIONS OF CONVERGENCE

Given a 2-D Z-transform $X(z_1, z_2)$, its region of convergence is given as the set of z for which

$$= \sum_{n_1=-\infty}^{+\infty} \sum_{n_2=-\infty}^{+\infty} |x(n_1, n_2)| |z_1|^{-n_1} |z_2|^{-n_2}$$

$$= \sum_{n_1=-\infty}^{+\infty} \sum_{n_2=-\infty}^{+\infty} |x(n_1, n_2)| r_1^{-n_1} r_2^{-n_2} < \infty,$$

where $r_1 \triangleq |z_1|$ and $r_2 \triangleq |z_2|$ are the moduli of the complex numbers z_1 and z_2. The region of convergence (ROC) can then be written in terms of such moduli values as

$$\mathcal{R}_x \triangleq \{(z_1, z_2) \mid |z_1| = r_1 \text{ and } |z_1| = r_2\}.$$

Since this specification depends only on magnitudes, we can plot ROCs in the convenient *magnitude plane*, shown in Figure 3.2.

EXAMPLE 3.3-1 (*Z-transform calculation*)
We consider the spatial, first-quadrant step function

$$x(n_1, n_2) = u_{-+}(n_1, n_2) = u(n_1, n_2).$$

Taking the Z-transform, we have from (3.2-1) that

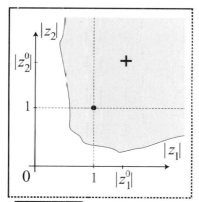

FIGURE 3.2   *The 2-D complex magnitude plane. Here · denotes the unit bi-circle and + denotes an arbitrary point at (z_1^0, z_2^0)*

$$X(z_1, z_2) = \sum_{n_1=-\infty}^{+\infty} \sum_{n_2=-\infty}^{+\infty} u(n_1, n_2) z_1^{-n_1} z_2^{-n_2}$$

$$= \sum_{n_1=0}^{+\infty} \sum_{n_2=0}^{+\infty} z_1^{-n_1} z_2^{-n_2}$$

$$= \sum_{n_1=0}^{+\infty} z_1^{-n_1} \cdot \sum_{n_2=0}^{+\infty} z_2^{-n_2}$$

$$= \frac{1}{1 - z_1^{-1}} \frac{1}{1 - z_2^{-1}} \quad \text{for } |z_1| > 1 \text{ and } |z_2| > 1,$$

$$= \frac{z_1}{z_1 - 1} \frac{z_2}{z_2 - 1}, \quad \text{with } \mathcal{R}_x = \{|z_1| > 1, |z_2| > 1\}.$$

We can plot this ROC on the complex z-magnitude plane as in Figure 3.3. Note that we have shown the ROC as the gray region and moved it slightly outside the lines $|z_1| = 1$ and $|z_2| = 1$, in order to emphasize that this open region does not include these lines. The zero loci for this separable signal are the manifold $z_1 = 0$ and the manifold $z_2 = 0$. These two distinct loci intersect at the complex point $z_1 = z_2 = 0$. The pole loci are also two in number and occur at the manifold $z_1 = 1$ and the manifold $z_2 = 1$. We note that these two pole loci intersect at the single complex point $z_1 = z_2 = 1$.

Next we consider how the Z-transform changes when the unit step switches to another quadrant.

EXAMPLE 3.3-2 (*unit step function in fourth quadrant*)
Here we consider a unit step function that has support on the fourth quad-

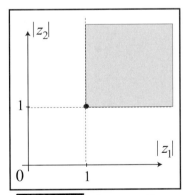

FIGURE 3.3   *Gray area illustrates the ROC for the Z-transform of the first-quadrant unit step function $u(n_1, n_2) = u_{++}(n_1, n_2)$*

rant. We denote it as $u_{+-}(n_1, n_2)$,

$$u_{+-}(n_1, n_2) \triangleq \begin{cases} 1, & n_1 \geqslant 0, \quad n_2 \leqslant 0 \\ 0, & \text{else.} \end{cases}$$

So setting $x(n_1, n_2) = u_{+-}(n_1, n_2)$, we next compute

$$X(z_1, z_2) = \sum_{n_1=-\infty}^{+\infty} \sum_{n_2=-\infty}^{-\infty} u_{+-}(n_1, n_2) z_1^{-n_1} z_2^{-n_2}$$

$$= \sum_{n_1=0}^{+\infty} z_1^{-n_1} \cdot \sum_{n_2=-\infty}^{0} z_2^{-n_2}$$

$$= \sum_{n_1=0}^{+\infty} z_1^{-n_1} \cdot \sum_{n_2'=0}^{+\infty} z_2^{n_2'} \quad \text{with } n_2' \triangleq -n_2$$

$$= \frac{1}{1 - z_1^{-1}} \frac{1}{1 - z_2} \quad \text{for } z_1| > 1 \quad \text{and} \quad |z_2| < 1,$$

$$= -\frac{z_1}{z_1 - 1} \frac{1}{z_2 - 1} \quad \text{with } \mathcal{R}_x = \{|z_1| > 1, |z_2| < 1\}.$$

The ROC is shown as the gray area in Figure 3.4.

All four quarter-plane support, unit step sequences have the special property of separability. Since the Z-transform is a separable operator, this makes the calculation split into the product of two 1-D transforms in the n_1 and n_2 directions, as we have just seen. The ROC then factors into the Cartesian product of the two 1-D ROCs. We look at a more general case next.

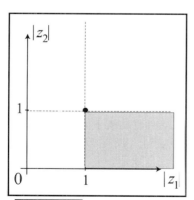

FIGURE 3.4 *ROC (gray area) for the fourth quadrant unit step function $u_{+-}(n_1, n_2)$*

3.3.1 MORE GENERAL CASE

In general, we have the Z-transform

$$X(z_1, z_2) = \frac{B(z_1, z_2)}{A(z_1, z_2)},$$

where both B and A are polynomials in the coefficients of some partial difference equation,

$$B(z_1, z_2) = \sum_{n_1=-N_1}^{+N_1} \sum_{n_2=-N_2}^{+N_2} b(n_1, n_2) z_1^{-n_1} z_2^{-n_2}$$

and

$$A(z_1, z_2) = \sum_{n_1=-N_1}^{+N_1} \sum_{n_2=-N_2}^{+N_2} a(n_1, n_2) z_1^{-n_1} z_2^{-n_2}.$$

To study the existence of this Z-transform, we focus on the denominator and rewrite A as

$$A(z_1, z_2) = z_1^{-N_1} z_2^{-N_2} \widetilde{A}(z_1, z_2),$$

where $\widetilde{A}$ is a *strict-sense polynomial* in z_1 and z_2, i.e., no negative powers of z_1 or z_2. Grouping together all terms in z_1^n, we can write

$$\widetilde{A}(z_1, z_2) = \sum_{n=0}^{\widetilde{N}_1} a_n(z_2) z_1^n,$$

yielding $\widetilde{N}_1$ poles (N_1 at most!) for each value of z_2,

$$z_1^i = f_i(z_2), \quad i = 1, \ldots, \widetilde{N}_1.$$

A sketch of such a pole surface is plotted in Figure 3.5. Note that we are plotting only the magnitude of one surface here, so that this plot does not tell the whole story. Also, there are $\widetilde{N}_1$ such sheets. Of course, there will be a similar number of zero loci or surfaces that come about from the numerator

$$\widetilde{B}(z_1, z_2) = \sum_{n=0}^{\widetilde{N}_1} b_n(z_2) z_1^n,$$

where $B(z_1, z_2) = z_1^{-N_1} z_2^{-N_2} \widetilde{B}(z_1, z_2)$. Note that these zero surfaces can intersect the pole surfaces (as well as each other) without being identical. Thus indeterminate $\frac{0}{0}$ situations can arise that cannot be simply canceled out. One classic

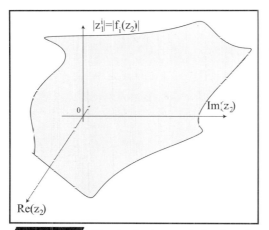

FIGURE 3.5 Sketch of pole magnitude $|z_1^i|$ as function of point in z_2 complex plane

example [4] is

$$\frac{z_1 + z_2 - 2}{(z_1 - 1)(z_2 - 1)},$$

which evaluates to $\frac{0}{0}$ at the point $(z_1, z_2) = (1, 1)$, and yet has no cancelable factors.

3.4 SOME Z-TRANSFORM PROPERTIES

Here we list some useful properties of the 2-D Z-transform that we will use in the sequel. Many are easy extensions of known properties of the 1-D Z-transform, but some are essentially new. In listing these properties, we introduce the symbol **Z** for the 2-D Z-transform operator.

Linearity property:

$$\mathbf{Z}\{ax(n_1, n_2) + by(n_1, n_2)\} = aX(z_1, z_2) + bY(z_1, z_2), \quad \text{with ROC} = \mathcal{R}_x \cap \mathcal{R}_y.$$

Delay property:

$$\mathbf{Z}\{x(n_1 - k_1, n_2 - k_2)\} = X(z_1, z_2)z_1^{-k_1}z_2^{-k_2}, \quad \text{with ROC} = \mathcal{R}_x.$$

Convolution property:

$$\mathbf{Z}\{x(n_1, n_2) * y(n_1, n_2)\} = X(z_1, z_2)Y(z_1, z_2) \quad \text{with ROC} = \mathcal{R}_x \cap \mathcal{R}_y.$$

Symmetry properties:

$$Z\{x^*(n_1, n_2)\} = X^*(z_1^*, z_2^*), \quad \text{with ROC} = \mathcal{R}_x.$$

$$Z\{x(-n_1, -n_2)\} = X(z_1^{-1}, z_2^{-1}), \quad \text{with ROC} = \{(z_1, z_2) \mid (z_1^{-1}, z_2^{-1}) \in \mathcal{R}_x\}.$$

3.4.1 LINEAR MAPPING OF VARIABLES

Consider two signals $x(n_1, n_2)$ and $y(n_1, n_2)$, which are related by the so-called linear mapping of variables

$$\begin{bmatrix} n_1' \\ n_2' \end{bmatrix} = \begin{bmatrix} l_{11} & l_{12} \\ l_{21} & l_{22} \end{bmatrix} \begin{bmatrix} n_1 \\ n_2 \end{bmatrix},$$

so that $x(n_1, n_2) = y(n_1', n_2') = y(l_{11}n_1 + l_{12}n_2, l_{21}n_1 + l_{22}n_2)$, where the matrix components l_{ij} are all integers, as required. At other points, y is taken as 0. The following relation then holds for the corresponding Z-transforms X and Y,

$$Y(z_1, z_2) = X(z_1^{l_{11}} z_2^{l_{21}}, z_1^{l_{12}} z_2^{l_{22}}), \tag{3.4-1}$$

with convergence region

$$\mathcal{R}_y = \{(z_1, z_2) \mid (z_1^{l_{11}} z_2^{l_{21}}, z_1^{l_{12}} z_2^{l_{22}}) \in \mathcal{R}_x\}.$$

This integer mapping of variables gives a warping of the spatial points (n_1, n_2), which is useful when we discuss stability tests and conditions for systems via their Z-transforms in a later section.

EXAMPLE 3.4-1 (*linear mapping of variables*)
In this example we use linear integer mapping of variables to map a signal from a general wedge support[3] to first-quadrant support. We set

$$x(n_1, n_2) = y(n_1 + n_2, n_2),$$

with x having the support indicated in Figure 3.6. Now, this transformation of variables also has an inverse with integer coefficients, so it is possible to also write y in terms of x,

$$y(n_1', n_2') = x(n_1' - n_2', n_2').$$

By the relevant Z-transform property we can say

$$Y(z_1, z_2) = X(z_1, z_1 z_2), \tag{3.4-2}$$

---

3. By wedge support we mean that the signal support is first quadrant plus a wedge from the second quadrant, with wedge indicated by a line at angle θ. For this example $\theta = 45°$.

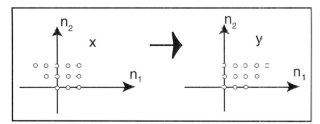

FIGURE 3.6   *Example of linear mapping of variables*

but also, because the inverse linear mapping is also integer valued, the same property says

$$X(z_1, z_2) = Y(z_1, z_1^{-1} z_2),$$

which is alternatively easily seen by solving (3.4-2) for $X(z_1, z_2)$.

In a later section we will develop Z-transform based stability tests for first-quadrant filters. Linear mapping of variables will be a way to extend these tests to other more general filters, i.e., those whose denominator coefficient support is wedge shaped.

3.4.2 INVERSE Z-TRANSFORM

The inverse Z-transform is given by the contour integral [6,7],

$$x(n_1, n_2) = \frac{1}{(2\pi j)^2} \oint_{C_1} \oint_{C_2} X(z_1, z_2) z_1^{n_1-1} z_2^{n_2-1} \, dz_1 \, dz_2,$$

where the integration path $C_1 \times C_2$ lies completely in $\mathcal{R}_x$ (the ROC of X), as it must. We can think of this 2-D inverse Z-transform as the concatenation of two 1-D inverse Z-transforms.

$$x(n_1, n_2) = \frac{1}{2\pi j} \oint_{C_2} \left[\frac{1}{2\pi j} \oint_{C_1} X(z_1, z_2) z_1^{n_1-1} \, dz_1 \right] z_2^{n_2-1} \, dz_2. \tag{3.4-3}$$

For a rational function X, the internal inverse Z-transform on the variable z_1 is straightforward, albeit with poles and zeros that are a function of the other variable z_2. For example, either partial fraction expansion or the residue method [7] could be used to evaluate the inner contour integral. Unfortunately, the second, or outer, inverse Z-transform over z_2 is often not of a rational function,[4] and, in

---

4. The reason it is not generally a rational function has to do with the formulas for the roots of a polynomial. In fact, it is known that above fourth order, these 1-D polynomial roots cannot be expressed in terms of any finite number of elementary functions of the coefficients [1].

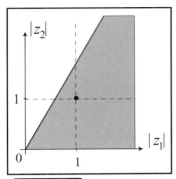

FIGURE 3.7   *Illustration of ROC (shaded area) of the example Z-transform*

general, is not amenable to closed-form expression. Some simple cases can be done, though.

EXAMPLE 3.4-2 (*simple example resulting in closed-form solution*)
A simple example where the preceding integrals can be easily carried out results from the Z-transform function,

$$X(z_1, z_2) = \frac{1}{1 - az_1^{-1}z_2}, \quad \text{with } \mathcal{R}_x = \left\{ |z_2| < |z_1|/|a| \right\},$$

where we take the case $|a| < 1$. We can illustrate this region of convergence as shown in Figure 3.7.

Proceeding with the inverse transform calculation (3.4-3), we get

$$x(n_1, n_2) = \frac{1}{2\pi j} \oint_{C_2} \left(\frac{1}{2\pi j} \oint_{C_1} \frac{1}{1 - az_1^{-1}z_2} z_1^{n_1-1} dz_1 \right) z_2^{n_2-1} dz_2.$$

Now, the inner integral corresponds to the first-order pole at $z_1 = az_2$, whose 1-D inverse Z-transform can be found using the residue method, or any other 1-D evaluation method, as

$$\frac{1}{2\pi j} \oint_{C_1} \frac{1}{1 - az_1^{-1}z_2} z_1^{n_1-1} dz_1 = (az_2)^{n_1} u(n_1).$$

Thus the overall 2-D inverse Z-transform reduces to

$$x(n_1, n_2) = \frac{1}{2\pi j} \oint_{C_2} (az_2)^{n_1} u(n_1) z_2^{n_2-1} dz_2$$

$$= a^{n_1} u(n_1) \frac{1}{2\pi j} \oint_{C_2} z_2^{n_1} z_2^{n_2-1} dz_2$$

$$= a^{n_1} u(n_1) \delta(n_1 + n_2),$$

with support on the diagonal set $\{n_1 \geqslant 0\} \cap \{n_2 = -n_1\}$. Rewriting, this result becomes

$$x(n_1, n_2) = \begin{cases} a^{n_1}, & n_1 \geqslant 0 \text{ and } n_2 = -n_1 \\ 0, & \text{else.} \end{cases}$$

There are other methods to invert the 2-D Z-transform:

- Direct long division of the polynomials.
- Known series expansion.
- Using Z-transform properties on known transform pairs.

EXAMPLE 3.4-3 (*long division method for inverse Z transform*)
We illustrate the long division method with the following example. Let

$$X(z_1, z_2) = \frac{1}{1 - z_1^{-1} z_2}, \quad \text{with } \mathcal{R}_x = \{ |z_2| < |z_1| \}.$$

Then we proceed to divide the denominator into the numerator as follows:

$$
\begin{array}{r}
1 + z_1^{-1} z_2 + z_1^{-2} z_2^2 + \cdots \\
1 - z_1^{-1} z_2 \enclose{longdiv}{} \\
\underline{1 - z_1^{-1} z_2} \\
z_1^{-1} z_2 \\
\underline{z_1^{-1} z_2 - z_1^{-2} z_2^2} \\
z_1^{-2} z_2^2 \\
\underline{z_1^{-2} z_2^2 - z_1^{-3} z_2^3} \\
+ \cdots
\end{array}
$$

The 1-D partial fraction expansion method for inverse Z-transform does not carry over to the 2-D case. This is a direct result of the nonfactorability of polynomials in more than one variable.

EXAMPLE 3.4-4 (*2-D polynomials do not factor*)
In the 1-D case, all high-order polynomials factor into first-order factors, and this property is used in the partial fraction expansion. In the multidimensional case, polynomial factorization cannot be guaranteed anymore. In fact, most of the time it is absent, as this example illustrates. Consider a signal x, with 3×3 support, corresponding to a 2×2 order polynomial in z_1 and z_2. If we could factor this polynomial into first-order factors, that would be the same as representing this signal as the convolution of two signals, say a and b of support 1×1. We would have

$$
\begin{array}{ccc}
x_{02} & x_{12} & x_{22} \\
x_{01} & x_{11} & x_{21} \\
x_{00} & x_{10} & x_{20}
\end{array}
=
\begin{array}{cc}
a_{01} & a_{11} \\
a_{00} & a_{10}
\end{array}
\circledast
\begin{array}{cc}
b_{01} & b_{11} \\
b_{00} & b_{10}
\end{array},
$$

or in Z-transforms (polynomials) we would have

$$
X(z_1, z_2) = A(z_1, z_2)B(z_1, z_2).
$$

The trouble here is that a general such x has nine degrees of freedom, while the total number of unknowns in a and b is only eight. If we considered factoring a general $N \times N$ array into two factors of order $N/2 \times N/2$, the deficiency in number of unknowns (variables) would be much greater, i.e., N^2 equations versus $2(N/2)^2 = N^2/2$ unknowns!

The fact that multidimensional polynomials do not factor has a number of other consequences, beyond the absence of a partial fraction expansion. It means that in filter design, one must take the implementation into account at the design stage. We will see in Chapter 5 that if one wants, say, the 2-D analog of second-order factors, then one has to solve the design approximation problem with this constraint built in. But also a factored form may have larger support than the corresponding nonfactored form, thus *possibly* giving a better approximation for the same number of coefficients. So this nonfactorability is not necessarily bad. Since we can always write the preceding first-order factors as separable, then 2-D polynomial factorability down to first-order factors would lead back to a finite set of isolated poles in each complex plane.

Generally we think of the Fourier transform as the evaluation of the Z-transform on the unit polycircle $\{|z_1| = |z_2| = 1\}$; however, this assumes the polycircle is in the region of convergence of $X(z_1, z_2)$, which is not always true. Example 3.4-5 demonstrates this exception.

EXAMPLE 3.4-5 (*comparison of Fourier and Z-transform*)
Consider the first-quadrant unit step function $u_{++}(n_1, n_2)$. Computing its

Z-transform, we earlier obtained

$$U_{++}(z_1, z_2) = \frac{1}{1 - z_1^{-1}} \frac{1}{1 - z_2^{-1}},$$

which is well behaved in its region of convergence $\mathcal{R}_x = \{|z_1| > 1, |z_2| > 1\}$. Note, however, that the corresponding Fourier transform has a problem here, and needs Dirac impulses. In fact, using the separability of $u_{++}(n_1, n_2)$ and noting the 1-D Fourier transform pair,

$$u(n) \Leftrightarrow U(\omega) = \pi \mathcal{E}(\omega) + \frac{1}{1 - e^{-j\omega}},$$

we obtain the 2-D Fourier transform,

$$U_{++}(\omega_1, \omega_2) = \left[\pi \delta(\omega_1) + \frac{1}{1 - e^{-j\omega_1}} \right] \left[\pi \mathcal{E}(\omega_2) + \frac{1}{1 - e^{-j\omega_2}} \right].$$

If we consider the convolution of two of these step functions, this should correspond to multiplying the transforms together. For the Z-transform, there is no problem, and the region of convergence remains unchanged. For the Fourier transform though, we would have to be able to interpret $(\delta(\omega_1))^2$ which is not possible, as powers and products of singularity functions are unstudied and so not defined.

Conversely, a sequence without a Z-transform but which possesses a Fourier transform is $\sin \omega n$, which in 2-D becomes the plane wave $\sin(\omega_1 n_1 + \omega_2 n_2)$. Thus, each transform has its own useful place. The Fourier transform is not strictly a subset of the Z-transform, because it can use impulses and other singularity functions, which are not permitted to Z-transforms.

We next turn to the stability problem for 2-D systems and find that the Z-transform plays a prominent role, just as in one dimension. However, the resulting 2-D stability tests are much more complicated, since the zeros and poles are functions, not points, in 2-D z space.

3.5 2-D FILTER STABILITY

Stability is an important concept for spatial filters as in the 1-D case. The FIR filters are automatically stable due to the finite number of presumably finite-valued coefficients. Basically stability means that the filter response will never get too large, if the input is bounded. For a linear filter, this means that nothing unpredictable or chaotic could be caused by extremely small perturbations in the input. A related notion is sensitivity to inaccuracies in the filter coefficients and computation, and stability is absolutely necessary to have the desired low sensitivity.

Stability is also necessary so that boundary conditions (often unknown in practice) will not have a big effect on the response far from the boundary. So, stable systems are preferred for a number of practical reasons. We start with the basic definition of stability of an LSI system.

DEFINITION 3.5-1 (*stability of an LSI system*)
We say a 2-D LSI system with impulse response $h(n_1, n_2)$ is bounded-input bounded-output (BIBO) stable if

$$\|h\|_1 \triangleq \sum_{k_1, k_2=-\infty}^{+\infty} \left| h(k_1, k_2) \right| < \infty.$$

The resulting linear space of signals is called l_1 and the norm $\|h\|_1$ is called the l_1 norm. Referring to this signal space, we can say that the impulse response is BIBO stable *if and only if* (iff) it is an element of the l_1 linear space, i.e.,

$$\text{system is stable} \quad \Longleftrightarrow \quad h \in l_1. \tag{3.5-1}$$

Clearly this means that for such a system described by convolution, the output will be uniformly bounded if the input is uniformly bounded. As we recall from Chapter 1, Section 1.1,

$$\left| y(n_1, n_2) \right| = \left| \sum_{k_1, k_2} x(n_1 - k_1, n_2 - k_2) h(k_1, k_2) \right|$$

$$\leqslant \max \left| x(k_1, k_2) \right| \cdot \left| \sum_{k_1, k_2} h(k_1, k_2) \right|$$

$$= M \cdot \|h\|_1$$

$$< \infty,$$

where $M < \infty$ is the assumed finite uniform bound on the input signal magnitude $|x|$, and the l_1 norm $\|h\|_1$ is assumed finite.

This BIBO stability condition is most easily expressed in terms of the Z-transform system function H. Because for Z-transforms convergence is really absolute convergence, the system is stable if

$$\text{the point} \quad (z_1, z_2) = (1, 1) \in \mathcal{R}_h \quad \text{the ROC of } H(z_1, z_2).$$

In general, for a rational system, we can write

$$H(z_1, z_2) = \frac{B(z_1, z_2)}{A(z_1, z_2)},$$

where A and B are 2-D polynomials in the variables z_1 and z_2. In one dimension, the corresponding statement would be $H(z) = B(z)/A(z)$, and stability would be determined by whether $1 \in$ ROC of $1/A$, i.e., we could ignore the numerator's presence, assuming no pole-zero cancellations. However, in the case of two and higher dimensions, this is no longer true, and it has been shown [4] that partial pole-zero cancellations can occur in such a way that no common factor can be removed, and so that the resulting filter is stable *because* of the effect of this numerator. Since this is a somewhat delicate situation, here we look for a more robust stability, and so look at just the *poles* of the system H, which are the zeros of the denominator polynomial $A(z_1, z_2)$. For such *robust stability* we must require $A(z_1, z_2) \neq 0$ in a region including $(z_1, z_2) = (1, 1)$, analogously to the 1-D case.

3.5.1 FIRST-QUADRANT SUPPORT

If a 2-D digital filter has first-quadrant support, then we have the following property for its Z-transform. If the Z-transform converges for some point (z_1^0, z_2^0), then it must also converge for all $(z_1, z_2) \in \{|z_1| \geqslant |z_1^0|, |z_2| \geqslant |z_2^0|\}$, because for such (z_1, z_2) we have

$$\sum_{k_1 \geqslant 0, k_2 \geqslant 0} |h(k_1, k_2)| |z_1|^{-k_1} |z_2|^{-k_2} \leqslant \sum_{k_1 \geqslant 0, k_2 \geqslant 0} |h(k_1, k_2)| |z_1^0|^{-k_1} |z_2^0|^{-k_2}.$$

Thus, since filters that are stable must have $(1, 1) \in \mathcal{R}_h$, then first-quadrant support filters that are stable must have an ROC that includes $\{|z_1| \geqslant 1, |z_2| \geqslant 1\}$, as sketched in Figure 3.8 as the gray region. Note that the region slightly overlaps the lines $|z_1| = 1$ and $|z_2| = 1$, in order to emphasize that these lines must be contained in the open region that is the region of convergence of a first-quadrant support and stable filter. We are not saying that the region of convergence of first-quadrant support stable filters will look like that sketched in Figure 3.8, just that the convergence region must include the gray area.

3.5.2 SECOND-QUADRANT SUPPORT

If a 2-D digital filter has second-quadrant support, then we have the following property for its Z-transform. If the Z-transform converges for some point (z_1^0, z_2^0), then it must also converge for all $(z_1, z_2) \in \{|z_1| \leqslant |z_1^0|, |z_2| \geqslant |z_2^0|\}$, because for such (z_1, z_2) we have

$$\sum_{k_1 \leqslant 0, k_2 \geqslant 0} |h(k_1, k_2)| |z_1|^{-k_1} |z_2|^{-k_2} \leqslant \sum_{k_1 \leqslant 0, k_2 \geqslant 0} |h(k_1, k_2)| |z_1^0|^{-k_1} |z_2^0|^{-k_2}.$$

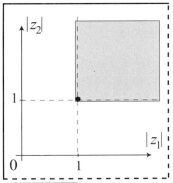

FIGURE 3.8 • *Region that must be included in the ROC of a first-quadrant support stable filter*

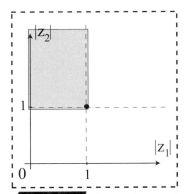

FIGURE 3.9 • *Region that must be included in the ROC of a second-quadrant support stable filter*

Thus, since filters that are stable must have $(1, 1) \in \mathcal{R}_h$, then second-quadrant support filters that are stable must have an ROC that includes $\{|z_1| \leqslant 1, |z_2| \geqslant 1\}$, as sketched in Figure 3.9 as the gray region. Note that the region slightly overlaps the lines $|z_1| = 1$ and $|z_2| = 1$, in order to emphasize that these lines must be contained in the open region that is the region of convergence of a second-quadrant support and stable filter.

EXAMPLE 3.5-1 (*a region of convergence*)
Consider the spatial filter with impulse response

$$h(n_1, n_2) = \begin{cases} a^{n_1}, & n_1 \leqslant 0 \text{ and } n_2 = -n_1 \\ 0, & \text{else.} \end{cases}$$

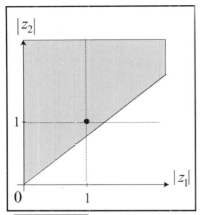

FIGURE 3.10 *Sketch of ROC (gray area) of the example Z-transform*

with $|a| > 1$. Note that the support of this h is in the second quadrant. Computing the Z-transform, we obtain

$$H(z_1, z_2) = 1 + a^{-1}z_1 z_2^{-1} + a^{-2}z_1^2 z_2^{-2} + \cdots$$

$$= \frac{1}{1 - a^{-1}z_1 z_2^{-1}}, \quad \text{with } \mathcal{R}_h = \left\{ |a|^{-1}|z_1||z_2|^{-1} < 1 \right\}.$$

We can sketch this ROC as shown in Figure 3.10.

If a 2-D filter has support in the remaining two quadrants, the ROC minimal size region can easily be determined similarly. We have thus proved our first theorems on spatial filter stability.

THEOREM 3.5-1 (*Shanks et al.* [9])

A 2-D or spatial filter with first-quadrant support impulse response $h(n_1, n_2)$ and rational system function $H(z_1, z_2) = B(z_1, z_2)/A(z_1, z_2)$ is bounded-input bounded-output stable if $A(z_1, z_2) \neq 0$ in a region including $\{|z_1| \geqslant 1, |z_2| \geqslant 1\}$. Ignoring the effect of the numerator $B(z_1, z_2)$, this is the same as saying that $H(z_1, z_2)$ is *analytic*, i.e., $H \neq \infty$, in this region.

For second-quadrant impulse response support, we can restate this theorem.

THEOREM 3.5-2 (*second quadrant stability*)

A 2-D or spatial filter with second-quadrant support impulse response $h(n_1, n_2)$ and rational system function $H(z_1, z_2) = B(z_1, z_2)/A(z_1, z_2)$ is bounded-input bounded-output stable if $A(z_1, z_2) \neq 0$ in a region including $\{|z_1| \leqslant 1, |z_2| \geqslant 1\}$.

Similarly, here are the theorem restatements for filters with impulse response support on either of the remaining two quadrants.

THEOREM 3.5-3 (*third quadrant stability*)
A 2-D or spatial filter with third-quadrant support impulse response $h(n_1, n_2)$ and rational system function $H(z_1, z_2) = B(z_1, z_2)/A(z_1, z_2)$ is bounded-input bounded-output stable if $A(z_1, z_2) \neq 0$ in a region including $\{|z_1| \leqslant 1, |z_2| \leqslant 1\}$.

THEOREM 3.5-4 (*fourth quadrant stability*)
A 2-D or spatial filter with fourth-quadrant support impulse response $h(n_1, n_2)$ and rational system function $H(z_1, z_2) = B(z_1, z_2)/A(z_1, z_2)$ is bounded-input bounded-output stable if $A(z_1, z_2) \neq 0$ in a region including $\{|z_1| \geqslant 1, |z_2| \leqslant 1\}$.

If we use the symbol $++$ to refer to impulse responses with first-quadrant support, $-+$ to refer to those with second-quadrant support, and so forth, then all four zero-free regions for denominator polynomials A can be summarized in the diagram of Figure 3.11. In words, we say "a $++$ support filter must have ROC including $\{|z_1| \geqslant 1, |z_2| \geqslant 1\}$," which is shown as the $++$ region in Figure 3.11. A general support spatial filter can be made up from these four components as either a sum or convolution product of these quarter-plane impulse responses,

$$h(n_1, n_2) = h_{++}(n_1, n_2) + h_{+-}(n_1, n_2) + h_{-+}(n_1, n_2) + h_{--}(n_1, n_2)$$

or

$$h(n_1, n_2) = h_{++}(n_1, n_2) * h_{+-}(n_1, n_2) * h_{-+}(n_1, n_2) * h_{--}(n_1, n_2).$$

Both of these general representations are useful for 2-D recursive filter design, as we will encounter in Chapter 5.

To test for stability using the preceding theorems would be quite difficult for all but the lowest order polynomials A, since we have seen previously that the zero loci of 2-D functions are continuous functions in one of the complex variables, hence requiring the evaluation of an infinite number of roots. Fortunately, it is possible to simplify these theorems. We will take the case of the $++$ or first-quadrant support filter here as a prototype.

THEOREM 3.5-5 (*simplified $++$ test*)
A $++$ filter with first-quadrant support impulse response $h(n_1, n_2)$ and rational system function $H(z_1, z_2) = B(z_1, z_2)/A(z_1, z_2)$ is bounded-input bounded-output (BIBO) stable if the following two conditions exist:

(a) $A(e^{j\omega_1}, z_2) \neq 0$ in a region including $\{$all $\omega_1, |z_2| \geqslant 1\}$.
(b) $A(z_1, e^{j\omega_2}) \neq 0$ in a region including $\{|z_1| \geqslant 1$, all $\omega_2\}$.

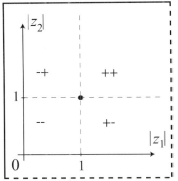

FIGURE 3.11 Illustration of necessary convergence regions for all four quarter-plane support filters

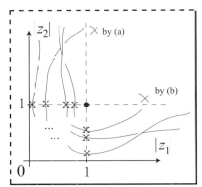

FIGURE 3.12 Figure used in proof of Theorem 3.5-5

Proof: We can construct a proof by relying on Figure 3.12. Here the $\times\times$ represent known locations of the roots when either $|z_1| = 1$ or $|z_2| = 1$. Now, since it is known in mathematics that the roots must be continuous functions of their coefficients [1], and since the coefficients, in turn, are simple polynomials in the other variable, it follows that each of these roots must be a continuous function of either z_1 or z_2. Now, as $|z_1| \nearrow$, beyond 1, the roots cannot cross the line $|z_2| = 1$ because of condition b. Also, as $|z_2| \nearrow$, beyond 1, the roots cannot cross the line $|z_1| = 1$ because of condition a. We thus conclude that the region $\{|z_1| \geqslant 1, |z_2| \geqslant 1\}$ will be zero-free by virtue of conditions a and b, as was to be shown. □

This theorem has given us considerable complexity reduction, in that the three-dimensional conditions a and b can be tested with much less work than can the original four-dimensional C^2 condition. To test condition a of the theorem, we must find the roots in z_2 of the indicated polynomial, whose coefficients are a function of the scalar variable ω_1, with the corresponding test for theorem condition b.

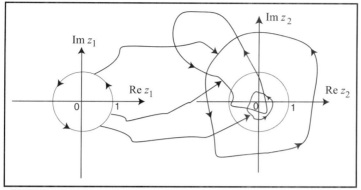

FIGURE 3.13   *Illustration of root map of condition a of Theorem 3.5-5*

3.5.3 ROOT MAPS

We can gain more insight into Theorem 3.5-5 by using the technique of root mapping. The idea is to fix the magnitude of one complex variable, say $|z_1| = 1$, and then find the location of the root locus in the z_2 plane as the phase of z_1 varies. Looking at condition **a** of Theorem 3.5-5, we see that as ω_1 varies from $-\pi$ to $+\pi$, the roots in the z_2 plane must stay inside the unit circle there, in order for the filter to be stable. In general, there will be N such roots or root maps. Because the coefficients of the z_2 polynomial are continuous functions of z_1, as mentioned previously, it follows from a theorem in algebra that the root maps will be continuous functions [1]. Since the coefficients are periodic functions of ω_1, the root maps will additionally be closed curves in the z_2 plane. This is illustrated in Figure 3.13. So a test for stability based on root maps can be constructed from Theorem 3.5-5, by considering both sets of root maps corresponding to condition a and condition b, and showing that all the root maps stay inside the unit circle in the target z plane.

Two further Z-transform stability theorem simplifications have been discovered, as indicated in the following two stability theorems.

THEOREM 3.5-6 (*Huang* [5])
The stability of a first-quadrant or $++$ quarter-plane filter of the form $H(z_1, z_2) = B(z_1, z_2)/A(z_1, z_2)$ is assured by the following two conditions:

(a) $A(e^{j\omega_1}, z_2) \neq 0$ in a region including $\{$all $\omega_1, |z_2| \geqslant 1\}$.
(b) $A(z_1, 1) \neq 0$ for all $|z_1| \geqslant 1$.

Proof: By condition **a**, the root maps in the z_2 plane are all inside the unit circle. But also, by condition **a**, none of the root maps in the z_1 plane can cross the unit circle. Then condition b states that one point on these z_1 plane root maps is inside the unit circle, i.e., the point corresponding to $\omega_2 = 0$. By the continuity of the root map, the whole z_1 plane root map must lie inside the unit circle. Thus by Theorem 3.5-5, the filter is stable. $\qquad\square$

A final simplification results in the next theorem.

THEOREM 3.5-7 (*DeCarlo–Strintzis* [11])
The stability of a first-quadrant or ++ quarter-plane filter of the form $H(z_1, z_2) = B(z_1, z_2)/A(z_1, z_2)$ is assured by the following three conditions:

(a) $A(e^{j\omega_1}, e^{j\omega_2}) \neq 0$ for all ω_1, ω_2.
(b) $A(z_1, 1) \neq 0$ for all $|z_1| \geqslant 1$.
(c) $A(1, z_2) \neq 0$ for all $|z_2| \geqslant 1$.

Proof: Here, condition a tells us that no root maps cross the respective unit circles. So, we know that the root maps are either completely outside the unit circles or completely inside them. Conditions b and c tell us that there is one point on each set of root maps that is inside the unit circles. We thus conclude that all root maps are inside the unit circles in their respective z planes. $\qquad\square$

A simple example of the use of these stability theorems in stability tests follows.

EXAMPLE 3.5-2 (*filter stability test*)
Consider the spatial filter with impulse response support on the first-quadrant and system function

$$H(z_1, z_2) = \frac{1}{1 - az_1^{-1} + bz_2^{-1}},$$

so that we have $A(z_1, z_2) = 1 - az_1^{-1} + bz_2^{-1}$. By Theorem 3.5-1, we must have $A(z_1, z_2) \neq 0$ for $\{|z_1| \geqslant 1, |z_2| \geqslant 1\}$. Thus all the roots $z_2 = f(z_1)$ must satisfy $|z_2| = |f(z_1)| < 1$ for all $|z_1| > 1$. In this case we have, by setting $A(z_1, z_2) = 0$, that

$$z_2 = f(z_1) = \frac{b}{1 - az_1^{-1}}.$$

Consider $|z_1| > 1$, and assuming that we must have $|a| < 1$, then we have

$$|z_2| = \frac{|b|}{|1 - az_1^{-1}|} \leqslant \frac{|b|}{1 - |a|},$$

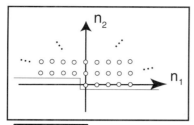

FIGURE 3.14   *Illustration of NSHP coefficient array support*

with the maximum value being achieved for some phase angle of z_1. Hence, for BIBO stability of this $++$ filter, we need

$$\frac{|b|}{1 - |a|} < 1 \quad \text{or equivalently} \quad |a| + |b| < 1.$$

The last detail now is to deal with the assumption $|a| < 1$. Actually, it is easy to see that this is necessary, by setting $|z_2| = \infty$, and noting the root at $z_1 = a$, which must be inside the unit circle.

3.5.4 STABILITY CRITERIA FOR NSHP SUPPORT FILTERS

The $++$ quarter-plane support filter is not the most general one that can follow a raster scan (horizontal across, and then down) and produce its output recursively, or what is called a recursively computable set of equations. We saw in Example 3.4-1 a wedge support polynomial that can be used in place of the quarter-plane $A(z_1, z_2)$ of the previous subsection. The most general wedge support, which is recursively computable in this sense, would have support restricted to a *nonsymmetric half-plane* (NSHP) defined as

$$\text{NSHP region} \triangleq \{n_1 \geqslant 0, n_2 \geqslant 0\} \cup \{n_1 < 0, n_2 > 0\},$$

as illustrated in Figure 3.14. With reference to this figure we can see that an NSHP filter makes wider use of the previously computed outputs, assuming a conventional raster scan, and hence is expected to have some advantages. We can see that this NSHP filter is a generalization of a first-quadrant filter, which includes some points from a neighboring quadrant, in this case the second. Extending our notation, we can call such a filter a $\oplus+$NSHP filter, with other types being denoted $\ominus+$, $+\ominus$, etc. We next present a stability test for this type of spatial recursive filter.

THEOREM 3.5-8 (*NSHP filter stability* [3])
A $\oplus+$NSHP support spatial filter is stable in the BIBO sense if its system function $H(z_1, z_2) = B(z_1, z_2)/A(z_1, z_2)$ satisfies the following conditions:

(a) $H(z_1, \infty)$ is analytic, i.e., free of singularities, on $\{|z_1| \geqslant 1\}$.

(b) $H(e^{+j\omega_1}, z_2)$ is analytic on $\{|z_2| \geqslant 1\}$, for all $\omega_1 \in [-\pi, +\pi]$.

Discussion: Ignoring possible effects of the numerator on stability, condition **a** states that $A(z_1, \infty) \neq 0$ on $\{|z_1| \geqslant 1\}$ and condition **b** is equivalently the condition $A(e^{+j\omega_1}, z_2) \neq 0$ on $\{|z_2| \geqslant 1\}$ for all $\omega_1 \in [-\pi, +\pi]$, which we have seen several times before. To see what the stability region should be for a $\oplus$+NSHP filter, we must realize that now we can no longer assume that $n_1 \geqslant 0$, so if the Z-transform converges for some point (z_1^0, z_2^0), then it must also converge for all $(z_1, z_2) \in \{|z_1| = |z_1^0|, |z_2| \geqslant |z_2^0|\}$. By filter stability we know that $h \in l_1$ so that the Z-transform H must converge for some region including $\{|z_1| = |z_2| = 1\}$, so the region of convergence for $H \in \oplus$+NSHP filter must only include $\{|z_1| = 1, |z_2| \geqslant 1\}$. To proceed further we realize that for any $\oplus$+NSHP filter $H = 1/A$, and we can write

$$A(z_1, z_2) = A(z_1, \infty) A_1(z_1, z_2),$$

with $A_1(z_1, z_2) \triangleq A(z_1, z_2)/A(z_1, \infty)$. Then the factor A_1 will not contain any coefficient terms $a_1(n_1, n_2)$ on the current line. As a result its stability can be completely described by the requirement that $A_1(e^{-j\omega_1}, z_2) \neq 0$ on $\{|z_2| \geqslant 1\}$ for all $\omega_1 \in [-\pi, +\pi]$.[5] Similarly, to have the first factor stable, in the $+n_1$ direction, we need $A(z_1, \infty) \neq 0$ on $\{|z_1| \geqslant 1\}$. Given both conditions **a** and **b** then, the $\oplus$+NSHP filter will be BIBO stable.

This stability test can be used on first-quadrant quarter-plane filters, since they are a subset of the NSHP support class. If we compare to Huang's Theorem 3.5-6, for example, we see that condition **b** here is like condition **a** there, but the 1-D test (condition **a** here) is slightly simpler than the 1-D test (condition **b** there). Here we just take the 1-D coefficient array on the horizontal axis for this test, while in Theorem 3.5-6, we must add the coefficients in vertical columns first.

EXAMPLE 3.5-3 *(stability test of $-+$ filter using NSHP test)*
Consider again the spatial filter with impulse response support on the first-quadrant, i.e., a ++ support filter, and system function

$$H(z_1, z_2) = \frac{1}{1 - az_1^{-1} + bz_2^{-1}}.$$

so that we have $A(z_1, z_2) = 1 - az_1^{-1} + bz_2^{-1}$. Since it is a subclass of the $\oplus$+NSHP filters, we can apply the theorem just presented. First we test condition **a**: $A(z_1, \infty) = 1 - az_1^{-1} = 0$ implies $z_1 = a$, and hence we need $|a| < 1$. Then testing condition **b**, we have $A(e^{+j\omega_1}, z_2) = 1 - ae^{-j\omega} - bz_2^{-1} = 0$, which

---

5. Filters with denominators solely of form A_1 are called *symmetric half-plane* (SHP).

implies

$$z_2 = \frac{b}{1 - ae^{-j\omega_1}}.$$

Since we need $|z_2| < 1$, we get the requirement $|a| + |b| < 1$, just as before.

The next example uses the NSHP stability test on an NSHP filter, where it is needed.

EXAMPLE 3.5-4 (*test of NSHP filter*)
Consider an impulse response with $\oplus$+NSHP support, given as

$$H(z_1, z_2) = \frac{1}{1 - az_1^{-1} + bz_1z_2^{-1}},$$

where we see that the recursive term in the previous line is now "above and to the right," instead of "above," as in a ++ quarter-plane support filter. First we test condition **a**: $A(z_1, \infty) = 1 - az_1^{-1} = 0$ implies $z_1 = a$, and hence we need $|a| < 1$. Then testing condition **b**, we have $A(e^{+j\omega_1}, z_2) = 1 - ae^{-j\omega_1} - be^{j\omega_1}z_2^{-1} = 0$, which implies

$$z_2 = \frac{be^{j\omega_1}}{1 - ae^{-j\omega_1}}.$$

Since we need $|z_2| < 1$, we get the requirement $|a| + |b| < 1$, same as before.

3.6 CONCLUSIONS

This chapter has looked at how the Z-transform generalizes to two dimensions. We have also looked at spatial difference equations and their stability in terms of the Z-transform. The main difference with the 1-D case is that 2-D polynomials do not generally factor into first-order, or even lower-order factors. As a consequence, we found that poles and zeros of Z-transforms were not isolated, and have turned out to be surfaces in the multidimensional complex space. Filter stability tests are much more complicated, although we have managed some simplifications that are computationally effective given today's computer power. Later, when we study filter design, we will incorporate the structure and a stability constraint into the formulation. We also introduced filters with the more general nonsymmetric half-plane support and briefly investigated their stability behavior in terms of Z-transforms.

3.7 PROBLEMS

1   Find the 2-D Z-transform and region of convergence of each of the following:

(a) $u_{++}(n_1, n_2)$

(b) $\rho^{(n_1+n_2)}, |\rho| < 1$

(c) $b^{(n_1+2n_2)}u(n_1, n_2)$

(d) $u_{-+}(n_1, n_2)$

2   Show that

$$y(n_1, n_2) = \binom{n_1 + n_2}{n_1} a^{n_1} b^{n_2} u_{++}(n_1, n_2)$$

satisfies the spatial difference equation

$$y(n_1, n_2) = ay(n_1 - 1, n_2) - by(n_1, n_2 - 1) + \delta(n_1, n_2),$$

over the region $n_1 \geqslant 0, n_2 \geqslant 0$. Assume initial rest, i.e., all boundary conditions on the top side and left-hand side are zero.

3   For the impulse response found in Example 3.1-2, use Stirling's formula[6] for the factorial to estimate whether this difference equation can be a stable filter or not.

4   In Section 3.1, it is stated that the solution of an LCCDE over a region can be written as

$$y(n_1, n_2) = y_{ZI}(n_1, n_2) + y_{ZS}(n_1, n_2)$$

where y_{ZI} is the solution to the boundary conditions with the input x set to zero, and y_{ZS} is the solution to the input x subject to zero boundary conditions. Show why this is true.

5   Consider the 2-D impulse response

$$h(n_1, n_2) = 5\rho_1^{n_1+n_2} u(n_1, n_2) * \rho_2^{n_1-n_2} u(n_1, n_2),$$

where u is the first-quadrant unit step function, and the ρ_i are real numbers that satisfy $-1 < \rho_i < +1$ for $i = 1, 2$. (Note the convolution, indicated by $*$.)

(a)   Find the Z-transform $H(z_1, z_2)$ along with its region of convergence.

(b)   Can h be the impulse response of a stable filter?

6   Find the inverse Z-transform of

$$X(z_1, z_2) = \frac{1}{1 - 0.9z_1^{-1}} \frac{1}{1 - 0.9z_2^{-1}} + \frac{0.9z_1}{1 - 0.9z_1} \frac{0.9z_2}{1 - 0.9z_2},$$

with $\text{ROC}_x \supset \{|z_1| = |z_2| = 1\}$. Is your resulting x absolutely summable?

---

6. A simple version of Stirling's formula (Stirling, 1730) is as follows: $n! \approx \sqrt{2\pi n}\, n^n e^{-n}$.

7   Prove the Z-transform relationship in (3.4-1) for linear mapping of variables.

8   Find the inverse Z-transform of

$$X(z_1, z_2) = \frac{1}{1 - (z_1^{-1} + z_2^{-1})},$$

with $\text{ROC} = \{(z_1, z_2) \mid |z_1|^{-1} + |z_2|^{-1} < 1\}$ via series expansion.

9   Show that an $N \times N$ support signal can always be written as the sum of N or fewer separable factors, by regarding the signal as a matrix and applying a singular value decomposition [10].

10   Use Z-transform root maps to numerically test the following filter for stability in the $++$ causal sense.

$$H(z_1, z_2) = \frac{1}{1 - \frac{1}{2}z_1^{-1} - \frac{1}{4}z_1^{-1}z_2^{-1} - \frac{1}{4}z_2^{-2}}.$$

Use MATLAB and the functions ROOTS and POLY.

11   Prove Theorem 3.5-3, the Z-transform stability condition for a spatial filter $H(z_1, z_2) = 1/A(z_1, z_2)$ with third-quadrant support impulse response $h(n_1, n_2)$.

12   Consider the following 2-D difference equation:

$$y(n_1, n_2) + 2y(n_1 - 1, n_2) + 3y(n_1, n_2 - 1) + 7y(n_1 - 1, n_2 - 1) = x(n_1, n_2).$$

In the following parts please use the standard image-processing coordinate system: n_1 axis horizontal, and n_2 axis downward.

(a)   Find a *stable* direction of recursion for this equation.

(b)   Sketch the impulse response support of the resulting system of part (a).

(c)   Find the Z-transform system function along with its associated ROC for the resulting system of part (a).

13   Consider the three-dimensional linear filter

$$H(z_1, z_2, z_3) = \frac{B(z_1, z_2, z_3)}{A(z_1, z_2, z_3)}, \quad \text{where } a(0, 0, 0) = 1,$$

where z_1 and z_2 correspond to spatial dimensions and z_3 corresponds to the time dimension. Assume that the impulse response $h(n_1, n_2, n_3)$ has *first-octant* support in parts (a) and (b) below. Note: Ignore any contribution of the numerator to system stability.

(a)   If this first-octant filter is stable, show that the region of convergence of $H(z_1, z_2, z_3)$ must include $\{|z_1| = |z_2| = |z_3| = 1\}$. Then extend this ROC as appropriate for the specified first-octant support of the impulse response $h(n_1, n_2, n_3)$. Hence conclude that *stability implies that*

$A(z_1, z_2, z_3)$ cannot equal zero in this region, i.e., this is a *necessary* condition for stability for first-octant filters. What is this stability region?

(b) Show that the condition that $A(z_1, z_2, z_3) \neq 0$ on the stability region of part (a) is a *sufficient* condition for stability of first-octant filters.

(c) Let's say that a 3-D filter is *causal* if its impulse response has support on $\{\mathbf{n} \mid n_3 \geq 0\}$. Note that there is now no restriction on the spatial support, i.e., in the n_1 and n_2 dimensions. What is the stability region of a *causal* 3-D filter?

14 Consider computing the output of a NSHP filter on a quarter-plane region. Specifically, the filter is

$$y(n_1, n_2) = 0.2y(n_1 - 1, n_2) + 0.4y(n_1, n_2 - 1) + 0.3y(n_1 - 1, n_2 - 1) + x(n_1, n_2)$$

and the solution region is $\{n_1 \geq 0, n_2 \geq 0\}$, with a boundary condition of zero along the two edges $n_1 = -1$ and $n_2 = -1$. This then specifies a system T mapping quarter-plane inputs x into quarter-plane outputs y, i.e., $y = T[x]$.

(a) Is T a linear operator?

(b) Is T a shift-invariant operator?

(c) Is T a stable operator?

REFERENCES

[1] G. A. Bliss, *Algebraic Functions*, Dover, New York, NY, 1966.

[2] D. E. Dudgeon and R. M. Mersereau, *Multidimensional Digital Signal Processing*, Prentice-Hall, Englewood Cliffs, NJ, 1983.

[3] M. P. Ekstrom and J. W. Woods, "Two-Dimensional Spectral Factorization with Applications in Recursive Digital Filtering," *IEEE Trans. Acoust., Speech, Signal Process.*, **ASSP-24**, 115–128, April 1976.

[4] D. M. Goodman, "Some Stability Properties of Two-Dimensional Liner Shift-Invariant Digital Filters," *IEEE Trans. Circ. Syst.*, **CAS-24**, 201–209, April 1977.

[5] T. S. Huang, "Stability of Two-Dimensional Recursive Filters," *IEEE Trans. Audio Electroacoust.*, **AU-20**, 158–163, June 1972.

[6] S. G. Krantz, *Function Theory of Several Complex Variables*, Wiley-Interscience, New York, NY, 1982.

[7] N. Levinson and R. M. Redheffer, *Complex Variables*, Holden-Day, San Francisco, CA, 1970.

[8] J. Lim, *Two-Dimensional Signal and Image Processing*, Prentice-Hall, Englewood Cliffs, NJ, 1990.

[9] J. L. Shanks, S. Treitel, and J. H. Justice, "Stability and Synthesis of Two-Dimensional Recursive Filters," *IEEE Trans. Audio Electroacoust.*, **AU-20**, 115–128, June 1972.

[10] G. Strang, *Linear Algebra and its Applications*, Academic Press, New York, NY, 1976.

[11] M. G. Strintzis, "Test of Stability of Multidimensional Filters," *IEEE Trans. Circ. Syst.*, **CAS-24**, 432–437, August 1977.

Two-Dimensional Discrete Transforms

4

In this chapter we look at discrete space transforms such as discrete Fourier series, discrete Fourier transform (DFT), and discrete cosine transform (DCT) in two dimensions. We also discuss fast and efficient realizations of the DFT and DCT. The DFT is a heavily used tool in image and multidimensional signal processing. Block transforms can be obtained from scanning the data into small blocks and then performing the DFT or DCT on each block. The block DCT is used extensively in image and video compression for transmission and storage. We also consider the *subband/wavelet transform* (SWT), which can be considered as a generalization of the block DCT wherein the basis functions can overlap from block-to-block. These SWTs can also be considered as a generalization of the Fourier transform wherein resolution in space can be traded off versus resolution in frequency.

4.1 DISCRETE FOURIER SERIES

The discrete Fourier series (DFS) has been called the "the Fourier transform for periodic sequences," in that it plays the same role for them that the Fourier transform plays for nonperiodic (ordinary) sequences. The DFS also provides a theoretical stepping stone toward the discrete Fourier transform, which has great practical significance in signal and image processing, as a Fourier transform for finite support sequences. Actually, we can take the Fourier transform of periodic sequences, but only with impulse functions. The DFS can give an equivalent representation without the need for Dirac impulses.

Since this section will mainly be concerned with periodic sequences, we establish the following convenient notation:

Notation: We write $\tilde{x}(n_1, n_2)$ to denote a sequence that is rectangularly periodic with period $N_1 \times N_2$ when only one period is considered. If two or more periods are involved in a problem, we will extend this notation and denote the periods explicitly.

DEFINITION 4.1-1 (*discrete Fourier series*)
For a periodic sequence $\tilde{x}(n_1, n_2)$, with rectangular period $N_1 \times N_2$ for positive integers N_1 and N_2, we define its DFS transform as

$$\widetilde{X}(k_1, k_2) \triangleq \sum_{n_1=0}^{N_1-1} \sum_{n_2=0}^{N_2-1} \tilde{x}(n_1, n_2) \exp -j2\pi \left(\frac{n_1 k_1}{N_1} + \frac{n_2 k_2}{N_2} \right),$$

for all integers k_1 and k_2 in the lattice Z^2, i.e., $-\infty < k_1, k_2 < +\infty$.[1]

Since k_1 and k_2 are integers, we can easily see that the DFS is itself periodic with period $N_1 \times N_2$, thus the DFS transforms or maps periodic sequences

---

1. Note that $-\infty < k_1, k_2 < +\infty$ is shorthand for $-\infty < k_1 < +\infty$ and $-\infty < k_2 < +\infty$.

into periodic sequences. It may be interesting to note that the Fourier transform for discrete space does not share the analogous property, since it maps discrete-space functions into continuous-space functions over $[-\pi, -\pi]^2$. Even though the Fourier transform over continuous space did map this space into itself, its discrete-space counterpart did not. In part because of this self-mapping feature of the DFS, the inverse DFS formula is very similar to the DFS formula.

THEOREM 4.1-1 (*inverse DFS*)
Given $\tilde{X}(k_1, k_2)$, the DFS of periodic sequence $\tilde{x}(n_1, n_2)$, the inverse DFS (or IDFS) is given as

$$\tilde{x}(n_1, n_2) = \frac{1}{N_1 N_2} \sum_{k_1=0}^{N_1-1} \sum_{k_2=0}^{N_2-1} \tilde{X}(k_1, k_2) \exp +j2\pi \left(\frac{n_1 k_1}{N_1} + \frac{n_2 k_2}{N_2} \right),$$

for all $-\infty < n_1, n_2 < +\infty$.

Proof: We start by inserting the DFS into this claimed inversion formula,

$$\frac{1}{N_1 N_2} \sum_{k_1=0}^{N_1-1} \sum_{k_2=0}^{N_2-1} \left[\sum_{n_1'=0}^{N_1-1} \sum_{n_2'=0}^{N_2-1} \tilde{x}(n_1', n_2') \exp -j2\pi \left(\frac{n_1' k_1}{N_1} + \frac{n_2' k_2}{N_2} \right) \right]$$

$$\times \exp +j2\pi \left(\frac{n_1 k_1}{N_1} + \frac{n_2 k_2}{N_2} \right)$$

$$= \sum_{n_1'=0}^{N_1-1} \sum_{n_2'=0}^{N_2-1} \tilde{x}(n_1', n_2') \left\{ \frac{1}{N_1 N_2} \sum_{k_1=0}^{N_1-1} \sum_{k_2=0}^{N_2-1} \exp +j2\pi \left[\frac{(n_1 - n_1')k_1}{N_1} + \frac{(n_2 - n_2')k_2}{N_2} \right] \right\}$$

$$= \sum_{n_1'=0}^{N_1-1} \sum_{n_2'=0}^{N_2-1} \tilde{x}(n_1', n_2') \left[\sum_{k_1=-\infty}^{+\infty} \varepsilon(n_1 - n_1' - k_1 N_1) \sum_{k_2=-\infty}^{+\infty} \delta(n_2 - n_2' + k_2 N_2) \right]$$

$$= \tilde{x}((n_1)_{N_1}, (n_2)_{N_2}),$$

i.e., a periodic extension of $\tilde{x}$ off its base period $[0, N_1 - 1] \times [0, N_2 - 1]$,

$$= \tilde{x}(n_1, n_2),$$

since $\tilde{x}$ is periodic with rectangular period $N_1 \times N_2$, thus completing the proof. $\square$

In fact, one could also take the DFS of the DFS. The reader may care to show that the result would be a scaled and reflected-through-the-origin version of the original periodic function $\tilde{x}$. The following example calculates and plots the DFS of a simple periodic discrete-space pulse function.

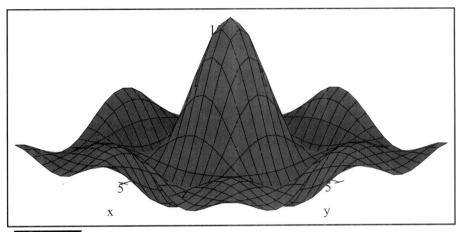

FIGURE 4.1 *Plot of the amplitude part of the DFS in Example 4.1-1*

EXAMPLE 4.1-1 (*calculation of the DFS*)

Let $\tilde{x}(n_1, n_2) = I_{4 \times 4}(n_1, n_2)$ for $0 \leqslant n_1, n_2 \leqslant 7$, where I_S is the *indicator function*[2] of the coordinate set S. Here N_1 and $N_2 = 8$. Considering the resulting periodic function we see a periodic pulse function on the plane with 25% ones and 75% zeros, which could serve as an optical image test pattern of sorts. Then

$$
\begin{aligned}
\widetilde{X}(k_1, k_2) &= \sum_{n_1=0}^{3} \sum_{n_2=0}^{3} W_8^{n_1 k_1} W_8^{n_2 k_2}, \quad \text{with } W_8 \triangleq \exp -j\frac{2\pi}{8} \\[1mm]
&= \sum_{n_1=0}^{3} W_8^{n_1 k_1} \sum_{n_2=0}^{3} W_8^{n_2 k_2} \\[1mm]
&= \frac{1 - W_8^{4k_1}}{1 - W_8^{k_1}} \frac{1 - W_8^{4k_2}}{1 - W_8^{k_2}} \\[1mm]
&= W_8^{\frac{3}{2}k_1} W_8^{\frac{3}{2}k_2} \frac{\sin(\frac{\pi k_1}{2})}{\sin(\frac{\pi k_1}{8})} \frac{\sin(\frac{\pi k_2}{2})}{\sin(\frac{\pi k_2}{8})}.
\end{aligned}
$$

2. The indicator function I_S has value 1 for a coordinate (n_1, n_2) contained in the set S and a 0 elsewhere, thereby *indicating* the set S.

We give a perspective plot of the amplitude part

$$\frac{\sin(\frac{\pi k_1}{2}) \sin(\frac{\pi k_2}{2})}{\sin(\frac{\pi k_1}{3}) \sin(\frac{\pi k_2}{3})},$$

in Figure 4.1.

4.1.1 PROPERTIES OF THE DFS TRANSFORM

The properties of the DFS are similar to those of the Fourier transform, but distinct in several important aspects, arising both from the periodic nature of $\tilde{x}$ as well as from the fact that the frequency arguments k_1 and k_2 are integers. We consider two sequences $\tilde{x}$ and $\tilde{y}$ with the same rectangular period $N_1 \times N_2$, having DFS transforms $\tilde{X}$ and $\tilde{Y}$, respectively. We offer some proofs below.

1 *Linearity*:

$$a\tilde{x} + b\tilde{y} \quad \Leftrightarrow \quad a\tilde{X} + b\tilde{Y},$$

when both sequences have the same period $N_1 \times N_2$.

2 *Periodic convolution*:
We define periodic convolution, with operator symbol $\otimes$, for two periodic sequences with the same 2-D period as

$$(\tilde{x} \otimes \tilde{y})(n_1, n_2) \triangleq \sum_{l_1=0}^{N_1-1} \sum_{l_2=0}^{N_2-1} \tilde{x}(l_1, l_2)\tilde{y}(n_1 - l_1, n_2 - l_2).$$

We then have the following transform pair:

$$(\tilde{x} \otimes \tilde{y})(n_1, n_2) \quad \Leftrightarrow \quad \tilde{X}(k_1, k_2)\tilde{Y}(k_1, k_2),$$

with proof below.

3 *Multiplication*:

$$\tilde{x}(n_1, n_2)\tilde{y}(n_1, n_2) \quad \Leftrightarrow \quad \frac{1}{N_1 N_2}\tilde{X}(k_1, k_2) \otimes \tilde{Y}(k_1, k_2),$$

4 *Separability*:

$$\tilde{x}_1(n_1)\tilde{x}_2(n_2) \quad \Leftrightarrow \quad \tilde{X}_1(k_1)\tilde{X}_2(k_2),$$

the separable product of a 1-D N_1-point and N_2-point DFS.

5 *Shifting (delay)*:

$$\tilde{x}(n_1 - m_1, n_2 - m_2) \quad \Leftrightarrow \quad \tilde{X}(k_1, k_2) \exp -j2\pi\left(\frac{m_1 k_1}{N_1} - \frac{m_2 k_2}{N_2}\right),$$

where the shift vector (m_1, m_2) is integer valued with proof below.

6 *Parseval's theorem*:

$$\sum_{n_1=0}^{N_1-1}\sum_{n_2=0}^{N_2-1} \tilde{x}(n_1,n_2)\tilde{y}^*(n_1,n_2) = \frac{1}{N_1 N_2}\sum_{k_1=0}^{N_1-1}\sum_{k_2=0}^{N_2-1} \tilde{X}(k_1,k_2)\tilde{Y}^*(k_1,k_2),$$

with special case for $x=y$, the "energy balance formula," since the left-hand side can be viewed as the "energy" in one period,

$$\sum_{n_1=0}^{N_1-1}\sum_{n_2=0}^{N_2-1} \left|\tilde{x}(n_1,n_2)\right|^2 = \frac{1}{N_1 N_2}\sum_{k_1=0}^{N_1-1}\sum_{k_2=0}^{N_2-1} \left|\tilde{X}(k_1,k_2)\right|^2,$$

with proof assigned as an end-of-chapter problem.

7 *Symmetry properties*:

(a) *Conjugation*:

$$\tilde{x}^*(n_1,n_2) \quad \Leftrightarrow \quad \tilde{X}^*(-k_1,-k_2),$$

(b) *Arguments reversed (reflection through origin)*:

$$\tilde{x}(-n_1,-n_2) \quad \Leftrightarrow \quad \tilde{X}(-k_1,-k_2).$$

(c) *Real-valued sequences (special case)*: By the conjugation property above, applied it to a real-valued sequence $\tilde{x}$, we have the *conjugate symmetry* property,

$$\tilde{X}(k_1,k_2) = \tilde{X}^*(-k_1,-k_2).$$

From this equation, the following four important symmetry properties of real periodic bi-sequences follow easily:

i. Re $\tilde{X}(k_1,k_2)$ is *even through the origin*, i.e.,

$$\mathrm{Re}\,\tilde{X}(k_1,k_2) = \mathrm{Re}\,\tilde{X}(-k_1,-k_2).$$

ii. Im $\tilde{X}(k_1,k_2)$ is *odd through the origin*, i.e.,

$$\mathrm{Im}\,\tilde{X}(k_1,k_2) = -\,\mathrm{Im}\,\tilde{X}(-k_1,-k_2).$$

iii. $|\tilde{X}(k_1,k_2)|$ is *even through the origin*, i.e.,

$$\left|\tilde{X}(k_1,k_2)\right| = \left|\tilde{X}(-k_1,-k_2)\right|.$$

iv. arg $\tilde{X}(k_1,k_2)$ is *odd through the origin*, i.e.,

$$\arg\tilde{X}(k_1,k_2) = -\arg\tilde{X}(-k_1,-k_2).$$

These last properties are used for reducing required data storage for the DFS by an approximate factor of $1/2$ in the real-valued image case.

4.1.2 PERIODIC CONVOLUTION

We choose to make some comments on periodic convolution. This property is very similar to the 1-D case for periodic sequences $\tilde{x}(n)$ of one variable. Rectangular periodicity, as considered here, allows an easy generalization. As in the 1-D case, we note that regular convolution is not possible with periodic sequences, since they have infinite energy. We offer next a proof of the periodic convolution property:

Proof of property 2: Let

$$\tilde{y}(n_1, n_2) \triangleq (\tilde{x}_1 \otimes \tilde{x}_2)(n_1, n_2)$$

$$= \sum_{l_1=0}^{N_1-1} \sum_{l_2=0}^{N_2-1} \tilde{x}_1(l_1, l_2)\tilde{x}_2(n_1 - l_1, n_2 - l_2).$$

Then

$$\widetilde{Y}(k_1, k_2) = \sum_{n_1=0}^{N_1-1} \sum_{n_2=0}^{N_2-1} \tilde{y}(n_1, n_2) \exp -j2\pi \left(\frac{n_1 k_1}{N_1} + \frac{n_2 k_2}{N_2} \right)$$

$$= \sum_{n_1,n_2} \sum_{l_1,l_2} \tilde{x}_1(l_1, l_2)\tilde{x}_2(n_1 - l_1, n_2 - l_2) W_{N_1}^{n_1 k_1} W_{N_2}^{n_2 k_2},$$

with $W_{N_i} \triangleq \exp -j2\pi/N_i = 1, 2$. Thus

$$\widetilde{Y}(k_1, k_2) = \sum_{n_1,n_2} \sum_{l_1,l_2} \tilde{x}_1(l_1, l_2) W_{N_1}^{l_1 k_1} W_{N_2}^{l_2 k_2} \left[\tilde{x}_2(n_1 - l_1, n_2 - l_2) W_{N_1}^{(n_1 - l_1)k_1} W_{N_2}^{(n_2 - l_2)k_2} \right]$$

$$= \sum_{l_1,l_2} \tilde{x}_1(l_1, l_2) W_{N_1}^{l_1 k_1} W_{N_2}^{l_2 k_2} \left[\sum_{n_1,n_2} \tilde{x}_2(n_1 - l_1, n_2 - l_2) W_{N_1}^{(n_1 - l_1)k_1} W_{N_2}^{(n_2 - l_2)k_2} \right]$$

$$= \sum_{l_1,l_2} \tilde{x}_1(l_1, l_2) W_{N_1}^{l_1 k_1} W_{N_2}^{l_2 k_2} \left[\widetilde{X}_2(k_1, k_2) \right],$$

by periodicity,

$$= \widetilde{X}_1(k_1, k_2)\widetilde{X}_2(k_1, k_2),$$

as was to be shown. $\square$

4.1.3 SHIFTING OR DELAY PROPERTY

According to property 5 in Section 4.1.1 $\tilde{x}(n_1 - m_1, n_2 - m_2)$ has DFS $\widetilde{X}(k_1, k_2) \exp -j2\pi(\frac{m_1 k_1}{N_1} + \frac{m_2 k_2}{N_2})$, which can be seen directly as follows: First, by definition,

$$\widetilde{X}(k_1, k_2) = \sum_{n_1=0}^{N_1-1} \sum_{n_2=0}^{N_2-1} \tilde{x}(n_1, n_2) \exp -j2\pi \left(\frac{n_1 k_1}{N_1} + \frac{n_2 k_2}{N_2} \right),$$

so the DFS of $\tilde{x}(n_1 - m_1, n_2 - m_2)$ is given by

$$\sum_{n_1=0}^{N_1-1} \sum_{n_2=0}^{N_2-1} \tilde{x}(n_1 - m_1, n_2 - m_2) \exp -j2\pi \left(\frac{n_1 k_1}{N_1} + \frac{n_2 k_2}{N_2} \right)$$

$$= \sum_{n_1'=-m_1}^{N_1-1-m_1} \sum_{n_2'=-m_2}^{N_2-1-m_2} \tilde{x}(n_1', n_2') \exp -j2\pi \left[\frac{(n_1' + m_1)k_1}{N_1} + \frac{(n_2' + m_2)k_2}{N_2} \right]$$

$$= \left[\sum_{n_1'=-m_1}^{N_1-1-m_1} \sum_{n_2'=-m_2}^{N_2-1-m_2} \tilde{x}(n_1', n_2') \exp -j2\pi \left(\frac{n_1' k_1}{N_1} + \frac{n_2' k_2}{N_2} \right) \right]$$

$$\times \exp -j2\pi \left(\frac{m_1 k_1}{N_1} + \frac{m_2 k_2}{N_2} \right)$$

$$= \left[\sum_{n_1'=0}^{N_1-1} \sum_{n_2'=0}^{N_2-1} \tilde{x}(n_1', n_2') \exp -j2\pi \left(\frac{n_1' k_1}{N_1} + \frac{n_2' k_2}{N_2} \right) \right] \exp -j2\pi \left(\frac{m_1 k_1}{N_1} + \frac{m_2 k_2}{N_2} \right)$$

$$= \widetilde{X}(k_1, k_2) \exp -j2\pi \left(\frac{m_1 k_1}{N_1} + \frac{m_2 k_2}{N_2} \right),$$

where we have substituted $n_1' \triangleq n_1 - m_1$ and $n_2' \triangleq n_1 - m_1$ in line 2 above, and where the second to the last line follows from the rectangular periodicity of $\tilde{x}$.

COMMENTS

1   If two rectangularly periodic sequences have different periods, we can use the DFS properties if we first find their common period, given as $N_i = \text{LCM}(N_i^x, N_i^y)$, where sequence $\tilde{x}$ is periodic $N_1^x \times N_2^x$ and similarly for $\tilde{y}$ (here LCM stands for *least common multiple*).

2   The separability property 4 in Section 4.1.1 comes about because the DFS operator is a separable operator and the operand $\tilde{x}(n_1, n_2)$ is assumed to be a separable function.

4.2 DISCRETE FOURIER TRANSFORM

The discrete Fourier transform (DFT) is "the Fourier transform for finite-length sequences" because, unlike the (discrete-space) Fourier transform, the DFT has a discrete argument and can be stored in a finite number of infinite word-length locations. Yet, still it turns out that the DFT can be used to exactly implement convolution for finite-size arrays. Following [5,8], our approach to the DFT will be through the DFS, which is made possible by the isomorphism between rectangular periodic and finite-length, rectangular-support sequences.

DEFINITION 4.2-1 (*discrete Fourier transform*)
For a finite-support sequence $x(n_1, n_2)$ with support $[0, N_1 - 1] \times [0, N_2 - 1]$, we define its DFT $X(k_1, k_2)$ for integers k_1 and k_2 as follows:

$$X(k_1, k_2) \triangleq \begin{cases} \sum_{n_1=0}^{N_1-1} \sum_{n_2=0}^{N_2-1} x(n_1, n_2) W_{N_1}^{n_1 k_1} W_{N_2}^{n_2 k_2}, \\ \quad (k_1, k_2) \in [0, N_1 - 1] \times [0, N_2 - 1] \\ 0, \quad \text{else.} \end{cases} \qquad (4.2\text{-}1)$$

We note that the DFT maps finite support rectangular sequences into themselves. Pictures of the DFT basis functions of size 8×8 are shown next, with real parts in Figure 4.2 and imaginary parts in Figure 4.3. The real part of the basis functions represent the components of x that are symmetric with respect to the 8×8 square, while the imaginary parts of these basis functions represent the nonsymmetric parts. In these figures, the color *white* is maximum positive $(+1)$, *mid-gray* is 0, and *black* is minimum negative (-1). Each basis function occupies a small square, all of which are then arranged into an 8×8 mosaic. Note that the highest frequencies are in the middle at $(k_1, k_2) = (4, 4)$ and correspond to the Fourier transform at $(\omega_1, \omega_2) = (\pi, \pi)$. So the DFT is seen as a projection of the finite-support input sequence $x(n_1, n_2)$ onto these basis functions. The DFT coefficients then are the representation coefficients for this basis. The inverse DFT (IDFT) exists and is given by

$$x(n_1, n_2) = \begin{cases} \frac{1}{N_1 N_2} \sum_{k_1=0}^{N_1-1} \sum_{k_2=0}^{N_2-1} X(k_1, k_2) W_{N_1}^{-n_1 k_1} W_{N_2}^{-n_2 k_2}, \\ \quad (n_1, n_2) \in [0, N_1 - 1] \times [0, N_2 - 1] \\ 0, \quad \text{else.} \end{cases} \qquad (4.2\text{-}2)$$

The DFT can be seen as a representation for x in terms of the basis functions $(1/N_1 N_2) W_{N_1}^{-n_1 k_1} W_{N_2}^{-n_2 k_2}$ and the expansion coefficients $X(k_1, k_2)$.

The correctness of this 2-D IDFT formula can be seen in a number of different ways. Perhaps the easiest, at this point, is to realize that the 2-D DFT is a separable transform, and as such, we can realize it as the concatenation of two 1-D DFTs.

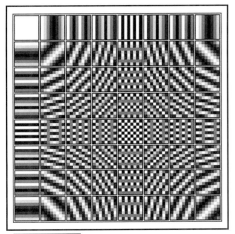

FIGURE 4.2   *Image of the real part of 8×8 DFT basis functions*

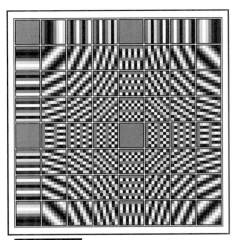

FIGURE 4.3   *Image of the imaginary part of 8×8 DFT basis functions*

Since we can see that the 2-D IDFT is just the inverse of each of these 1-D DFTs in the given order, say row first and then by column, we have the desired result based on the known validity of the 1-D DFT/IDFT transform pair, applied twice.

A second method of proof is to rely on the DFS that we have established in the previous section. Indeed, this is the main value of the DFS, i.e., as an aid in understanding DFT properties. The key concept is that rectangular periodic and rectangular finite-support sequences are isomorphic to one another, i.e., given $\tilde{x}$, we can define a finite-support x as $x \triangleq \tilde{x} I_{N_1 \times N_2}$, and given a finite-support x, we can find the corresponding $\tilde{x}$ as $\tilde{x} = x[(n_1)_{N_1}, (n_2)_{N_2}]$, where we use the notation

$(n)_N$, meaning "$n \mod N$." Still a third method is to simply insert (4.2-1) into (4.2-2) and perform the 2-D proof directly.

EXAMPLE 4.2-1 (*1's on diagonal of square*)

Consider a spatial sequence of finite extent $x(n_1, n_2) = \delta(n_1 - n_2)I_{N \times N}$, as illustrated in the left-hand side of Figure 4.4. Proceeding to take the DFT, we obtain for $(k_1, k_2) \in [0, N-1] \times [0, N-1]$:

$$X(k_1, k_2) = \sum_{n_1=0}^{N-1} \sum_{n_2=0}^{N-1} \delta(n_1 - n_2) W_N^{n_1 k_1} W_N^{n_2 k_2}$$

$$= \sum_{n_1=0}^{N-1} W_N^{n_1(k_1+k_2)}$$

$$= N\delta(k_1 + k_2) \quad \text{for } (k_1, k_2) \in [0, N-1] \times [0, N-1]$$

and, of course, $X(k_1, k_2) = 0$ elsewhere, as illustrated on the right-hand side of Figure 4.4. Note that this is analogous to the case of the continuous-parameter Fourier transform of an impulse line (see Chapter 1). This ideal result occurs only when the line is exactly at 0°, 45°, or 90°, or at 135° and the DFT size is square. For other angles, we get a $\sin Nk/N \sin k$ type of approximation that tends to the impulse line solution more and more as N grows larger without bound. Of course, as was true for the continuous-space (parameter) Fourier transform, the line of impulses in frequency is perpendicular or at 90° to the impulse line in space.

4.2.1 DFT PROPERTIES

The properties of the DFT are similar to those of the DFS. The key difference is that the support of the sequence x and of the DFT X is finite. We consider

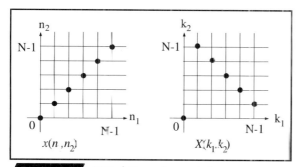

FIGURE 4.4 *DFT of 1's on diagonal of square*

two sequences x and y with the same rectangular $N_1 \times N_2$ support, having DFT transforms X and Y, respectively. We then offer proofs of some of these properties below:

1 *Linearity*:

$$ax + by \quad \Leftrightarrow \quad aX + bY,$$

when both sequences have the same support $[0, N_1 - 1] \times [0, N_2 - 1]$.

2 *Circular convolution*:

We define circular convolution for two finite support sequences with the same period as

$$(x \otimes y)(n_1, n_2) \triangleq \sum_{l_1=0}^{N_1-1} \sum_{l_2=0}^{N_2-1} x(l_1, l_2) y\big[(n_1 - l_1)_{N_1}, (n_2 - l_2)_{N_2}\big] I_{N_1 \times N_2}(n_1, n_2),$$

(4.2-3)

using the operator symbol $\otimes$. We then have the following transform pair:

$$(x \otimes y)(n_1, n_2) \quad \Leftrightarrow \quad X(k_1, k_2) Y(k_1, k_2),$$

with proof below.

3 *Multiplication*:

$$x(n_1, n_2) y(n_1, n_2) \quad \Leftrightarrow \quad \frac{1}{N_1 N_2} X(k_1, k_2) \otimes Y(k_1, k_2).$$

4 *Separability*:

$$x_1(n_1) x_2(n_2) \quad \Leftrightarrow \quad X_1(k_1) X_2(k_2),$$

the separable product of a 1-D N_1-point and N_2-point DFT.

5 *Circular shifting*:

$$x\big[(n_1 - m_1)_{N_1}, (n_2 - m_2)_{N_2}\big] I_{N_1 \times N_2}(n_1, n_2)$$

$$\Leftrightarrow \quad X(k_1, k_2) \exp -j2\pi \left(\frac{m_1 k_1}{N_1} + \frac{m_2 k_2}{N_2} \right),$$

where the shift vector (m_1, m_2) is integer valued (proof given below).

6 *Parseval's theorem*:

$$\sum_{n_1=0}^{N_1-1} \sum_{n_2=0}^{N_2-1} x(n_1, n_2) y^*(n_1, n_2) = \frac{1}{N_1 N_2} \sum_{k_1=0}^{N_1-1} \sum_{k_2=0}^{N_2-1} X(k_1, k_2) Y^*(k_1, k_2),$$

with special case for $x = y$, the "energy balance formula," since the left-hand side then becomes the energy of the signal

$$\mathcal{E}_x = \sum_{n_1=0}^{N_1-1} \sum_{n_2=0}^{N_2-1} |x(n_1, n_2)|^2 = \frac{1}{N_1 N_2} \sum_{k_1=0}^{N_1-1} \sum_{k_2=0}^{N_2-1} |X(k_1, k_2)|^2,$$

with proof assigned as an end-of-chapter problem.

7 *Symmetry properties:*

(a) *Conjugation:*

$$x^*(n_1, n_2) \quad \Leftrightarrow \quad X^*\big[(-k_1)_{N_1}, (-k_2)_{N_2}\big] I_{N_1 \times N_2}(k_1, k_2).$$

(b) *Arguments reversed (modulo reflection through origin):*

$$x\big[(-n_1)_{N_1}, (-n_2)_{N_2}\big] I_{N_1 \times N_2}(n_1, n_2)$$
$$\Leftrightarrow \quad X\big[(-k_1)_{N_1}, (-k_2)_{N_2}\big] I_{N_1 \times N_2}(k_1, k_2).$$

(c) *Real-valued sequences (special case):* By the conjugation property above, applying it to a real-valued sequence x, we have the *conjugate symmetry* property

$$X(k_1, k_2) = X^*\big[(-k_1)_{N_1}, (-k_2)_{N_2}\big] I_{N_1 \times N_2}(k_1, k_2). \qquad (4.2\text{-}4)$$

From this equation, the following four properties follow easily:

i. $\operatorname{Re} X(k_1, k_2)$ is *even*, i.e.,

$$\operatorname{Re} X(k_1, k_2) = \operatorname{Re} X\big[(-k_1)_{N_1}, (-k_2)_{N_2}\big] I_{N_1 \times N_2}(k_1, k_2).$$

ii. $\operatorname{Im} X(k_1, k_2)$ is *odd*, i.e.,

$$\operatorname{Im} X(k_1, k_2) = -\operatorname{Im} X\big[(-k_1)_{N_1}, (-k_2)_{N_2}\big] I_{N_1 \times N_2}(k_1, k_2).$$

iii. $|X(k_1, k_2)|$ is *even*, i.e.,

$$|X(k_1, k_2)| = X\big[(-k_1)_{N_1}, (-k_2)_{N_2}\big] |I_{N_1 \times N_2}(k_1, k_2).$$

iv. $\arg \widetilde{X}(k_1, k_2)$ is *odd*, i.e.,

$$\arg \widetilde{X}(k_1, k_2) = -\arg \widetilde{X}(-k_1, -k_2).$$

These last properties are used for reducing required data storage for the DFT by an approximate factor of $1/2$ in the real-valued image case.

PROOF OF DFT CIRCULAR CONVOLUTION PROPERTY 2

Key property 2 says that multiplication of 2-D DFTs corresponds to the defined circular convolution in the spatial domain, in a manner very similar to that in one dimension. In fact, we can see from (4.2-3) that this operation is separable into a 1-D circular convolution along the rows, followed by a 1-D circular convolution over the columns. The correctness of this property can then be proved by making use of the 1-D proof twice, once for the rows and once for the columns. Another way to see the result is to consider the corresponding periodic sequences $\tilde{x}$ and $\tilde{y}$. Then (4.2-3) can be seen as the first period from their periodic convolution, since $\tilde{x}(n_1, n_2) = x[(n_1)_{N_1}, (n_2)_{N_2}]$ and $\tilde{y} = y[(n_1)_{N_1}, (n_2)_{N_2}]$. So the first period of their DFS product must be the corresponding DFT result, and so by property 2 of DFS, we have the result here, since by the correspondence $\tilde{x} \sim x$, it must be that $\tilde{X} \sim X$.

PROOF OF DFT CIRCULAR SHIFT PROPERTY 5

Since $\tilde{x}(n_1, n_2) = x[(n_1)_{N_1}, (n_2)_{N_2}]$, it follows that the periodic shift of $\tilde{x}$ agrees with the circular shift of x,

$$\tilde{x}(n_1 - m_1, n_2 - m_2) = x\big[(n_1 - m_1)_{N_1}, (n_2 - m_2)_{N_2}\big],$$

for $(n_1, n_2) \in [0, N_1 - 1] \times [0, N_2 - 1]$, so the DFS of the left-hand side must equal the DFT of the right-hand side, over the fundamental period in frequency, i.e., $(k_1, k_2) \in [0, N_1 - 1] \times [0, N_2 - 1]$. We thus have the DFT

$$\tilde{X}(k_1, k_2) \exp -j2\pi \left(\frac{m_1 k_1}{N_1} + \frac{m_2 k_2}{N_2} \right) I_{N_1 \times N_2}(k_1, k_2)$$

$$= X(k_1, k_2) \exp -j2\pi \left(\frac{m_1 k_1}{N_1} + \frac{m_2 k_2}{N_2} \right).$$

An end-of-chapter problem asks you to use this circular shift property to prove the 2-D DFT circular convolution property directly in the 2-D DFT domain.

EXAMPLE 4.2-2 (*circular convolution*)

Let $N_1 = N_2 = 4$. The diagram in Figure 4.5 shows an example of the 2-D circular convolution of two small arrays x and y. In this figure, the two top plots show the arrays $x(n_1, n_2)$ and $y(n_1, n_2)$, where the open circles indicate zero values of these 4×4 support signals. The non-zero values are denoted by filled-in circles in the 3×3 triangle-support x, and various filled-in shapes on the square 2×2 support y. To perform their convolution, over dummy variables (l_1, l_2), the middle plots of the figure show the inputs as functions of these dummy variables. The bottom two plots then show y circularly shifted to correspond to the output points $(n_1, n_2) = (0, 0)$ on the left and $(n_1, n_2) =$

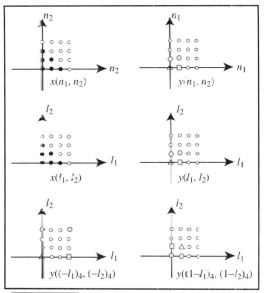

FIGURE 4.5 Example of 2-D circular convolution of small triangle support x and small square support y, both considered as $N_1 \times N_2 = 4 \times 4$ support

(1, 1) on the right. Continuing in this way, we can see that for all (n_1, n_2), we get the linear convolution result. This has happened because the circular wrap-around points in y occur only at zero values for x. We see that this was due to the larger value we took for N_1 and N_2, i.e., larger than the necessary value for the DFT.

We have seen in this example that the circular or wrap-around does not affect all output points. In fact, by enclosing the small triangular support signal and the small pulse in the larger 4×4 square, we have avoided the wrap-around for all $(n_1 \geq 1, n_2 \geq 1)$. We can generalize this result as follows. Let the supports of signals x and y be given as

$$\text{supp}\{x\} = [0, M_1 - 1] \times [0, M_2 - 1] \quad \text{and} \quad \text{supp}\{y\} = [0, L_1 - 1] \times [0, L_2 - 1].$$

Then, if the DFT size is taken as $N_1 = M_1 + L_1 - 1$ or larger, and $N_2 = M_2 + L_2 - 1$, or larger, we get the linear convolution result. Thus to avoid the circular or spatial-aliasing result, we simply have to pad the two signals with zeros out to a DFT size that is large enough to contain the *linear convolution* result, i.e., the result of ordinary noncircular convolution.

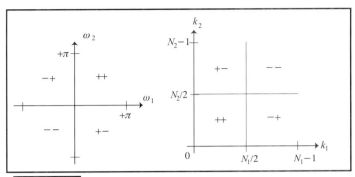

FIGURE 4.6   *Mapping of FT samples to DFT locations*

4.2.2 RELATION OF DFT TO FOURIER TRANSFORM

For a finite support sequence x, we can compute both the Fourier transform $X(\omega_1, \omega_2)$ as well as the DFT $X(k_1, k_2)$. We now answer the question of the relation of these two. Comparing (1.2-1) of Chapter 1 to (4.2-1), we have

$$X(k_1, k_2) = X(\omega_{k_1}, \omega_{k_2}), \quad \text{where } \omega_{k_i} \triangleq 2\pi k_i / N_i, \ i = 1, 2, \qquad (4.2\text{-}5)$$

for $(k_1, k_2) \in [0, N_1 - 1] \times [0, N_2 - 1]$. This means that the central or primary period of the Fourier transform $X(\omega_1, \omega_2)$ is not what is sampled, but rather an area of equal size in its first quadrant. Of course, by periodicity this is equivalent to sampling in the primary period. Based on such considerations, we can construct the diagram of Figure 4.6 indicating where the DFT samples come from in the Fourier transform plane. In particular, all the FT samples along the ω_i axes map into $k_i = 0$ in the DFT, and the possible samples (they occur only for N_i even) at $\omega_i = \pi$ map into $k_i = N_i/2$. Also note the indicated mapping of the four quadrants of $[-\pi, +\pi]^2$ in the $\omega_1 \times \omega_2$ plane onto $[0, N_1 - 1] \times [0, N_2 - 1]$ in the first quadrant of the $k_1 \times k_2$ plane.

EXAMPLE 4.2-3 (*DFT symmetry in real-valued signal case*)
When the image (signal) is real valued, we have the following DFT symmetry property $X(k_1, k_2) = X^*[(-k_1)_{N_1}, (-k_2)_{N_2}]I_{N_1 \times N_2}(k_1, k_2)$. When the DFT is stored, we only need to consider the locations $[0, N_1 - 1] \times [0, N_2 - 1]$. For these locations, we can then write

$$X(k_1, 0) = X^*(N_1 - k_1, 0) \quad (k_2 = 0),$$

$$X(0, k_2) = X^*(0, N_2 - k_2) \quad (k_1 = 0),$$

$$X(k_1, k_2) = X^*(N_1 - k_1, N_2 - k_2) \quad \text{otherwise.}$$

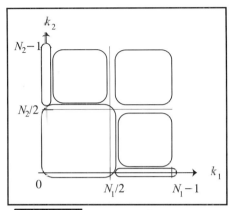

Illustration of the conjugate symmetry in DFT storage, for the real-valued image case

This then gives us the conjugate symmetry through the point $(k_1, k_2) = (N_1/2, N_2/2)$ as shown in Figure 4.7. The big square with round corners shows an essential part of the DFT coefficients and comprises locations $\{0 \leqslant k_1 \leqslant N_1/2, 0 \leqslant k_2 \leqslant N_2/2\}$. The other essential part is one of the two smaller round-cornered squares that share a side with this large square—for example, the upper square, which comprises points $\{1 \leqslant k_1 < N_1/2, N_2/2 < k_2 \leqslant N_2 - 1\}$. The other smaller squares are not needed by the symmetry condition above. Neither are the two narrow regions along the axes. Such symmetry can be used to reduce storage by approximately one-half. We need only to store the coefficients for the resulting *nonsymmetric half-size* region.

4.2.3 EFFECT OF SAMPLING IN FREQUENCY

Let $x(n_1, n_2)$ be a general signal with Fourier transform $X(\omega_1, \omega_2)$, then we can sample it as in (4.2-5) If we then take the IDFT of the function $X(k_1, k_2) I_{N_1 \times N_2}(k_1, k_2)$, we obtain the signal $y(n_1, n_2)$,

$$y(n_1, n_2) = \frac{1}{N_1 N_2} \sum_{k_1=0}^{N_1-1} \sum_{k_2=0}^{N_2-1} X\left(\frac{2\pi k_1}{N_1}, \frac{2\pi k_2}{N_2}\right) W_{N_1}^{-n_1 k_1} W_{N_2}^{-n_2 k_2} \qquad (4.2\text{-}6)$$

$$= \frac{1}{N_1 N_2} \sum_{k_1=0}^{N_1-1} \sum_{k_2=0}^{N_2-1} \left[\sum_{\text{all } m_1, m_2} x(m_1, m_2) W_{N_1}^{m_1 k_1} W_{N_2}^{m_2 k_2} \right] W_{N_1}^{-n_1 k_1} W_{N_2}^{-n_2 k_2}$$

$$(4.2\text{-}7)$$

$$\times \sum_{\text{all } m_1, m_2} x(m_1, m_2) \left[\frac{1}{N_1 N_2} \sum_{k_1=0}^{N_1-1} \sum_{k_2=0}^{N_2-1} W_{N_1}^{(m_1-n_1)k_1} W_{N_2}^{(m_2-n_2)k_2} \right] \quad (4.2\text{-}8)$$

$$= \sum_{\text{all } m_1, m_2} x(m_1, m_2) \left[\sum_{\text{all } l_1, l_2} \delta(m_1 - n_1 + l_1 N_1, m_2 - n_2 + l_2 N_2) \right] \quad (4.2\text{-}9)$$

$$= \sum_{\text{all } l_1, l_2} x(n_1 - l_1 N_1, n_2 - l_2 N_2), \quad (4.2\text{-}10)$$

which displays *spatial domain aliasing* caused by sampling in frequency. If the original signal x had a finite support $[0, M_1 - 1] \times [0, M_2 - 1]$ and we took dense enough samples of its Fourier transform, satisfying $N_1 \geqslant M_1$ and $N_2 \geqslant M_2$, then we would have no overlap in (4.2-10), or equivalently,

$$x(n_1, n_2) = y(n_1, n_2) I_{N_1 \times N_2}(n_1, n_2),$$

and will have avoided the aliased terms from coming into this spatial support region. One interpretation of (4.2-6) is as a numerical approximation to the IFT whose exact form is not computationally feasible. We thus see that the substitution of an IDFT for the IFT can result in spatial domain aliasing, which, however, can be controlled by taking the uniformly spaced samples at high enough rates is spatial to avoid significant (or any) spatial overlap (alias).

4.2.4 INTERPOLATING THE DFT

Since the DFT consists of samples of the FT as in (4.2-5), for an $N_1 \times N_2$ support signal, we can take the inverse or IDFT of these samples to express the original signal x in terms of $N_1 N_2$ samples of its Fourier transform $X(\omega_1, \omega_2)$. Then taking one more FT, we can write $X(\omega_1, \omega_2)$ in terms of its samples, i.e.,

$$X(\omega_1, \omega_2) = \text{FT}\{\text{IDFT}\{X(\omega_{k_1}, \omega_{k_2})\}\},$$

thus constituting a 2-D sampling theorem for (rectangular) samples in frequency. Actually performing the calculation, we proceed

$$X(\omega_1, \omega_2) = \sum_{n_1=0}^{N-1} \sum_{n_2=0}^{N-1} \left[\frac{1}{N_1 N_2} \sum_{k_1=0}^{N_1-1} \sum_{k_2=0}^{N_2-1} X(2\pi k_1/N_1, 2\pi k_2/N_2) W_{N_1}^{-n_1 k_1} W_{N_2}^{-n_2 k_2} \right]$$

$$\times \exp -j(\omega_1 n_1 + \omega_2 n_2)$$

$$= \sum_{k_1=0}^{N_1-1} \sum_{k_2=0}^{N_2-1} X(2\pi k_1/N_1, 2\pi k_2/N_2)$$

$$\times \left[\frac{1}{N_1 N_2} \sum_{n_1=0}^{N-1} \sum_{n_2=0}^{N-1} W_{N_1}^{-n_1 k_1} W_{N_2}^{-n_2 k_2} \exp -j(\omega_1 n_1 + \omega_2 n_2) \right]$$

$$= \sum_{k_1=0}^{N_1-1} \sum_{k_2=0}^{N_2-1} X(2\pi k_1/N_1, 2\pi k_2/N_2) \left(\frac{1}{N_1} \sum_{n_1=0}^{N-1} W_{N_1}^{-n_1 k_1} \exp -j\omega_1 n_1 \right)$$

$$\times \left(\frac{1}{N_2} \sum_{n_2=0}^{N-1} W_{N_2}^{-n_2 k_2} \exp -j\omega_2 n_2 \right)$$

$$= \sum_{k_1=0}^{N_1-1} \sum_{k_2=0}^{N_2-1} X(2\pi k_1/N_1, 2\pi k_2/N_2) \frac{1}{N_1} \frac{\sin \frac{N_1}{2}(\omega_1 - \frac{2\pi k_1}{N_1})}{\sin \frac{1}{2}(\omega_1 - \frac{2\pi k_1}{N_1})}$$

$$\times \frac{1}{N_2} \frac{\sin \frac{N_2}{2}(\omega_2 - \frac{2\pi k_2}{N_2})}{\sin \frac{1}{2}(\omega_2 - \frac{2\pi k_2}{N_2})} \exp -j\frac{N_1-1}{2}\left(\omega_1 - \frac{2\pi k_1}{N_1}\right)$$

$$\times \exp -j\frac{N_2-1}{2}\left(\omega_2 - \frac{2\pi k_2}{N_2}\right).$$

Upon definition of the 1-D interpolation functions

$$\Phi_i(\omega) \triangleq \frac{1}{N_i} \frac{\sin \frac{N_i}{2}\omega}{\sin \frac{1}{2}\omega} \exp -j\frac{N_i-1}{2}\omega, \quad \text{for } i=1,2,$$

we can write

$$X(\omega_1, \omega_2)$$
$$= \sum_{k_1=0}^{N_1-1} \sum_{k_2=0}^{N_2-1} X(2\pi k_1/N_1, 2\pi k_2/N_2) \Phi_1\left(\omega_1 - \frac{2\pi k_1}{N_1}\right) \Phi_2\left(\omega_2 - \frac{2\pi k_2}{N_2}\right).$$

This is analogous to the 1-D case [8]; as there, we must note that the key is that we have taken a number of samples $N_1 \times N_2$ that is consistent with the spatial support of the signal x. Of course, larger values of the N_i are permissible, but smaller values would create spatial aliasing. In which case, this 2-D frequency-domain sampling theorem would not be valid.

4.3 2-D DISCRETE COSINE TRANSFORM

One disadvantage to the DFT for some applications is that the transform X is complex valued, even for real data. A related transform, the discrete cosine trans-

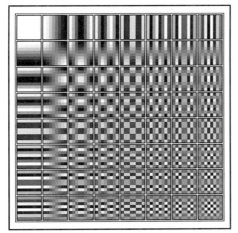

FIGURE 4.8   *Image of the basis functions of the* 8×8 *DCT*

form (DCT), does not have this problem. The DCT is a separate transform and not the real part of the DFT. It is widely used in image and video compression applications, e.g., JPEG and MPEG. It is also possible to use DCT for filtering using a slightly different form of convolution called *symmetric convolution*, cf., end-of-chapter problem 14.

DEFINITION 4.3-1 (*2-D DCT*)
Assume that the data array has finite rectangular support on $[0, N_1 - 1] \times [0, N_2 - 1]$, then the 2-D DCT is given as

$$X_C(k_1, k_2) \triangleq \sum_{n_1=0}^{N_1-1} \sum_{n_2=0}^{N_2-1} 4x(n_1, n_2) \cos \frac{\pi k_1}{2N_1}(2n_1 + 1) \cos \frac{\pi k_2}{2N_2}(2n_2 + 1),$$

(4.3-1)

for $(k_1, k_2) \in [0, N_1 - 1] \times [0, N_2 - 1]$. Otherwise, $X_C(k_1, k_2) \triangleq 0$.

The DCT basis functions for size 8×8 are shown in Figure 4.8. The mapping between the mathematical values and the colors (gray levels) is the same as in the DFT figures earlier. Each basis function occupies a small square; the squares are then arranged into an 8×8 mosaic. Note that unlike the DFT, where the highest frequencies occur near $(N_1/2, N_2/2)$, the highest frequencies of the DCT occur at the highest indices $(k_1, k_2) = (7, 7)$.

The inverse DCT exists and is given for $(n_1, n_2) \in [0, N_1 - 1] \times [0, N_2 - 1]$ as

$$x(n_1, n_2) = \frac{1}{N_1 N_2} \sum_{k_1=0}^{N_1-1} \sum_{k_2=0}^{N_2-1} w(k_1) w(k_2) X_C(k_1, k_2)$$

$$\times \cos \frac{\pi k_1}{2N_1} (2n_1 + 1) \cos \frac{\pi k_2}{2N_2} (2n_2 + 1),$$

where the weighting function $w(k)$ is given just as in the case of the 1-D DCT by

$$w(k) \triangleq \begin{cases} 1/2, & k = 0 \\ 1, & k \neq 0. \end{cases}$$

By observation of (4.3-1), we see that the 2-D DCT is a separable operator. As such it can be applied to the rows and then the columns, or *vice versa*. Thus the 2-D theory can be developed by repeated application of the 1-D theory. In the following subsections we relate the 1-D DCT to the 1-D DFT of a symmetrically extended sequence. This not only provides an understanding of the DCT but also enables its fast calculation via methods intended for the DFT. Later, we present a fast DCT calculation that can avoid the use of complex arithmetic in the usual case where z is a real-valued signal, e.g., an image. (Note: The next two subsections can be skipped by the reader familiar with the 1-D DCT theory.)

4.3.1 Review of 1-D DCT

In the 1-D case the DCT is defined as

$$X_C(k) \triangleq \begin{cases} \sum_{n=0}^{N-1} 2x(n) \cos \frac{\pi k}{2N} (2n + 1), & k \in [0, N - 1] \\ 0, & \text{else,} \end{cases} \qquad (4.3\text{-}2)$$

for every N point signal x having support $[0, N - 1]$. The corresponding inverse transform, or IDCT, can be written as

$$x(n) = \begin{cases} \frac{1}{N} \sum_{k=0}^{N-1} w(k) X_C(k) \cos \frac{\pi k}{2N} (2n + 1), & n \in [0, N - 1] \\ 0, & \text{else.} \end{cases} \qquad (4.3\text{-}3)$$

It turns out that this 1-D DCT can be understood in terms of the DFT of a *symmetrically extended sequence*[3]

$$y(n) \triangleq x(n) + x(2N - 1 - n), \qquad (4.3\text{-}4)$$

---

3. This is not the only way to symmetrically extend x, but this method results in the most widely used DCT sometimes called DCT-2 [8].

with support $[0, 2N - 1]$. In fact, on defining the $2N$ point DFT $Y(k) \triangleq$ $\mathrm{DFT}_{2N}\{y(n)\}$, we will show that the DCT can be alternatively expressed as

$$
X_C(k) = \begin{cases} W_{2N}^{k/2} Y(k), & k \in [0, N - 1] \\ 0, & \text{else.} \end{cases}
\tag{4.3-5}
$$

Thus the DCT is just the DFT analysis of the symmetrically extended signal (4.3-4). Looking at this equation, we see that there is no overlap in its two components, which fit together without a gap. We can see that right after $x(N-1)$ comes $x(N-1)$ at position $n = N$, which is then followed by the rest of the nonzero part of x in reverse order, out to $n = 2N - 1$, where sits $x(0)$. We can see a point of symmetry midway between $n = N - 1$ and N, i.e., at $n = N - \frac{1}{2}$. If we consider its periodic extension $\tilde{y}(n)$, we will also see a symmetry about the point $n = -\frac{1}{2}$. We thus expect that the $2N$ point DFT $Y(k)$ will be real valued except for the phase factor $W_{2N}^{-k/2}$. So the phase factor in (4.3-5) is just what is needed to cancel out the phase term in Y and make the DCT real, as it must if the two equations, (4.3-1) and (4.3-5), are to agree for real valued inputs x.

To reconcile these two definitions, we start out with (4.3-5), and proceed as follows:

$$
\begin{aligned}
Y(k) &= \sum_{n=0}^{N-1} x(n) W_{2N}^{nk} + \sum_{n=N}^{2N-1} x(2N - 1 - n) W_{2N}^{nk} \\
&= \sum_{n=0}^{N-1} x(n) W_{2N}^{nk} + \sum_{n'=0}^{N-1} x(n') W_{2N}^{-(n'+1)k}, \quad \text{with } n' \triangleq 2N - 1 - n \\
&= W_{2N}^{-k/2} \sum_{n=0}^{N-1} x(n) W_{2N}^{(n+0.5)k} + W_{2N}^{-k/2} \sum_{n=0}^{N-1} x(n) W_{2N}^{-(n+.5)k} \\
&= W_{2N}^{-k/2} \sum_{n=0}^{N-1} 2x(n) \cos \frac{\pi k}{2N}(2n + 1) \quad \text{for } k \in [0, 2N - 1],
\end{aligned}
$$

the last line following from $W_{2N}^{-(n+.5)k} = \exp(j2\pi(n + 0.5)k/2N)$ and Euler's relation, which agrees with the original definition, (4.3-2).

The formula for the inverse DCT, (4.3-3), can be established similarly, starting out from

$$
x(n) = \left[\frac{1}{2N} \sum_{k=0}^{2N-1} Y(k) W_{2N}^{-nk} \right] I_N(n).
$$

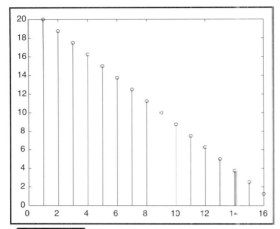

FIGURE 4.9   *MATLAB plot of x(n) over its support*

EXAMPLE 4.3-1 (*1-D DCT*)

In this example, we use MATLAB to take the DCT of the finite support function x given on its support region $[0, N-1]$ as

$$x(n) = 20(1 - n/N).$$

We choose $N = 16$ and call the MATLAB function dct using the .m file DCTeg2.m contained on the included CD-ROM,

```
clear
N=16;
for i1=1:N,
    n=i1-1;
    signal(i1)=20*(1-n/N);
end
stem(signal)
figure
x=fft(signal);
xm=abs(x);
stem(xm)
figure
xc=dct(signal);
stem(xc),
```

where we also calculate the FFT magnitude for comparison purposes. The resulting plots are shown in Figures 4.9–4.11. We see that for this highly asymmetric signal, the DCT is a much more efficient representation than is the DFT.

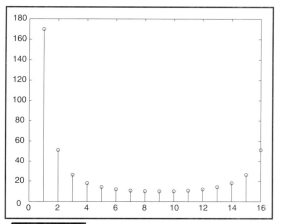

FIGURE 4.10   *MATLAB plot of DFT magnitude* $|X(k)|$

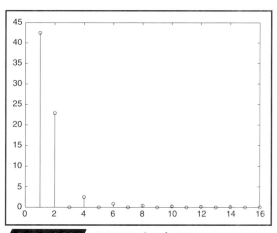

FIGURE 4.11   *MATLAB plot of DCT* $X_C(k)$

4.3.2 SOME 1-D DCT PROPERTIES

1   *Linearity*:

$$ax + by \quad \Leftrightarrow \quad aX_C + bY_C.$$

2   *Energy conservation*:

$$\sum_{n=0}^{N-1} |x(n)|^2 = \frac{1}{2N} \sum_{k=0}^{N-1} w(k) |X_C(k)|^2. \qquad (4.3\text{-}6)$$

3 *Symmetry:*

(a) *General case:*

$$x^*(n) \quad \Leftrightarrow \quad X_C^*(k).$$

(b) *Real-valued case:*

$$\text{real}\, x(n) \quad \Rightarrow \quad \text{real}\, X_C(k).$$

4 *Eigenvectors of unitary DCT:* Define the column vector $\mathbf{x} \triangleq [x(0), x(1), \dots, x(N-1)]^T$ and define the matrix $\mathbf{C}$ with elements:

$$C_{k',n'} \triangleq \begin{cases} \sqrt{\dfrac{1}{N}}, & k' = 1 \\ \sqrt{\dfrac{2}{N}} \cos \dfrac{\pi}{2N}(k'-1)(2n'-1), & 1 < k' \leqslant N, \end{cases}$$

then the vector $\mathbf{y} = \mathbf{Cx}$ contains the *unitary* DCT, whose elements are given as

$$\mathbf{y} \triangleq \left[\sqrt{\frac{1}{N}} X_C(0), \sqrt{\frac{2}{N}} X_C(1), \dots, \sqrt{\frac{2}{N}} X_C(N-1) \right]^T.$$

A *unitary matrix* is one whose inverse is the same as the transpose $\mathbf{C}^{-1} = \mathbf{C}^T$. For the unitary DCT, we have

$$\mathbf{x} = \mathbf{C}^T \mathbf{y},$$

and energy balance equation,

$$\mathbf{x}^T \mathbf{x} = \mathbf{y} \mathbf{C} \mathbf{C}^T \mathbf{y}$$
$$= \mathbf{y}^T \mathbf{y},$$

a slight modification on the DCT Parseval equation of (4.3-6). So the unitary DCT preserves the energy of the signal x.

It turns out that eigenvectors of the unitary DCT are the same as those of the symmetric tridiagonal matrix [4]

$$\mathbf{Q} = \begin{bmatrix} 1-\alpha & -\alpha & 0 & \cdot\,\cdot & \cdots & 0 \\ -\alpha & 1 & -\alpha & 0 & \cdots & 0 \\ 0 & -\alpha & 1 & -\alpha & \ddots & \vdots \\ \vdots & 0 & -\alpha & \cdot & \ddots & 0 \\ \vdots & \ddots & \ddots & \cdot & 1 & -\alpha \\ 0 & \cdots & \cdots & 0 & -\alpha & 1-\alpha \end{bmatrix}$$

and this holds true for arbitrary values of the parameter α.

We can relate this matrix $\mathbf{Q}$ to the inverse covariance matrix of a 1-D first-order stationary Markov random sequence [10], with correlation coefficient ρ, necessarily satisfying $|\rho| < 1$.

$$\mathbf{R}^{-1} = \frac{1}{\beta^2} \begin{bmatrix} 1-\rho\alpha & -\alpha & 0 & \cdots & \cdots & 0 \\ -\alpha & 1 & -\alpha & 0 & \cdots & 0 \\ 0 & -\alpha & 1 & -\alpha & \ddots & \vdots \\ \vdots & 0 & -\alpha & \ddots & \ddots & 0 \\ \vdots & \ddots & \ddots & \ddots & 1 & -\alpha \\ 0 & \cdots & \cdots & 0 & -\alpha & 1-\rho\alpha \end{bmatrix},$$

where $\alpha \triangleq \rho/(\rho+1)$ and $\beta^2 \triangleq (1-\rho^2)/(1+\rho^2)$. The actual covariance matrix of the Markov random sequence is

$$\mathbf{R} = \begin{bmatrix} 1 & \rho & \rho^2 & \rho^3 & \cdots & \rho^{N-1} \\ \rho & 1 & \rho & \rho^2 & \ddots & \vdots \\ \rho^2 & \rho & 1 & \rho & \ddots & \rho^3 \\ \rho^3 & \rho^2 & \rho & \ddots & \ddots & \rho^2 \\ \vdots & \ddots & \ddots & \ddots & \ddots & \rho \\ \rho^{N-1} & \cdots & \rho^3 & \rho^2 & \rho & 1 \end{bmatrix},$$

with corresponding, first-order difference equation,

$$x(n) = \rho x(n-1) + w(n).$$

It can further be shown that when $\rho \simeq 1$, the matrix $\mathbf{Q} \simeq \beta^2 \mathbf{R}^{-1}$, so that their eigenvectors approximate each other too. Because the eigenvectors of a matrix and its inverse are the same, we then have the fact that the unitary DCT basis vectors approximate the Karhunen–Loeve expansion [10], with basis vectors given as the solution to the matrix-vector equation,

$$\mathbf{R}\boldsymbol{\Phi} = \boldsymbol{\Lambda}\boldsymbol{\Phi},$$

and corresponding Karhunen–Loeve transform (KLT) given by

$$\mathbf{y} = \boldsymbol{\Phi}^\dagger \mathbf{x}.$$

Thus the 1-D DCT of a first-order Markov random vector of dimension N should be close to the KLT of x when its correlation coefficient $\rho \simeq 1$. This ends the review of the 1-D DCT.

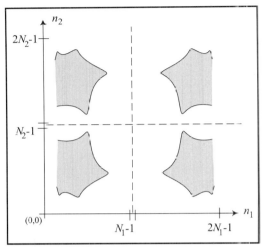

FIGURE 4.12   *Illustration of 2-D symmetric extension used in the DCT*

4.3.3 SYMMETRIC EXTENSION IN 2-D DCT

Since the 2-D DCT

$$X_C(k_1, k_2) = \sum_{n_1=0}^{N_1-1} \sum_{n_2=0}^{N_2-1} 4x(n_1, n_2) \cos \frac{\pi k_1}{2N_1}(2n_1 + 1) \cos \frac{\pi k_2}{2N_2}(2n_2 + 1),$$

is just the separable operator resulting from application of the 1-D DCT along first one dimension and then the other, the order being immaterial, we can easily extend the 1-D DCT properties to the 2-D case. In terms of the connection of the 2-D DCT with the 2-D DFT, we thus see that we must symmetrically extend in, say, the horizontal direction and then symmetrically extend that result in the vertical direction. The resulting symmetric function (extension) becomes

$$y(n_1, n_2) \triangleq x(n_1, n_2) + x(n_1, 2N_2 - 1 - n_2) + x(2N_1 - 1 - n_1, n_2)$$
$$+ x(2N_1 - 1 - n_1, 2N_2 - 1 - n_2),$$

which is sketched in Figure 4.12, where we note that the symmetry is about the lines $N_1 - \frac{1}{2}$ and $N_2 - \frac{1}{2}$. Then from (4.3-5), it follows that the 2-D DCT is given in terms of the $N_1 \times N_2$ point DFT as

$$X_C(k_1, k_2) = \begin{cases} W_{2N_1}^{k_1/2} W_{2N_2}^{k_2/2} Y(k_1, k_2), & (k_1, k_2) \in [0, N_1 - 1] \times [0, N_2 - 1] \\ 0, & \text{else} \end{cases}$$

COMMENTS

1 We see that both the 1-D and 2-D DCTs involve only real arithmetic for real-valued data, and this may be important in some applications.

2 The symmetric extension property can be expected to result in fewer high-frequency coefficients in DCT with respect to DFT. Such would be expected for lowpass data, since there would often be a jump at the four edges of the $N_1 \times N_2$ period of the corresponding periodic sequence $\tilde{x}(n_1, n_2)$, which is not consistent with small high-frequency coefficients in the DFS or DFT. Thus the DCT is attractive for lossy data storage applications, where the exact value of the data is not of paramount importance.

3 The DCT can be used for a symmetrical type of filtering with a symmetrical filter (see end-of-chapter problem 14).

4 2-D DCT properties are easy generalizations of 1-D DCT properties in Section 4.3.2.

4.4 SUBBAND/WAVELET TRANSFORM (SWT)[4]

The DCT transform is widely used for compression, in which case it is almost always applied as a *block transform*. In a block transform, the data are scanned, a number of lines at a time, and the data are then mapped into a sequence of blocks. These blocks are then operated upon by 2-D DCT. In the image compression standard JPEG, the DCT is used with a block size of 8×8. SWTs can be seen as a generalization of such block transforms, a generalization that allows the blocks to overlap. SWTs make direct use of decimation and expansion and LSI filters. Recalling Example 2.3-1 of Chapter 2, we saw that ideal rectangular filters could be used to decompose the $2\pi \times 2\pi$ unit cell of the frequency domain into smaller regions. Decimation could then be used on these smaller regions to form component signals at lower spatial sample rates. The collection of these lower sample rate signals then would be equivalent to the original high-rate signal. However, being separate signals, now they are often amenable to processing targeted to their reduced frequency range, and this has been found useful in several to many applications.

4.4.1 IDEAL FILTER CASE

To be specific, consider the four-subband decomposition shown in Figure 4.13.

Using ideal rectangular filters, we can choose H_{00} to pass the LL subband $\{|\omega_1| \leqslant \pi/2, |\omega_2| \leqslant \pi/2\}$ and reject other frequencies in $[-\pi, +\pi]^2$. Then, after 2×2 decimation, set H_{10} to pass the HL subband $\{\pi/2 < |\omega_1|, |\omega_2| \leqslant \pi/2\}$, set

4. What we here call the subband/wavelet transform (SWT) is usually referred to as either the discrete wavelet transform (DWT) or as the subband transform.

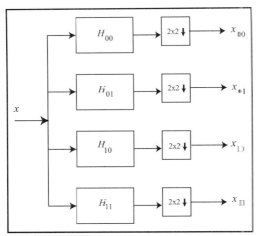

FIGURE 4.13   *Illustration of 2×2 rectangular subband/wavelet transform*

H_{01} to pass the LH subband $\{|\omega_1| \leqslant \pi/2, \pi/2 < |\omega_2|\}$, and, finally, H_{11} to pass the HH subband $\{\pi/2 < |\omega_1|, \pi/2 < |\omega_2|\}$, all defined as subsets of $[-\pi, +\pi]^2$. Then the four subband signals x_{00}, x_{10}, x_{01}, and x_{11} will contain all the information from the input signal x, each with a four-times lower spatial sample rate after 2×2 decimation. For example, using (2.3-1) from Chapter 2, Section 3, we have for the 2×2 decimation case, where $M_1 = M_2 = 2$,

$$X_{ij}(\omega_1, \omega_2) = \frac{1}{4} \sum_{k_1=0}^{1} \sum_{k_2=0}^{1} H_{ij}\left(\frac{\omega_1 - 2\pi k_1}{2}, \frac{\omega_2 - 2\pi k_2}{2}\right) X\left(\frac{\omega_1 - 2\pi k_1}{2}, \frac{\omega_2 - 2\pi k_2}{2}\right).$$

$$(4.4-1)$$

For the case $i = j = 0$, and noting the filter H_{00} has unity gain in its $[-\pi/2, +\pi/2]^2$ passband, we have the LL subband

$$X_{00}(\omega_1, \omega_2) = \frac{1}{4} X\left(\frac{\omega_1}{2}, \frac{\omega_2}{2}\right) \quad \text{on } [-\pi, +\pi]^2, \qquad (4.4-2)$$

because the other three terms in (4.4-1), with i and j not both 0, are all 0 in the frequency domain unit cell $[-\pi, +\pi]^2$. Thus we can write the LL subband signal x_{00} as[5]

$$x_{00}(n_1, n_2) = \text{IFT}\left\{\tfrac{1}{4} X(\omega_1/2, \omega_2/2) \text{ on } [-\pi, +\pi]^2\right\}.[6]$$

5. Note that $\frac{1}{4} X(\omega_1/2, \omega_2/2)$ is not periodic with period $2\pi \times 2\pi$ and so is not a complete expression for the FT. Still, over the fundamental period $[-\pi, -\pi]^2$, it is correct. The complete expression for the FT just repeats this expression, periodically, outside $[-\pi, +\pi]^2$.

6. We write the function to be inverse Fourier transformed in the unconventional way, $\frac{1}{4} X(\omega_1/2, \omega_2/2)$ on $[-\pi, +\pi]^2$, to emphasize that $\frac{1}{4} X(\omega_1/2, \omega_2/2)$ by itself is not a Fourier transform, as mentioned in the preceding note.

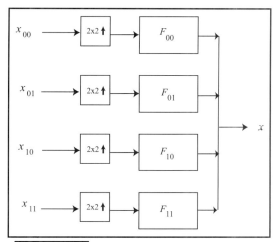

FIGURE 4.14   *Illustration of 2×2 rectangular subband/wavelet inverse transform*

Similarly, regarding the LH subband x_{01}, we can say that

$$x_{01}(n_1, n_2) = \mathbf{IFT}\left\{ \tfrac{1}{4} X(\omega_1/2, \omega_2/2 - \pi) \text{ on } [-\pi, +\pi]^2 \right\}$$

and so-forth for x_{10} and x_{11}. Thus, we have the following frequency domain expressions, each valid only in the unit frequency cell $[-\pi, +\pi]^2$,

$$X_{00}(\omega_1, \omega_2) = \frac{1}{4} X(\omega_1/2, \omega_2/2),$$

$$X_{01}(\omega_1, \omega_2) = \frac{1}{4} X(\omega_1/2, \omega_2/2 - \pi),$$

$$X_{10}(\omega_1, \omega_2) = \frac{1}{4} X(\omega_1/2 - \pi, \omega_2/2),$$

and

$$X_{11}(\omega_1, \omega_2) = \frac{1}{4} X(\omega_1/2 - \pi, \omega_2/2 - \pi).$$

To complete this 2×2 rectangular SWT, we must show the inverse transform that reconstructs the original signal x from these four subband signals x_{00}, x_{01}, x_{10}, and x_{11}. The inverse transform system diagram is shown in Figure 4.14. With reference to (2.3-4) from Chapter 2, we can write the Fourier transform output of each filter as

$$F_{kl}(\omega_1, \omega_2) X_{kl}(2\omega_1, 2\omega_2), \quad \text{for } k, l = 0, 1.$$

Combining these results, we obtain the overall equation for transform followed by inverse transform as

$$X(\omega_1, \omega_2) = \sum_{k,l=0,1} F_{kl}(\omega_1, \omega_2) X_{kl}(2\omega_1, 2\omega_2)$$

$$= \sum_{k,l=0,1} F_{kl}(\omega_1, \omega_2) \frac{1}{4} X(\omega_1 - k\pi, \omega_2 - l\pi). \qquad (4.4\text{-}3)$$

Now, if we set the gains of the ideal synthesis filters $F_{kl} = 4$, and define their frequency domain regions of support to be the same as those of the analysis filters H_{kl}, we obtain

$$X(\omega_1, \omega_2) = \begin{cases} X(\omega_1, \omega_2), & |\omega_1| \leqslant \pi/2, |\omega_2| \leqslant \pi/2 \\ X(\omega_1, \omega_2), & |\omega_1 - \pi| < \pi/2, |\omega_2| \leqslant \pi/2 \\ X(\omega_1, \omega_2), & |\omega_1| \leqslant \pi/2, |\omega_2 - \pi| < \pi/2 \\ X(\omega_1, \omega_2), & |\omega_1 - \pi| < \pi/2, |\omega_2 - \pi| < \pi/2, \end{cases}$$

$$= X(\omega_1, \omega_2) \quad \text{for } |\omega_1| \leqslant \pi, |\omega_2| \leqslant \pi.$$

We can regard this as an ideal SWT in the sense that the Fourier transform is an ideal frequency transform, while the DFT and DCT are more practical ones, i.e., they involve a finite amount of computation and do not have ideal filter shapes. We can substitute nonideal filters for the preceding ideal filter case and get a more practical type of subband/wavelet transform. The resulting transform, while not avoiding aliasing error in the subband signals, does manage to cancel out this aliasing error in the *reconstructed* or inverse transformed signal x. To distinguish the two, we can call the preceding case an *ideal SWT*, and the practical type, with finite-order filters, just an SWT. Of course, there is no unique SWT, as it will depend on the filters used.

4.4.2 1-D SWT with Finite-Order Filter

We can avoid the need for ideal filters if we can accept some aliasing error in the subbands. We first review the 1-D case for a two-channel filter bank. Then we will construct the required 2-D filters as separable. If we construct the 1-D filters to cancel out any aliasing in the reconstruction, then the same will be true of the separable 2-D transform. Using the 1-D theory we find

$$X_0(\omega) = \frac{1}{2} \big[H_0(\omega/2) X(\omega/2) + H_0(\omega/2 - \pi) X(\omega/2 - \pi) \big]$$

and

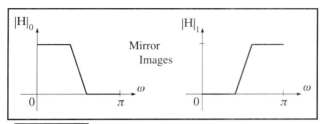

FIGURE 4.15   *Illustration of magnitude responses of a quadrature magnitude filter (QMF) pair*

$$X_1(\omega) = \frac{1}{2}\big[H_1(\omega/2)X(\omega/2) + H_1(\omega/2 - \pi)X(\omega/2 - \pi)\big].$$

Similarly, the reconstructed signal (inverse transform) is given as

$$\widehat{X}(\omega) = F_0(\omega)X_0(2\omega) + F_1(\omega)X_1(2\omega)$$

$$= \frac{1}{2}\big[F_0(\omega)H_0(\omega) + F_1(\omega)H_1(\omega)\big]X(\omega)$$

$$+ \frac{1}{2}\big[F_0(\omega)H_0(\omega - \pi) + F_1(\omega)H_1(\omega - \pi)\big]X(\omega - \pi).$$

Therefore, to cancel out aliasing, we need

$$F_0(\omega)H_0(\omega - \pi) + F_1(\omega)H_1(\omega - \pi) = 0, \tag{4.4-4}$$

and then, to achieve perfect reconstruction, we additionally need

$$F_0(\omega)H_0(\omega) + F_1(\omega)H_1(\omega) = 2.$$

The first solution to this problem used the quadrature mirror condition [1,3],

$$F_0(\omega) = H_1(\omega - \pi) \quad \text{and} \quad F_1(\omega) = -H_0(\omega - \pi),$$

which satisfies (4.4-4). In order to get a lowpass–highpass filter pair, they then set

$$H_1(\omega) = H_0(\omega - \pi),$$

where we see the "mirror symmetry" as sketched in Figure 4.15.

We can see that if H_0 is chosen as a lowpass filter, then H_1 will be highpass, and similarly the reconstruction filter F_0 will be lowpass and F_1 will be highpass. To get perfect reconstruction we then need

$$H_0^2(\omega) - H_0^2(\omega - \pi) = 2.$$

Since the 2-D separable case can be thought of as the sequential application of the above 1-D subsampling to the rows and columns, we have immediately that the corresponding separable 2-D filters will work to achieve alias cancellation and perfect reconstruction also.

$$H_{ij}(\omega_1, \omega_2) = H_i(\omega_1)H_j(\omega_2) \quad \text{and} \quad F_{ij}(\omega_1, \omega_2) = F_i(\omega_1)F_j(\omega_2).$$

An alternative formulation of the alias cancellation is due to Mintzer [7], and Simoncelli and Adelson [9]. In their approach,

$$F_0(\omega) = H_0(-\omega) \quad \text{and} \quad F_1(\omega) = H_1(-\omega),$$

which results in the reconstruction formula

$$\widehat{X}(\omega) = \frac{1}{2}\big[H_0(\omega)H_0(-\omega) - H_1(\omega)H_1(-\omega)\big]X(\omega)$$

$$+ \frac{1}{2}\big[H_0(\omega - \pi)H_0(-\omega) + H_1(\omega - \pi)H_1(-\omega)\big]X(\omega - \pi)$$

$$= \frac{1}{2}\big[|H_0(\omega)|^2 + |H_1(\omega)|^2\big]X(\omega) + \text{alias terms.}$$

Upon setting

$$H_0(\omega) = H(\omega) \quad \text{and} \quad H_1(\omega) = e^{+j\omega}H(-\omega + \pi),$$

again the alias terms cancel out. Here $H(\omega)$ is chosen as a lowpass filter designed to achieve

$$|H_0(\omega)|^2 + |H_1(\omega)|^2 = 2,$$

$$|H(\omega)|^2 + |H(-\omega + \pi)|^2 = 2.$$

Some methods of subband filter design will be discussed in Chapter 5.

4.4.3 2-D SWT with FIR Filters

We can apply the just developed 1-D theory separably to the horizontal and vertical axes, resulting in four subbands $X_{ij}(\omega_1, \omega_2)$, just as in (4.4-1). Due to finite-order filters though, equations like (4.4-2) will not hold and alias error will be present in the subbands X_{ij}. However, if we use the separable filters we just discussed, which have the property of canceling out aliasing in the reconstruction, then an *inverse* SWT (ISWT) exists and is given by (4.4-3). Depending on the filters used and their length, the amount of aliasing in the subbands may be significant or not. Filter designs for SWT ameliorate this problem mainly by going to longer filters.

4.4.4 Relation of SWT to DCT

Consider an 8×8 rectangular SWT. It would have 64 subbands, with center frequencies evenly distributed over the frequency unit cell. Now, consider applying the DCT to the same data, but blockwise, using 8×8 DCTs. We can easily identify each set of DCT coefficients with one of the subbands. In fact, if we were to use the DCT basis functions, reflected through the origin as the subband/wavelet filters, i.e.,

$$h_{kl}(n_1, n_2) = 4 \cos \frac{\pi k}{2N_1}(-2n_1 + 1) \cos \frac{\pi l}{2N_2}(-2n_2 + 1),$$

then the subband values would be exactly the sequence of DCT coefficients at that frequency k, l. We can thus see the SWT as a generalization of the block DCT transform, wherein the basis functions (subband filter impulse responses) are allowed to overlap in space.

4.4.5 Relation of SWT to Wavelets

Wavelet theory is essentially the continuous-time theory that corresponds to dyadic subband transforms, i.e., those where only the L (LL) subband is recursively split over and over. Wavelet analysis of a continuous-time signal begins as follows. Let $f(t) \in L^2$ (L^2 being the space of square integrable functions $\int_{-\infty}^{+\infty} |f(t)|^2 \, dt < \infty$), and specify a *mother wavelet* $\psi(t)$ as some

$$\psi(t) \in L^1 \quad \text{satisfying} \quad \int_{-\infty}^{+\infty} \psi(t) \, dt = 0,$$

i.e., a specified highpass function. We then define a *continuous wavelet transform* (CWT) of f as [6,11]

$$\gamma(s, \tau) \triangleq \int_{-\infty}^{+\infty} f(t) \psi_{s,\tau}(t) \, dt,$$

with *wavelets* $\psi_{s,\tau}(t) \triangleq \frac{1}{\sqrt{s}} \psi(\frac{t-\tau}{s})$. The positive parameter s is called the *scale*, and τ is called the *delay*. We can see that, if ψ is concentrated in time, a CWT gives a time-varying analysis of f into scales or resolutions, where increasing the scale parameter s provides a lower resolution analysis. You might think that f is being overanalyzed here, and indeed it is. Wavelet theory shows that a discrete version of this analysis suffices to describe the function. A *discrete wavelet transform* DWT (still for continuous-time functions) is defined via the discrete family of wavelets [6,11]:

$$\psi_{k,l}(t) \triangleq \frac{1}{\sqrt{s_0^k}} \psi\left(\frac{t - l\tau_0 s_0^k}{s_0^k}\right),$$

$$= 2^{-k/2}\psi(2^{-k}t - l), \quad \text{with } s_0 = 2 \text{ and } \tau_0 = 1.$$

as

$$\gamma(k,l) \triangleq \int_{-\infty}^{+\infty} f(t)\psi_{k,l}(t)\,dt.$$

This analysis can be inverted as follows:

$$f(t) = \sum_{k,l} \gamma(k,l)\psi_{k,l}(t), \tag{4.4-5}$$

which is called an *inverse discrete wavelet transform* (IDWT), thus creating the analysis–synthesis pair that justifies the use of the word "transform." Now, the k index denotes scale, so that as k increases the scale gets larger and larger, while the l index denotes delay or position of the scale information on the time axis. The values of the DWT $\gamma(k,l)$ are also called the *wavelet coefficients* of the continuous-time function f. Here both indices k,l are doubly infinite, i.e., $-\infty < k,l < +\infty$, in order to span the full range of scales and delays.

Rewriting (4.4-5) as a sum over scales,

$$f(t) = \sum_k \left(\sum_l \gamma(k,l)\psi_{k,l}(t)\right), \tag{4.4-6}$$

we can break the scales up into two parts,

$$f(t) = \sum_{k < k_C} \left[\sum_l \gamma(k,l)\psi_{k,l}(t)\right] - \sum_{k \geq k_0} \left[\sum_l \gamma(k,l)\psi_{k,l}(t)\right],$$

$$\underbrace{\qquad\qquad\qquad}_{\text{low-scale (frequency) part}} \qquad \underbrace{\qquad\qquad\qquad}_{\text{fine-scale (frequency) part}}$$

which is reminiscent of an SWT that we have defined only for discrete-time functions earlier in this chapter. In fact, wavelet theory shows that the two theories fit together in the following sense. If a continuous-time function is wavelet analyzed with a DWT at a fine scale, then the wavelet coefficients at the different scales are related by an SWT. Going a bit further, to each DWT there corresponds a dyadic SWT that recursively calculates the wavelet coefficients. Unfortunately, there are SWTs that do not correspond to wavelet transforms due to lack of convergence as the scale gets larger and larger. This is the so-called *regularity* problem. However, it can be generally said that most all SWTs currently being used have good enough regularity properties, given the rather small number of scales that are used both for image analysis and for image coding, typically less than 7. More information on wavelets and wavelet transforms can be obtained from many sources, including the classic sources [6,11].

Before leaving the topic of wavelets, notice that the scale parameter k in (4.4-5) never quite gets to $-\infty$, as the infinite sum is really just the limit of finite sums. However, this is OK, since there cannot be any "dc" value or constant in any function $f(t) \in L^2$ over the infinite interval $(-\infty, +\infty)$, remember we require that the mother wavelet and, as a result, the actual wavelet basis functions have zero mean, i.e., $\int_{-\infty}^{+\infty} \psi(t)\, dt = 0$.

4.5 FAST TRANSFORM ALGORITHMS

The DFT and DCT are widely used in image and video signal processing, due in part to the presence of fast algorithms for their calculation. These fast algorithms are largely inherited from the 1-D case due to the separable nature of these transforms.

4.5.1 FAST DFT ALGORITHM

Here we briefly look at efficient algorithms to calculate the DFT. We cover the so-called row–column approach, which expresses the 2-D DFT as a series of 1-D transforms. To understand this method we start out with

$$X(k_1, k_2) = \sum_{n_1=0}^{N_1-1} \sum_{n_2=0}^{N_2-1} x(n_1, n_2) W_{N_1}^{n_1 k_1} W_{N_2}^{n_2 k_2},$$

valid for $(k_1, k_2) \in [0, N_1 - 1] \times [0, N_2 - 1]$, and bring the sum over n_1 inside, to obtain

$$= \sum_{n_2=0}^{N_2-1} \left[\sum_{n_1=0}^{N_1-1} x(n_1, n_2) W_{N_1}^{n_1 k_1} \right] W_{N_2}^{n_2 k_2}$$

$$= \sum_{n_2=0}^{N_2-1} X(k_1; n_2) W_{N_2}^{n_2 k_2} \tag{4.5-1}$$

upon defining the row-transforms $X(k_1; n_2)$ as

$$X(k_1; n_2) \triangleq \sum_{n_1=0}^{N_1-1} x(n_1, n_2) W_{N_1}^{n_1 k_1},$$

for each row n_2 of the original data array x. If we store these values back in the same row, then we see that (4.5-1) is then just the 1-D column DFTs of this intermediate data array. If the data are stored sequentially on a tape or disk, it may be advantageous to perform a data transposition step between the column and row transforms, to minimize access to this secondary storage [5]. When there is sufficient random-access storage, such an intermediate step is not necessary. If we use an in-place 1-D Fourier transform (FFT) routine to implement the row and column transforms, then the amount of random-access memory (RAM) needed would be determined by the size of the data array x with support $N_1 \times N_2$.

COMPUTATION

Here we assume that both N_1 and N_2 are powers of 2:

Step 1: We use the Cooley–Tukey 1-D FFT algorithm [8] to do the row transforms, either *decimation in time* (DIT) or *decimation in frequency* (DIF) with $N_2 \cdot \frac{N_1}{2} \log_2 N_1$ complex multiplies and $N_2 \cdot N_1 \log_2 N_1$ complex additions.

Step 2: We do any necessary data mappings due to insufficient RAM, e.g., matrix transposition.

Step 3: We use the Cooley–Tukey 1-D FFT algorithm again to do the column transforms in $N_1 \cdot \frac{N_2}{2} \log_2 N_2$ complex multiplies and $N_1 \cdot N_2 \log_2 N_2$ complex additions.

The total amount of computation then comes to $\frac{N_1 N_2}{2} \log_2 N_1 N_2$ complex multiplications and $N_1 N_2 \log_2 N_1 N_2$ complex additions. These are the so-called *radix-2* algorithms. There are also *radix-4* algorithms for 1-D FFT and vector radix algorithms for 2-D FFT [5]. They both can offer some modest improvement over the radix-2 case considered here. For example, if N is a power of 4, then the radix-4 approach can save about 25% in computation.

4.5.2 FAST DCT METHODS

We can employ the fast method of 1-D DFT calculation to the 1-D DCT in implementing the 2-D DCT by a row–column method. We would simply calculate the 2N-point DFT of the symmetrically extended sequence y. This would be an efficient $N \log_2 N$ method but would involve complex arithmetic. An alternative method involving only real arithmetic for a real-valued $x(n)$ has been obtained [12] as follows: They define the real and imaginary parts of the N-point DFT as

$$\text{cos-DFT}_N(k) \triangleq \sum x(n) \cos \frac{2\pi nk}{N}$$

and

$$\text{sin-DFT}_N(k) \triangleq \sum x(n) \sin \frac{2\pi nk}{N}.$$

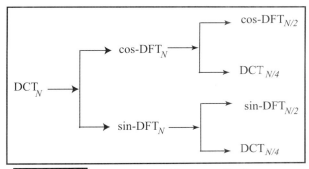

FIGURE 4.16   *Tree diagram of fast DCT of* [12]

Then they find

$$\text{DFT}_N(k) = \text{cos-DFT}_N(k) - j\,\text{sin-DFT}_N(k),$$

$$\text{DCT}_N(N) = 0,$$

$$\text{DCT}_N(-k) = \text{DCT}_N(+k),$$

$$\text{DCT}_N(2N - k) = -\text{DCT}_N(k),$$

$$\text{cos-DFT}_N(N - k) = \text{cos-DFT}_N(k),$$

and

$$\text{sin-DFT}_N(N - k) = -\text{sin-DFT}_N(k),$$

where in these equations, and in others of this section, the DCT and DFT transforms are not automatically set to zero outside their fundamental period. Using these relations and some others, the dependency tree of Figure 4.16 is developed [12]. The resulting computation for N a positive integral power of two, becomes approximately $\frac{N}{2}\log_2 N$ real multiplications and $\frac{N}{2}(3\log_2 N - 2) + 1$ real additions (please see [12] for the details). The 2-D $N_1 \times N_2$ DCT, via a row–column method, would then require $N_1 \cdot \frac{N_2}{2}\log_2 N_2 + N_2 \cdot \frac{N_2}{2}\log_2 N_2$ real multiplications.

4.6 SECTIONED CONVOLUTION METHODS

Consider an image of size $M_1 \times M_2$, to be convolved with an FIR filter of size $L_1 \times L_2$, where usually we have $L_i \ll M_i$, $i = 1, 2$. In such a case, it is not efficient to directly implement the 2-D convolution in the DFT domain. However, we can first section the input into adjoining rectangular sections $x_{i,j}(n_1, n_2)$ so that

$$x(n_1, n_2) = \sum_{i,j} x_{i,j}(n_1, n_2).$$

Then we can write the convolution $y = h * x$ as

$$y(n_1, n_2) = \sum_{i,j} h(n_1, n_2) * x_{i,j}(n_1, n_2),$$

where each section output $y_{i,j}(n_1, n_2) \triangleq h(n_1, n_2) * x_{i,j}(n_1, n_2)$ is of much smaller extent, and so can be efficiently calculated via DFT methods. In the 2-D *Overlap-and-Add* (O&A) method [8], we select the section size as $(N_1 - L_1 + 1) \times (N_2 - L_2 + 1)$ and the DFT size as $N_1 \times N_2$ so as to avoid spatial aliasing. The section outputs $y_{i,j}$ then overlap in vertical strips of width L_1 and horizontal strips of height L_2.

The total number of sections for the image N_s is given as

$$N_s = \left\lceil \frac{M_1}{N_1 - L_1 + 1} \right\rceil \left\lceil \frac{M_2}{N_2 - L_2 + 1} \right\rceil.$$

where $\lceil \cdot \rceil$ indicates the *greatest integer function*. The computation needed per section is two DFTs and $N_1 N_2$ complex multiplies and additions. Using row–column FFTs and assuming that N_1 and N_2 are integral powers of 2, we obtain the total requirement as

$$\left\lceil \frac{M_1}{N_1 - L_1 + 1} \right\rceil \left\lceil \frac{M_2}{N_2 - L_2 + 1} \right\rceil (2N_1 N_2 \log_2(N_1 N_2) + 2N_1 N_2)$$

real multiplications.

Now the direct computation, assuming x and h are real valued, is $M_1 M_2 L_1 L_2$. We can offer the following comparisons: For $L_1 = L_2 \geqslant 5$, the O&A method has less computation according to these formulas, and for $L_1 = L_2 \geqslant 10$, the O&A method may be actually preferable, considering the overhead of implementing the required DFTs. In the literature on such block filtering, typically $N_1 = N_2 = 64$ or 128.

4.7 CONCLUSIONS

The DFS, DFT, and DCT can easily be extended to two dimensions as separable operators. The familiar concepts of circular convolution and now spatial aliasing are visited again, due to the sampling in frequency space. We investigated the SWT and interpreted it as an extension to overlapping base functions of the DCT. Fast algorithms for both the DFT and DCT were presented relying on the so-called row–column method that exploits operator separability. Sectioned convolution allows "linear convolution" results to be achieved via DFT-based circular convolution. More on the 2-D DFT, including a special DFT for general sampling lattices, can be found in Chapter 2 of Dudgeon and Mersereau [2].

4.8 PROBLEMS

1 Consider the rectangularly periodic sequence $\tilde{x}(n_1, n_2)$, with period $N_1 \times N_2$, and express its Fourier transform in terms of impulse functions. Relate your result to the corresponding DFS $\tilde{X}(k_1, k_2)$.

2 Show that the sum and product of two periodic functions $\tilde{x}$ and $\tilde{y}$ that have the same periods are periodic with the common period.

3 Find the N-point DFT of $x(n) = u(n) - u(n - N)$ and then the same size DFT of $x(n) = u(n) - u(n - N + 1)$.

4 Let the signal $x(n_1, n_2)$ have support $[0, N_1 - 1] \times [0, N_2 - 1]$. Express the $N_1 \times N_2$-point DFS or DFT, *as appropriate*, of each of the following signals in terms of $X(\omega_1, \omega_2)$, the Fourier transform of x:

   (a) $x((n_1)_{N_1}, (n_2)_{N_2})$

   (b) $\sum_{\text{all } k_1} \sum_{\text{all } k_2} x(n_1 - k_1 N_1, n_2 - k_2 N_2)$. (remember, k_1 and k_2 are integers)

   (c) $x((N_1 - n_1)_{N_1}, (N_2 - n_2)_{N_2})$

   (d) $x(N_1 - 1 - n_1, N_2 - 1 - n_2)$.

5 Find the DFT of $x(n_1, n_2) = 1\delta(n_1, n_2) + 2\delta(n_1 - 1, n_2) + 1\delta(n_1 - 2, n_2)$.

6 Let an image $x(n_1, n_2)$ have support $n_1, n_2 := 0, \ldots, N - 1$. Assuming the image is real-valued, its discrete Fourier transform $X(k_1, k_2)$ will display certain symmetry properties. In particular, please show that $X(k_1, k_2) = X^*(N - k_1, N - k_2)$ for $k_1, k_2 := 0, \ldots, N - 1$. Please prove this directly without appealing to the discrete Fourier series.

7 Prove Parseval's theorem for the $N_1 \times N_2$-point DFT,

$$\sum_{n_1=0}^{N_1-1} \sum_{n_2=0}^{N_2-1} x(n_1, n_2) y^*(n_1, n_2) = \frac{1}{N_1 N_2} \sum_{k_1=0}^{N_1-1} \sum_{k_2=0}^{N_2-1} X(k_1, k_2) Y^*(k_1, k_2).$$

Your proof is to be direct from the definitions of DFT and IDFT and should not use any other properties, unless you prove them too.

8 Find the DFT of the finite support signal x, with support $[0, N_1 - 1] \times [0, N_2 - 1]$, given on its support as

$$x(n_1, n_2) = 10 + 2\cos(2\pi 5 n_1 / N_1) + 5\sin(2\pi 8 n_2 / N_2).$$

Assume that $N_1 > 5$ and $N_2 > 8$.

9 Prove the correctness of DFT property 2 directly in the 2-D DFT transform domain, making use of the DFT circular shift or delay property 5 in Section 4.2.1.

10 Start with a general signal $s(n_1, n_2)$ and take its Fourier transform to get $S(\omega_1, \omega_2)$. Then sample this Fourier transform at $\omega_1 = \frac{2\pi k_1}{256}$ and $\omega_2 = \frac{2\pi k_2}{256}$, where $k_1, k_2 : 0, \ldots, 255$ to get $S(k_1, k_2)$. Find an expression for **IDFT**$\{S(k_1, k_2)\}$ in terms of the original $s(n_1, n_2)$. This is called *spatial aliasing*.

11 Fill in the details in going from (4.2-3) to (4.2-10), thus deriving the consequences of sampling in the 2-D frequency domain.

12 Let the 1-D finite-support signal x be given over its support region $[0, N-1]$ as

$$x(n) = 10 + 8\cos\left[\frac{\pi}{2N}(2n + 1)\right] + 2\cos\left[\frac{10\pi}{2N}(2n + 1)\right].$$

Assume $N > 10$.

(a) Sketch the plot of $x(n)$.
(b) Find the DFT of x.
(c) Find the DCT of x.
(d) Use MATLAB to perform a 2-D version of this problem for the finite-support function x given over its support region $[0, N-1]^2$ as

$$x(n_1, n_2) = 10 + 8\cos\left[\frac{\pi}{2N}(2n_1 + 1)\right]\cos\left[\frac{\pi}{2N}(2n_2 + 1)\right]$$
$$+ 2\cos\left[\frac{10\pi}{2N}(2n_1 + 1)\right]\cos\left[\frac{10\pi}{2N}(2n_2 + 1)\right].$$

Please provide plots of $x(n_1, n_2)$, $X(k_1, k_2)|$, and $X_C(k_1, k_2)$.

13 This problem concerns DCT properties, both in 1-D and 2-D cases.

(a) Let $X_C(k)$ be the DCT of $x(n)$ with support $[0, N-1]$. Find a simple expression for the DCT of $x(N-1-n)$. Express your answer directly in terms of $X_C(k)$.
(b) Repeat for a 2-D finite support sequence $x(n_1, n_2)$ with support on $[0, N-1]^2$, and find a simple expression for the 2-D DCT of $x(N-1-n_1, N-1-n_2)$ in terms of $X_C(k_1, k_2)$. Please do not slight this part. It is the general case, and therefore does not follow directly from result of part a.

14 This problem concerns filtering with the discrete cosine transform (DCT). Take the version of DCT defined in Section 4.3 with N points and consider only the one-dimensional (1-D) case.

(a) If possible, define a type of *symmetric convolution* so that zero-phase filtering may be done with the DCT. Justify your definition of symmetric convolution.

(b) Consider a symmetric zero-phase FIR filter with support $[-M, +M]$ and discuss whether *linear convolution* can be obtained for a certain relation between M and N. Please justify carefully.

(c) Is there a version of *sectioned convolution* that can be done using the DCT? Explain.

15 Let the signal (image) $x(n_1, n_2)$ have finite support $[0, M_1 - 1] \times [0, M_2 - 1]$. Then we have seen it is sufficient to sample its Fourier transform $X(\omega_1, \omega_2)$ on a uniform Cartesian grid of $N_1 \times N_2$ points, where $N_1 \geqslant M_1$ and $N_2 \geqslant M_2$. Here we consider two such DFTs, i.e., $DFT_{N_1 \times N_2}$ and $DFT_{M_1 \times M_2}$

(a) Express $DFT_{N_1 \times N_2}$ in terms of $DFT_{M_1 \times M_2}$ as best as you can.

(b) Consider the case where $N_i = L_i M_i$ for $i = 1, 2$, where the L_i are integers, and find an amplitude and phase closed-form representation for the resulting interpolator function.

16 We have been given an FIR filter h with support on the square $[0, N - 1] \times [0, N - 1]$. We must use a 512×512-point row–column FFT to approximately implement this filter for a 512×512 pixel image. Assume $N \ll 512$.

(a) First, evaluate the number of complex multiplies used to implement the filter as a function of M for an $M \times M$ image, assuming the component 1-D FFT uses the Cooley–Tukey approach (assume M is a power of 2). Please then specialize your result to $M = 512$. Assume that the FIR filter is provided as H in the DFT domain.

(b) What portion of the output will be the correct (*linear convolution*) result?

17 This problem concerns *2-D vectors* that are rectangular $N_1 \times N_2$ arrays of values that correspond to finite sets of 2-D data, ordinarily thought of as matrices. We call them 2-D vectors because we want to process them by linear operators to produce output 2-D vectors. Such operators can be thought of as 4-D arrays, the first two indices corresponding to a point in the output 2-D vector, and the last two indices corresponding to a point in the input 2-D vector. We will call these linear operators *2-D matrices*, and so notationally we can write

$$\mathbf{Y} = \mathcal{H}\mathbf{X},$$

where values $\mathbf{X}, \mathbf{Y}$ denote 2-D vectors and value $\mathcal{H}$ denotes the 2-D matrix.

(a) Show that the set of 2-D $N_1 \times N_2$ vectors constitutes a finite-dimensional vector space.

(b) Define the appropriate addition and 2-D matrix multiplication in this vector space.

(c) Show that this notationally simple device is equivalent to *stacking*,[7] wherein the 2-D vectors $\mathbf{X}, \mathbf{Y}$ are scanned in some way into 1-D vec-

---

7. Often the scanning used is referred to as *lexicographical order*, meaning left-to-right, and then advance one line and repeat, also called *raster scan*.

tors **x**, **y** and then the general linear operator becomes a 1-D or ordinary matrix **H**, which has a block structure. Finally relate the blocks in **H** in terms of the elements of $\mathcal{H}$.

(d)   How can we find the determinant, $\det \mathcal{H}$, and then, if $\det H \neq 0$, how can we define and then find $\mathcal{H}^{-1}$?

REFERENCES

[1] A. Croisier, D. Esteban, and C. Galand, "Perfect Channel Splitting by use of Interpolation, Decimation, and Tree Decomposition Techniques," *Proc. Int. Conf. Inform. Sciences/Systems*, Patras, Greece, Aug. 1976. pp. 443–446.

[2] D. E. Dudgeon and R. M. Mersereau, *Multidimensional Digital Signal Processing*, Prentice-Hall, Englewood Cliffs, NJ, 1983.

[3] D. Esteban and C. Galand, "Application of Quadrature Mirror Filters to Split Band Voice Coding Schemes," *Proc. Int. Conf. Acoust., Speech, Signal. Proc., (ICASSP)*, May 1977. pp. 191–195.

[4] A. K. Jain, *Fundamentals of Digital Image Processing*, Prentice-Hall, Englewood Cliffs, NJ, 1989.

[5] J. Lim, *Two-Dimensional Signal and Image Processing*, Prentice-Hall, Englewood Cliffs, NJ, 1990.

[6] S. Mallat, "A Theory for Multiresolution Signal Decomposition: the Wavelet Representation," *IEEE Trans. Pattern Anal. Machine Intel'l.*, **11**, 674–693, Nov. 1989.

[7] F. Mintzer, "Filters for Distortion-Free Two-Bond Multirate Filter Banks," *IEEE Trans. Acoust., Spech, and Signal Process*, **ASSP-33**, 526–630, June 1985.

[8] A. V. Oppenheim, R. W. Schafer, and J. R. Buck, *Discrete-Time Signal Processing*, 2nd edn., Prentice-Hall, Englewood Cliffs, NJ, 1999.

[9] E. P. Simoncelli and E. H. Adelson, "Subband Transforms," In *Subband Image Coding*, J. W. Woods, ed. Kluwer Academic Publ., Norwell, MA, 1991. pp. 143–192.

[10] H. Stark and J. W. Woods, *Probability and Random Processes with Application in Signal Processing*, 3rd edn., Prentice-Hall, Englewood Cliffs, NJ, 2002.

[11] M. Vetterli and C. Herley, "Wavelets and Filter Banks: Theory and Design," *IEEE Trans. Signal Process.*, **40**, 2207–2232, Sept. 1992.

[12] M. Vetterli and H. J. Nussbaumer, "Simple FFT and DCT Algorithms with Reduced Number of Operations," *Signal Processing*, **6**, 267–278, 1984.

TWO-DIMENSIONAL FILTER DESIGN

5

This chapter concerns the design of spatial and other two-dimensional digital filters. We will introduce both finite and infinite impulse response duration (FIR and IIR) filter design. In the FIR case we first present window function design of rectangular support filters. We then briefly present a 1-D to 2-D transformation method due to McClellan and a method based on successive projections onto convex sets of frequency domain specifications. In the IIR case, we look at both space-domain and frequency-domain design methods. We consider both conventional spatial recursive filters as well as a more general class called fully recursive. Finally, we discuss subband/wavelet filter designs for use in the SWT.

5.1 FIR FILTER DESIGN

Finite impulse response support (FIR) filters are often preferred in applications because of their ease of implementation and freedom from instability worries. However, they generally call for a greater number of multiplies and adds per output point. Also, they typically require more temporary storage than the infinite impulse response designs that are covered in Section 5.2.

5.1.1 FIR WINDOW FUNCTION DESIGN

The method of window function design carries over simply from the 1-D case. We start out with an ideal, but infinite-order, impulse response $h_I(n_1, n_2)$, which may have been obtained by inverse Fourier transform of an ideal frequency response $H_I(\omega_1, \omega_2)$. We then proceed to generate our FIR filter impulse response $h(n_1, n_2)$ by simply multiplying the ideal impulse response by a prescribed *window function* $w(n_1, n_2)$ that has finite rectangular support,

$$h(n_1, n_2) \triangleq w(n_1, n_2) h_I(n_1, n_2),$$

with corresponding multiplication in the frequency domain,

$$H(\omega_1, \omega_2) = W(\omega_1, \omega_2) \circledast H_I(\omega_1, \omega_2),$$

where $\circledast$ indicates periodic continuous-parameter convolution with fundamental period $[-\pi, +\pi] \times [-\pi, +\pi]$, specifically

$$H(\omega_1, \omega_2) = \frac{1}{(2\pi)^2} \iint_{[-\pi,+\pi] \times [-\pi,+\pi]} W(\phi_1, \phi_2) H_I(\omega_1 - \phi_1, \omega_2 - \phi_2) \, d\phi_1 \, d\phi_2.$$

In applying this method it is important that the window function support supp(w) coincide with the largest coefficients of the ideal impulse response h_I, and such can be accomplished by shifting one or the other till they best align.

We would like the window function w and its Fourier transform W to have the following properties:

- w should have rectangular $N_1 \times N_2$ support.
- w should approximate a circularly symmetric function (about its center point) and be real-valued.
- The volume of w should be concentrated in the space domain.
- The volume of W should be concentrated in the frequency domain.

Ideally we would like W to be an impulse in frequency, but we know even from 1-D signal processing that this will not be possible for a space-limited function w. The desired symmetry in the window w will permit the window-designed impulse response h to inherit any symmetries present in the ideal impulse response h_I, which can be useful for implementation. In the 2-D case, we can consider two classes of window functions: *separable windows* and *circular windows*.

SEPARABLE (RECTANGULAR) WINDOWS

Here we simply define the 2-D window function $w(n_1, n_2)$ as the product of two 1-D window functions,

$$w_s(n_1, n_2) \triangleq w_1(n_1)w_2(n_2).$$

This approach generally works well when the component functions are good 1-D windows.

CIRCULAR (ROTATED) WINDOWS

The circular window is defined in terms of a 1-D continuous-time window function $w(t)$, as

$$w_c(n_1, n_2) \triangleq w\left(\sqrt{n_1^2 + n_2^2}\right),$$

which has the potential, at least, of offering better circular symmetry in the designed filter.

COMMON 1-D CONTINUOUS-TIME WINDOWS

We define the following windows as centered on $t = 0$, but you should keep in mind that when they are applied in filter design, they must be shifted to line up with the significant coefficients in the ideal impulse response.

- *Rectangular* window:

$$w(t) \triangleq \begin{cases} 1, & |t| < T \\ 0, & \text{else} \end{cases}$$

- *Bartlett (triangular)* window:

$$w(t) \triangleq \begin{cases} 1 - t/T, & 0 \leqslant t \leqslant T \\ 1 + t/T, & -T \leqslant t \leqslant 0 \\ 0, & \text{else.} \end{cases}$$

- *Hanning* window:

$$w(t) \triangleq \begin{cases} \frac{1}{2}(1 + \cos \pi t/T), & |t| < T \\ 0, & \text{else.} \end{cases}$$

- *Kaiser* window:

$$w(t) \triangleq \begin{cases} I_0\big(\beta\sqrt{1 - (t/T)^2}\big)/I_0(\beta), & |t| < T \\ 0, & \text{else,} \end{cases}$$

where $I_0(t)$ is the modified Bessel function of the first kind and of zero order [9]. The free parameter β in this definition means that the Kaiser window is actually a family of windows. As β is varied over a range, typically [0, 8], the transition bandwidth goes up while the filter attenuation in the stopband goes down, in such a manner as to better or equal the performance of the other known 1-D window functions. We thus concentrate on Kaiser window function designs for our 2-D filters.

Approximate design formulas have been developed for 2-D circular symmetric lowpass filters [4] to estimate the required values of the filter size N_1 and N_2 and the Kaiser parameter[1] β. They are based on the desired filter transition bandwidth $\Delta\omega$ and stopband attenuation ATT, defined as follows:

$$\Delta\omega \triangleq \omega_s - \omega_p,$$

where ω_s and ω_p denote the circular radii of the desired stopband and passband, respectively, and each with maximum value of π. The attenuation parameter ATT is given as,

$$\text{ATT} \triangleq -20 \log_{10}\big(\sqrt{\delta_s \delta_p}\big),$$

where δ_s and δ_p denote the desired peak frequency domain errors in the stopband and passband, respectively. The estimated filter orders are given for the square support case $N_1 = N_2 = N_s$ *separable*,

$$N_s \approx \frac{\text{ATT} - 8}{2.10\Delta\omega},$$

---

1. This β parameter is denoted as α in the Oppenheim, Schafer, and Buck text.

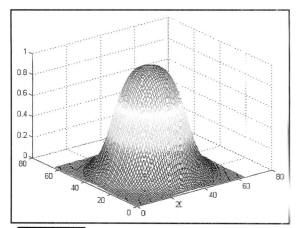

FIGURE 5.1   *Magnitude response of separable Kaiser window FIR filter design with $\beta = 8$*

and N_c *circular,*

$$N_c \approx \frac{\text{ATT} - 7}{2.18 \Delta \omega}.$$

The estimate of β is then given as β_s *separable,*

$$\beta_s \approx \begin{cases} 0.42(\text{ATT} - 19.3)^{0.4} + 0.089(\text{ATT} - 19.3), & 20 < \text{ATT} < 60 \\ 0, & \text{else,} \end{cases}$$

and β_c *circular,*

$$\beta_c \approx \begin{cases} 0.56(\text{ATT} - 20.2)^{0.4} + 0.083(\text{ATT} - 20.2), & 20 < \text{ATT} < 60 \\ 0, & \text{else.} \end{cases}$$

EXAMPLE 5.1-1 (*window design using* MATLAB)

We have designed an 11×11 lowpass filter with filter cutoff $f_c = 0.3$ ($\omega_c = 0.6\pi$) and using two Kaiser windows, with $\beta = 8$, both separable and circular. Figure 5.1 shows the magnitude response of the filter obtained using the separable Kaiser window. The contour plot of the magnitude data is shown in Figure 5.2. Figure 5.3 shows the 11×11 impulse response of the designed FIR filter. Figure 5.4 shows a contour plot of the impulse response, to see its degree of "circularity."

The next set of figures shows corresponding results for a circular Kaiser window. Figure 5.5 shows the magnitude response, followed by Figure 5.6 showing its contour plot in frequency. Figure 5.7 shows the corresponding impulse response plot, and then Figure 5.8 shows the contour plot. Note that both the circular and the separable designed filters display a lot of circular

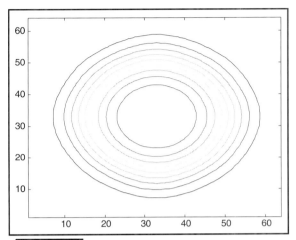

FIGURE 5.2   *Contour plot of separable Kaiser filter magnitude response*

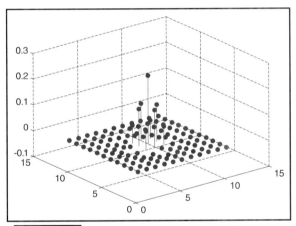

FIGURE 5.3   *Plot of* 11 × 11 *separable impulse response*

symmetry, which they inherit from the exact circular symmetry of the ideal response H_I. Both designs are comparable. A lower value of β would result in lower transition bandwidth, but less attenuation in the stopband and more ripple in the passband.

The MATLAB .m files WinDes permit easy experimentation with Kaiser window design of spatial FIR filters. The program WinDesS.m uses separable window design, while WinDesC.m uses the circular window method. Both programs make use of the MATLAB image processing toolbox function fwind1.m, which uses 1-D windows. The other toolbox function fwind2.m does window design

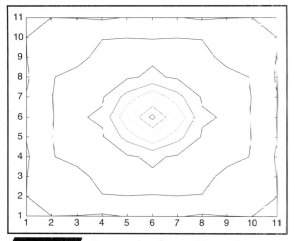

FIGURE 5.4   *Contour plot of separable impulse response*

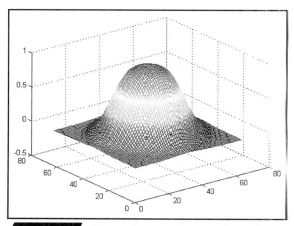

FIGURE 5.5   *Magnitude response, $\beta = 8$, but with circular Kaiser window (see color insert)*

with a 2-D window that you supply. It is not used in either WinDes program. As an exercise, try using WinDesC with the parameters of Example 5.1-1, except that $\beta = 2$. These programs are on the enclosed CD-ROM.

EXAMPLE 5.1-2 (*filter Eric image*)
We take the filter designed with the circular window and apply it to filtering the *Eric* image from Figure 1.6 of Chapter 1. The result is shown in Figure 5.9, where we note the visual blurring effect of the lowpass filter.

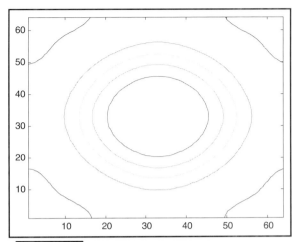

FIGURE 5.6   *Contour plot of magnitude response of Kaiser circular window designed filter*

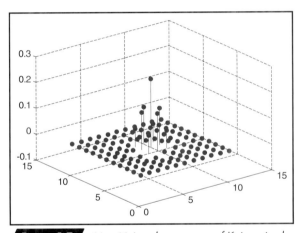

FIGURE 5.7 11×11 *impulse response of Kaiser circular window designed filter*

We now turn to an FIR design technique that transforms 1-D filters into 2-D filters.

5.1.2 Design by Transformation of 1-D Filter

Another common and powerful design method is the transformation of 1-D filters to 2-D filters. The basic idea starts with a 1-D filter and then transforms it in such a way that 2-D filter response is constant along contours specified by the transformation. Thus for contours with an approximate circular shape, and 1-D lowpass filter, we can achieve approximate circular symmetry in a 2-D lowpass

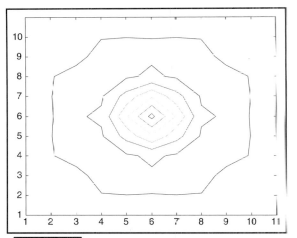

FIGURE 5.8   *Contour plot of impulse response for circular Kaiser window designed filter*

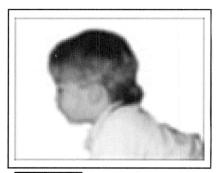

FIGURE 5.9   *Lowpass filtering of Eric image of Chapter 1*

filter. The basic idea is due to McClellan [12] and it is developed much more fully in Lim's text [11]; here, we present only the basics. The method is also available in the MATLAB image processing toolbox.

We start with a 1-D FIR filter, of type I or II, with linear phase, so that we can write the frequency response as the real function (neglecting delay),

$$H(\omega) = \sum_{n=0}^{M} a(n) \cos(n\omega), \tag{5.1-1}$$

where the coefficients $h(n)$ can be expressed in terms of the $a(n)$. Next, we rewrite this equation in terms of Chebyshev polynomials $\cos(n\omega) = T_n[\cos \omega]$ as

$$H(\omega) = \sum_{n=0}^{M} a'(n) \cos^n \omega, \tag{5.1-2}$$

where the $a'(n)$ can be expressed in terms of the $a(n)$ (see end-of-chapter problem 7). Note that the first few Chebyshev polynomials are given as

$$T_0(x) = 1, \qquad T_1(x) = x, \qquad T_2(x) = 2x^2 - 1,$$

and, in general,

$$T_n(x) = 2xT_{n-1}(x) - T_{n-2}(x), \quad n > 1.$$

Next we introduce a transformation from the 2-D frequency plane to $\cos \omega$, given as [12]

$$F(\omega_1, \omega_2) = A + B\cos\omega_1 + C\cos\omega_2 + D\cos(\omega_1 - \omega_2) + E\cos(\omega_1 + \omega_2),$$

where the constants A, B, C, D, and E are adjusted to the constraint $|F(\omega_1, \omega_2)| \leqslant 1$ so that F will equal $\cos\omega$ for some ω.

At this point we parenthetically observe that when the 1-D filter is expressed in terms of a Z-transform, the term $\cos\omega$ will map over to $\frac{1}{2}(z + z^{-1})$, and that the transformation F will map into the 2-D Z-transform of the 1×1 order symmetric FIR filter f given as

$$f(n_1, n_2) = \left\{ \begin{matrix} \frac{1}{2}D & \frac{1}{2}C & \frac{1}{2}E \\ \frac{1}{2}B & A & \frac{1}{2}B \\ \frac{1}{2}E & \frac{1}{2}C & \frac{1}{2}D \end{matrix} \right\},$$

with axes n_1 horizontal and n_2 upward, with 0 and the center. Thus we can take a realization of the 1-D filter and, replacing the delay boxes z^{-1} with the filter $F(z_1, z_2)$, obtain a realization of the 2-D filter. Of course, this can be generalized in various ways, including the use of higher-order transformations and the other types (II and IV) of 1-D FIR filters.

If we take the values $A = -0.5, B = C = 0.5$, and $D = E = 0.25$, we get contours $F(\omega_1, \omega_2)$ with a near-circular shape, as sketched in Figure 5.10, which is very nearly circular in shape out beyond midfrequency $\approx \pi/2$.

If we start out with a 1-D filter that is optimal in the Chebyshev sense, i.e., l^∞ optimal in the magnitude response, then it has been shown [11] that the preceding 1×1-order transformation preserves optimality in this sense for the 2-D filter, *if* the passband and stopband contours of the desired filter are isopotentials of the transformation.

The following examples show the transformation of 1-D optimal Chebyshev lowpass filters into 2-D filters via this method.

EXAMPLE 5.1-3 (*9 × 9 lowpass filter*)
We start out with the design of a 9-tap (eighth-order) FIR lowpass filter with passband edge $f_p = 0.166$ ($\omega_p = 0.332\pi$) and stopband edge $f_s = 0.333$ ($\omega_s = 0.666\pi$), using the Remez (aka Parks–McClellan) algorithm. The result

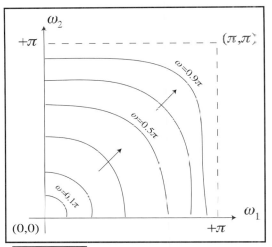

FIGURE 5.10   Illustration of 1 × 1-order McClellan transformation for near-circular symmetric contours

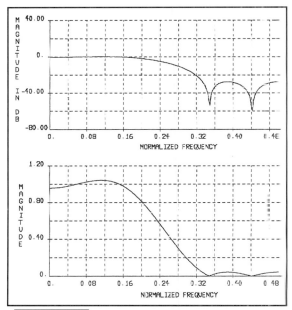

FIGURE 5.11   Magnitude response of 9-tap FIR type I filter

is a passband ripple of $\delta = 0.05$ and a stopband attenuation ATT = 30 dB, and resulting linear and logarithmic magnitude frequency response shown in Figure 5.11. When this 9-tap filter is transformed with the near-circular 1×1-order McClellan transform, we obtain the 2-D amplitude response shown in

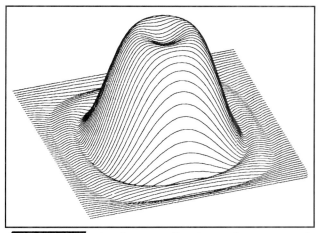

FIGURE 5.12   *Plot of 9 × 9 FIR filter designed as a 2-D transform of a 9-tap lowpass filter*

Figure 5.12, where the zero frequency point $\omega = 0$ is in the middle of this 3-D perspective plot.

We note that this design has resulted in a very good approximation of circular symmetry as we expect from the contour plot of Figure 5.10 for the transformation used. For this particular transformation, at $\omega_2 = 0$, we get $\cos\omega = \cos\omega_1$, so that the passband and stopband edges are located at the same frequencies, on the axes at least, as those of the 1-D prototype filter, i.e., $\omega_p = 0.332\pi$ and $\omega_s = 0.666\pi$, respectively.

EXAMPLE 5.1-4 (*application of transform lowpass filter*)

In this example, we design and apply a transformation-designed lowpass filter to a digital image by use of MATLAB functions. Using the transformation of the previous example, we convert a lowpass 1-D filter with passband edge $\omega_p = 0.025$ and stopband edge $\omega_p = 0.250$ into an 11×11 near-circular symmetric lowpass filter. The resulting 2-D frequency response is shown in Figure 5.13. The input image is shown in Figure 5.14. The output image is shown in Figure 5.15. The difference image (biase up by $+128$ for display on $[0, 255]$) is shown in Figure 5.16. The actual MATLAB .m file `McDesFilt.m` is given on the book's CD-ROM. You can experiment with changing the passband, filter type, image, etc. (Note: this routine requires MATLAB with the image processing toolbox.) Similarly, using an approximately circular highpass filter, we can get the output shown in Figure 5.17; note the similarity to Figure 5.16.

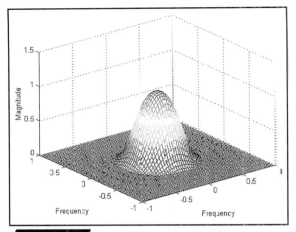

FIGURE 5.13   *Lowpass filter for image filtering*

FIGURE 5.14   *Input image*

5.1.3 PROJECTION-ONTO-CONVEX-SETS METHOD

The method of *projection onto convex sets* (PCCS) is quite general and has been applied to several image processing optimization problems [16] Here we follow an application to 2-D FIR filter design due to Abo-Taleb and Fahmy [1]. Using an assumed symmetry in the ideal function and the desired FIR filter, we rewrite the filter response in terms of a set of frequency-domain basis functions as

$$H(\omega_1, \omega_2) = \sum_{k=1}^{M} a(k)\phi_k(\omega_1, \omega_2),$$

FIGURE 5.15   *Output of lowpass filter*

FIGURE 5.16   *Corresponding difference image (biased up for display by +128)*

where the filter coefficients $h(n_1, n_2)$ can be simply expressed in terms of the $a(k)$, using the built-in symmetries assumed for h. Next we densely and uniformly discretize the basic frequency cell with N^2 points as $\mathbf{x}_i = (\omega_1^i, \omega_2^i)$, for $i = 1, \ldots, N^2$. Then we can express the filter frequency response at grid point $\mathbf{x}_i$ as

$$H(\mathbf{x}_i) = \langle \mathbf{a}, \phi(\mathbf{x}_i) \rangle,$$

where $\mathbf{a}$ is a column vector of the $a(k)$ and ϕ is a vector of the ϕ_k, each of dimension M, and $\langle \cdot, \cdot \rangle$ denotes the conventional vector-space inner product. The

FIGURE 5.17   *Output of McClellan-transformed near-circular highpass filter*

frequency-domain error at the point $\mathbf{x}_i$ then becomes

$$e(\mathbf{x}_i) = I(\mathbf{x}_i) - \langle \mathbf{a}, \phi(\mathbf{x}_i) \rangle.$$

We can state the approximation problem as the equations

$$\left| I(\mathbf{x}_i) - \langle \mathbf{a}, \phi(\mathbf{x}_i) \rangle \right| \leqslant \delta \quad \text{for all } i = 1, \ldots, N^2.$$

To apply the POCS method, we must first observe that the sets

$$Q_i \triangleq \left\{ \mathbf{a} : \left| I(\mathbf{x}_i) - \langle \mathbf{a}, \phi(\mathbf{x}_i) \rangle \right| \leqslant \delta \right\}$$

are *convex*[2] and closed. We note that we want to find a vector a that is contained in the set intersection

$$Q \triangleq \bigcap_{i=1}^{N^2} Q_i,$$

if such a point exists. In that case, the POCS algorithm is guaranteed to converge to such a point(s) [1]. Note that the existence of a solution will depend on the tolerance δ, but that we do not know how small the tolerance should be. What this means in practice is that a sequence of problems will have to be solved, for a decreasing set of δ's, until the algorithm fails to converge, in which case the last viable solution is taken as the result. Of course, one could also specify a needed value δ and then increase the filter order, in terms of M, until a solution was obtained. The basic POCS algorithm can be given as follows:

2. A convex set Q is one for which, f points $\mathbf{p}_1$ and $\mathbf{p}_2$ are in the set, then so must be $\mathbf{p} = \alpha\mathbf{p}_1 + (1-\alpha)\mathbf{p}_2$ for all $0 \leqslant \alpha \leqslant 1$.

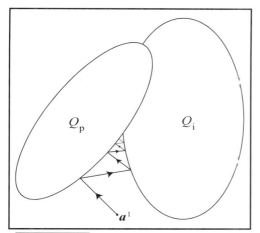

FIGURE 5.18   *Illustration of the convergence based on orthogonal projection onto convex sets*

BASIC POCS ALGORITHM

1   Scan through the frequency grid points x_i to find the frequency x_p of maximum error $e(x_p) = e_p$.

2   If $e_p \leqslant \delta$, stop.

3   Else, update the coefficient vector as the orthogonal projection onto Q_p [1,16],

$$\mathbf{a}^{n+1} = \begin{cases} \mathbf{a}^n - (\delta - |e_p|)\phi(x_p)\frac{\text{sgn}(e_p)}{||\phi(x_p)||}, & \text{if } e_p > \delta \\ \mathbf{a}^n, & \text{else.} \end{cases}$$

4   Return to 1.

The key step of this algorithm is step 3, which performs an orthogonal projection of the M-dimensional filter coefficient vector $\mathbf{a}^n$ onto the constraint set Q_p. As the algorithm progresses, the solution point $\mathbf{a}$ may move into and out of any given set many times, but is guaranteed to converge finally to a point in the intersection of all the constraint sets, *if such a point exists*. An illustrative diagram of POCS convergence is shown in Figure 5.18.

The references [1,16] contain several examples of filters designed by this method. The obtained filters appear to be quite close to the Chebyshev optimal linear phase FIR designs of Harris and Mersereau [6], which are not covered in this text. The POCS designs appear to have a significant computational advantage, particularly for larger filter supports.

5.2 IIR FILTER DESIGN

In this section, we first look at conventional 2-D recursive filter design. This is followed by a discussion of a generalization called "fully recursive filters," which offers the potential of even better performance

5.2.1 2-D RECURSIVE FILTER DESIGN

The Z-transform system function of a spatial IIR or recursive filter is given as

$$H(z_1, z_2) = \frac{B(z_1, z_2)}{A(z_1, z_2)},$$

where both B and A are polynomial[3] functions of z_1, z_2, which in the spatial domain yields the computational procedure, assuming $a_{0,0} = 1$,

$$y(n_1, n_2) = - \sum_{(k_1, k_2) \in \mathcal{R}_a - (0,0)} a_{k_1, k_2} y(n_1 - k_1, n_2 - k_2)$$
$$+ \sum_{(k_1, k_2) \in \mathcal{R}_b} b_{k_1, k_2} x(n_1 - k_1, n_2 - k_2),$$

where the coefficient support regions of the denominator and numerator are denoted as $\mathcal{R}_a$ and $\mathcal{R}_b$, respectively. Here we consider a so-called *direct form* or unfactored design, but factored designs are possible, and even preferable for implementation in finite-wordlength arithmetic, as would be expected from the 1-D case. However, as we have seen earlier, various factored forms will not be equivalent due to the lack of a finite-order factorization theorem in two and higher dimensions.

TYPICAL DESIGN CRITERIA

The choice of error criteria is important and related to the expected use of the filter. In addition to the choice of either spatial-domain or frequency-domain error criteria or a combination of both, there is the choice of error norm. The following error criteria are often used:

- *Space-domain* design: Often a least-squares design criteria is chosen,

$$\left\| h_1(n_1, n_2) - k(n_1, n_2) \right\|_2,$$

---

3. Strictly speaking, the "polynomial" may include both positive and negative powers of the z_i, but we do not make the distinction here. We call both polynomials if they are just finite order.

via use of the l^2 error norm

$$\|f\|_2 \triangleq \sqrt{\sum \sum |f|^2(n_1, n_2)}.$$

- *Magnitude-only* approximation:

$$\big\||H_I|(\omega_1, \omega_2) - |H|(\omega_1, \omega_2)\big\|,$$

where $|H_I|$ denotes an ideal magnitude function and $|H|$ is the magnitude of the filter being designed. The most common choice for the norm $\|\cdot\|$, is the L^2 norm

$$\|F\|_2 \triangleq \sqrt{\int_{-\pi}^{+\pi} \int_{-\pi}^{+\pi} |F|^2(\omega_1, \omega_2)\, d\omega_1\, d\omega_2}.$$

Magnitude-only design is generally a carryover from 1-D filter design, where the common IIR design methods are based on transforming an analog filter via the bilinear transform. These methods lead to very good filter magnitude response, but with little control over the phase. For single-channel audio systems, these filter phase distortions can be unimportant. For image processing, though, where 2-D filters see a big application, phase is very important. So a magnitude-only design of an image processing filter may not be adequate.

- *Zero-phase* design:

$$\big\|H_I(\omega_1, \omega_2) - |H|^2(\omega_1, \omega_2)\big\|,$$

where the ideal function H_I is assumed to be *real-valued and nonnegative*, often the case in image processing. The filter is then realized by two passes. In the first pass the image data is filtered by $h(n_1, n_2)$ in its stable direction, generally from right-to-left and then top-to-bottom, the so-called *raster scan* of image processing. The output data from this step is then filtered by $h^*(-n_1, -n_2)$ in its stable direction, from left-to-right, and bottom-to-top. The overall realized frequency response is then $|H|^2(\omega_1, \omega_2)$, which is guaranteed zero-phase.

- *Magnitude and phase* (delay) design:

$$\big\|H_I(\omega_1, \omega_2) - H(\omega_1, \omega_2)\big\|$$

or

$$\big\||H_I|(\omega_1, \omega_2) - |H|(\omega_1, \omega_2)\big\| + \lambda \big\|\arg H_I(\omega_1, \omega_2) - \arg H(\omega_1, \omega_2)\big\|,$$

where the Lagrange parameter λ controls the weight given to phase error versus magnitude error. A particular choice of λ will give a prescribed amount of phase error. This critical value of λ, though, will usually only be approximately determined after a series of designs for a range of λ values.

SPACE-DOMAIN DESIGN

We look at two methods, the first being the design method called Padé approximation, which gives exact impulse response values, but over a usually small range of points. Then we look at least-squares extensions, which minimize a modified error.

PADÉ APPROXIMATION   This simple method carries over from the 1-D case. Let

$$e(n_1, n_2) = h_I(n_1, n_2) - b(n_1, n_2).$$

Then the squared l^2 norm becomes

$$\|e\|_2^2 = \sum e^2(n_1, n_2)$$
$$\triangleq f(\mathbf{a}, \mathbf{b}),$$

a nonlinear function of the denominator and numerator coefficient vectors $\mathbf{a}$ and $\mathbf{b}$, respectively. The Padé approximation method gives a linear closed-form solution that achieves zero-impulse response error over a certain support region, whose number of points equals $\dim(\mathbf{a}) + \dim(\mathbf{b})$.

Let $x = \delta$ so that $y = h$, then

$$h(n_1, n_2) = - \sum_{(k_1,k_2)>(0,0)} a(k_1, k_2)h(n_1 - k_1, n_2 - k_2) + b(n_1, n_2)$$
$$\neq h_I(n_1, n_2),$$

where we have again assumed that coefficient $a(0, 0) = 1$ without loss of generality. The error function then becomes

$$e(n_1, n_2) = h_I(n_1, n_2) + \sum_{(k_1,k_2)>(0,0)} a(k_1, k_2)h(n_1 - k_1, n_2 - k_2) - b(n_1, n_2),$$

which is nonlinear in the parameters a and b thru the function h. To avoid the nonlinearities, we now define the *modified error*

$$e_M(n_1, n_2) \triangleq h_I(n_1, n_2) + \sum_{(k_1,k_2)>(0,0)} a(k_1, k_2)h_I(n_1 - k_1, n_2 - k_2) - b(n_1, n_2)$$
$$= a(n_1, n_2) * h_I(n_1, n_2) - b(n_1, n_2). \tag{5.2-1}$$

Comparison of these two equations reveals that we have substituted h_I for h inside the convolution with a, which can be a reasonable approximation given a presumed smoothing effect from coefficient sequence a. To this extent we can hope that $e_M \approx e$. Note in particular, though, that the modified error e_M is linear in the coefficient vectors $\mathbf{a}$ and $\mathbf{b}$.

If $\dim(\mathbf{a}) = p$ and $\dim(\mathbf{b}) = q + 1$, then there are $N = p + q + 1$ unknowns. We can then set $e_M = 0$ on $\mathcal{R}_{\text{Padé}} = \{$a connected N point region$\}$. We have to choose the region so that the filter masks slide over the set, rather than bringing new points (pixels) into the equations. A simple example suffices to illustrate this linear FIR design method.

EXAMPLE 5.2-1 (*Padé approximation*)
Consider the 1×1-order system function

$$H(z_1, z_2) = \frac{b_{00} + b_{01}z_2^{-1}}{1 + a_{10}z_1^{-1} + a_{01}z_2^{-1}},$$

where we have $p = 2$ and $q + 1 = 2$, equivalent to $N = 4$ unknowns. We look for a first-quadrant support for the impulse response h and set $\mathcal{R}_{\text{Padé}} = \{(0, 0), (1, 0), (0, 1), (1, 1)\}$. Setting $e_M = 0$, we have the following recursion

$$h_I(n_1, n_2) = -a_{10}h_I(n_1 - 1, n_2) - a_{01}h_I(n_1, n_2 - 1)$$
$$+ b_{00}\delta(n_1, n_2) + b_{01}\delta(n_1, n_2 - 1).$$

Assuming zero initial (boundary conditions), we evaluate this equation on the given region $\mathcal{R}_{\text{Padé}}$ to obtain the four equations,

$$h_I(0, 0) = b_{00},$$
$$h_I(1, 0) = -a_{10}h_I(0, 0),$$
$$h_I(0, 1) = -a_{01}h_I(0, 0) + b_{01},$$
$$h_I(1, 1) = -a_{10}h_I(0, 1) - a_{01}h_I(1, 0),$$

which can be easily solved for the four coefficient unknowns.

We have not yet mentioned that the filter resulting from Padé approximation may not be stable! Unfortunately this is so. Still, if it is stable, and if the region is taken large enough, good approximations can result. If the ideal impulse response is stable, i.e., $h_I \in l^1$, then the Padé approximate filter *should* also be stable, again if the region is taken large enough. We also note that, because of the shape of the Padé region $\mathcal{R}_{\text{Padé}}$, effectively honoring the output mask of the filter, $e_M = e$ there, and so the obtained impulse response values are exact *on that region*. So the key is to include in the region $\mathcal{R}_{\text{Padé}}$ all the "significant values" of the ideal impulse response. Of course, this is not always easy and can lead to rather large filter orders.

We next turn to an extension of Padé approximation called Prony's method which minimizes the l^2 spatial domain modified error over (in theory) its entire support region, rather than just setting it to zero over the small region $\mathcal{R}_{\text{Padé}}$.

PRONY'S METHOD (SHANK'S)   This is a linear least-squares method that minimizes the modified error, (5.2-1), but again ignores stability. We write

$$\mathcal{E}_{\text{M}} = \sum e_{\text{M}}^2(n_1, n_2)$$

$$= \sum_{\mathcal{R}_{\text{b}}} e_{\text{M}}^2(n_1, n_2) + \sum_{\mathcal{R}_{\text{b}}^c} e_{\text{M}}^2(n_1, n_2)$$

$$\triangleq \mathcal{E}_{\text{M},\mathcal{R}_{\text{b}}} + \mathcal{E}_{\text{M}\mathcal{R}_{\text{b}}^c},$$

making the obvious definitions, and note that the second error term will be independent of coefficient vector **b**. This because on the complimentary set $\mathcal{R}_{\text{b}}^c$ we can write

$$e_{\text{M}}(n_1, n_2) = a(n_1, n_2) * h_{\text{I}}(n_1, n_2)$$

$$= f(\mathbf{a}).$$

so we can perform linear least squares to minimize $\mathcal{E}_{\text{M}\mathcal{R}_{\text{b}}^c}$ over **a**. Again we emphasize that we cannot just choose $\mathbf{a} = 0$ to make the modified error $e_{\text{M}} = 0$, since we have the constraint $a(0, 0) = 1$ such that $a(0, 0)$ is not an element in the design vector **a**.

Considering the practical case where there are a finite number of points to consider, we can order the values $e_{\text{M}}(n_1, n_2)$ on $\mathcal{R}_{\text{b}}^c$ onto a vector, say $\mathbf{e}_{\text{M}}$, and then minimize $\mathbf{e}_{\text{M}}^T\mathbf{e}_{\text{M}}$ over the denominator coefficient vector a. A typical term in $\mathbf{e}_{\text{M}}^T\mathbf{e}_{\text{M}}$ would look like

$$e_{\text{M}}^2(n_1, n_2) = \left[h_{\text{I}}(n_1, n_2) + a(1, 0)h_{\text{I}}(n_1 - 1, n_2) - a(0, 1)h_{\text{I}}(n_1, n_2 - 1) + \cdots\right]^2,$$

since $a(0, 0) = 1$. Thus, taking partials with respect to the $a(k_1, k_2)$, we obtain

$$\partial\left(\mathbf{e}_{\text{M}}^T\mathbf{e}_{\text{M}}\right) / \partial a(k_1, k_2)$$

$$= 2 \sum_{n_1, n_2} h_{\text{I}}(n_1 - k_1, n_2 - k_2)$$

$$\times \left[h_{\text{I}}(n_1, n_2) + a(1, 0)h_{\text{I}}(n_1 - 1, n_2) + a(0, 1)h_{\text{I}}(n_1, n_2 - 1) + \cdots\right]$$

$$= 2 \sum_{n_1, n_2} h_{\text{I}}(n_1 - k_1, n_2 - k_2)\left[h_{\text{I}}(n_1, n_2) + \sum_{l_1, l_2} a(l_1, l_2)h_{\text{I}}(n_1 - l_1, n_2 - l_2)\right],$$

and setting these partial derivatives to zero, we obtain the so-called *normal equations*

$$\sum_{l_1, l_2} R_I(k_1 - l_1, k_2 - l_2) a(l_1, l_2) = -R_I(k_1, k_2) \quad \text{for } (k_1, k_2) \in \text{supp}\{a\}, \quad (5.2\text{-}2)$$

with the definition of the "correlation terms"

$$R_I(k_1, k_2) \triangleq \sum_{n_1, n_2} h_I(n_1 - k_1, n_2 - k_2) h_I(n_1, n_2).$$

Finally, (5.2-2) can be put into matrix-vector form and solved for the denominator coefficient vector **a** in the case when the matrix corresponding to the "correlation function" is nonnegative definite—fortunately, most of the time. Turning now to the first error term $\mathcal{E}_{M,\mathcal{R}_b}$, we can write

$$e_M(n_1, n_2) = a(n_1, n_2) * h_I(n_1, n_2) - b(n_1, n_2)$$

$$= 0, \quad (5.2\text{-}3)$$

upon setting $b(n_1, n_2) = a(n_1, n_2) * h_I(n_1, n_2)$, using $a(0, 0) = 1$ and the values of **a** obtained in the least-squares solution. We then obtain $\mathcal{E}_{M,\mathcal{R}_b} = 0$, and hence have achieved a minimum for the total modified error $\mathcal{E}_M$.

EXAMPLE 5.2-2 (1×1-*order case*)

Consider the design problem with 1×1-order denominator with variable coefficients $\{a(1, 0), a(0, 1), a(1, 1)\}$ and numerator consisting of coefficients $\{b(0, 0), b(1, 0)\}$. Then $\mathcal{R}_b^c = \{n_1 \geqslant 2, n_2 = 0\} \cup \{n_1 \geqslant 0, n_2 > 0\}$ for a designed first-quadrant support, since $\mathcal{R}_b = \{(0, 0), (1, 0)\}$. For a given ideal impulse response, with $\text{supp}\{h_I\} = \{n_1 \geqslant 0, n_2 \geqslant 0\}$, we next compute $R_I(k_1, k_2)$ for $(k_1, k_2) \in \mathcal{R}_b^c$. Then we can write the normal equations as

$$R_I(0, 0)a(1, 0) + R_I(1, -1)a(0, 1) + R_I(0, -1)a(1, 1) = -R_I(1, 0),$$

$$R_I(-1, 1)a(1, 0) + R_I(0, 0)a(0, 1) + R_I(-1, 0)a(1, 1) = -R_I(0, 1),$$

$$R_I(0, 1)a(1, 0) + R_I(1, 0)a(0, 1) + R_I(0, 0)a(1, 1) = -R_I(1, 1),$$

which can be put into matrix form

$$\mathbf{R}_I \mathbf{a} = -\mathbf{r}_I$$

and solved for vector $\mathbf{a} = [a(1, 0), a(0, 1), a(1, 1)]^T$. Finally we go back to (5.2-3) and solve for $b(0, 0)$ and $b(1, 0)$ as $b(n_1, n_2) = a(n_1, n_2) * h_I(n_1, n_2)$ for $(n_1, n_2) = (0, 0)$ and $(1, 0)$.

As mentioned previously, there is no constraint of filter stability here. The resulting filter may be stable or it may not be. Fortunately, the method has been found useful in practice, depending on the stability of the ideal impulse response h_1 that is being approximated. More on Prony's method, and an iterative improvement extension, is contained in Lim [11].

5.2.2 FULLY RECURSIVE FILTER DESIGN

The Z-transform system function of a *fully recursive filter* (FRF) is given as

$$H(z_1, z_2) = \frac{B(z_1, z_2)}{A(z_1, z_2)},$$

where both B and A are *rational functions* of z_1 and polynomial functions of z_2. The 2-D difference equation is given by

$$\sum_{k_1=-\infty}^{+\infty} \sum_{k_2=0}^{L_D} a(k_1, k_2)y(n_1 - k_1, n_2 - k_2) = \sum_{k_1=-\infty}^{+\infty} \sum_{k_2=0}^{L_N} b(k_1, k_2)x(n_1 - k_1, n_2 - k_2),$$

where L_D and L_N are positive integers. It can be expressed in *row-operator* form as

$$a_0(n_1) * y_{n_2}(n_1) = -\sum_{k_2=1}^{L_D} a_{k_2}(k_1) * y_{n_2-k_2}(n_1) + \sum_{k_2=0}^{L_N} b_{k_2}(k_1) * x_{n_2-k_2}(n_1), \quad (5.2\text{-}4)$$

where the 1-D coefficient sequences are given as

$$a_0(n_1) \triangleq a(n_1, 0), \qquad b_0(n_1) \triangleq b(n_1, 0),$$
$$a_{n_2}(n_1) \triangleq a(n_1, n_2), \qquad b_{n_2}(n_1) \triangleq b(n_1, n_2),$$

and the convolution symbol $*$ indicates a 1-D operation on a row, whether input or output. Also the input and output *row sequences* are denoted as

$$y_{n_2}(n_1) \triangleq y(n_1, n_2) \quad \text{and} \quad x_{n_2}(n_1) \triangleq x(n_1, n_2).$$

If the coefficient row sequences are 1-D FIR, then we have a conventional 2-D recursive filter, but here we consider the possibility that they may be of infinite support. Note that even 2-D FIR filters are included in the FRF class, by setting $a_0(n_1) = \delta(n_1)$, other $a_{n_2}(n_1) = 0$, and $b_{n_2}(n_1) = b(n_1, n_2)$. Upon taking the 2-D Z-transform of (5.2-4), we obtain

$$A_0(z_1)Y(z_1, z_2) = -\sum_{k_2=1}^{L_D} A_{k_2}(z_1)Y(z_1, z_2)z_2^{-k_2} + \sum_{k_2=0}^{L_N} B_{k_2}(z_1)X(z_1, z_2)z_2^{-k_2}.$$

Now we introduce the following rational forms for the 1-D coefficient Z-transforms A_{k_2} and B_{k_2},

$$A_{k_2}(z_1) = \frac{N^a_{k_2}(z_1)}{D^a_{k_2}(z_1)} \qquad \text{and} \qquad B_{k_2}(z_1) = \frac{N^b_{k_2}(z_1)}{D^b_{k_2}(z_1)},$$

where the numerators and denominators are finite-order polynomials in z_1. They thus constitute rational row operators that can be implemented via recursive filters separately processing the present and most recent L_N input rows and L_D previous output rows. The row coefficients $a_{n_2}(n_1)$ and $b_{n_2}(n_1)$ are then just the impulse responses of these 1-D row operators. The row operator on the current output row A_0 can then be implemented via inverse filtering, assuming A_0 has a stable inverse, after completion of the sums indicated on the right-hand side of (5.2-4).

The overall FRF system function can then be written as

$$H(z_1, z_2) = \frac{\sum_{k_2=0}^{L_N} \frac{N^b_{k_2}(z_1)}{D^b_{k_2}(z_1)} z_2^{-k_2}}{\sum_{k_2=0}^{L_D} \frac{N^a_{k_2}(z_1)}{D^a_{k_2}(z_1)} z_2^{-k_2}}.$$

We pause now for an example.

EXAMPLE 5.2-3 (*first-order FRF with NSHP impulse response support*)
Let $L_N = 1$ and $L_D = 1$. Take the feedback row operators A_0 and A_1 with the following system functions

$$A_0(z_1) = \frac{1 + 0.8z_1^{-1}}{1 + 0.9z_1^{-1}} \quad \text{and} \quad A_1(z_1) = 0.8\frac{2 + z_1^{-1}}{1 + 0.6z_1^{-1}},$$

with ROC containing $\{|z_1| \leqslant 1\}$ and input row operators B_0 and B_1 with system functions

$$B_0(z_1) = \frac{1 + z_1^{-1}}{1 + 0.7z_1^{-1}} \quad \text{and} \quad B_1(z_1) = 0.7\frac{1 + z_1^{-1}}{1 + 0.8z_1^{-1}},$$

also with ROC containing $\{|z_1| \leqslant 1\}$. Then the overall FRF system function is

$$H(z_1, z_2) = \frac{B_0(z_1) + B_1(z_1)z_2^{-1}}{A_0(z_1) + A_1(z_1)z_2^{-1}}$$

$$= \frac{\frac{1+z_1^{-1}}{1+0.7z_1^{-1}} + \frac{1+z_1^{-1}}{1+0.8z_1^{-1}}z_2^{-1}}{\frac{1+0.8z_1^{-1}}{1+0.9z_1^{-1}} + \frac{2+z_1^{-1}}{1+0.6z_1^{-1}}z_2^{-1}}.$$

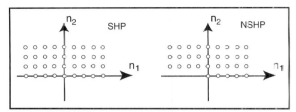

FIGURE 5.19   *Illustration of FRF support options. Note that FRF impulse response support extends to the complete SHP or NSHP region*

OBSERVATIONS

1   Each 1-D row operator can be factored into poles inside and poles outside the unit circle and then split into $\oplus$ and $\ominus$ components, where the $\oplus$ component is recursively stable as a right-sided (causal) sequence and the $\ominus$ component is recursively stable as a left-sided (anticausal) sequence.

2   If the inverses $A_0^{-1}(z_1)$ and $B_0^{-1}(z_1)$ are stable and right-sided, then we get NSHP support for the overall FRF impulse response. We get SHP response support as shown in Figure 5.19 for A_0^{-1} and B_0^{-1} stable and two-sided.

3   It is important to note, in either case, that impulse responses are not restricted to wedge support as would be the case for a conventional NSHP-recursive filter.

The stability of FRF filters is addressed in [10]. The filter design in the next Example 5.2-4 uses a numerical measure of stability as an additional design constraint, thus ensuring that the designed filter satisfies a numerical version of the FRF stability test [10]. The example concerns an FRF design for the ideal 2-D Wiener filter, an optimal linear filter for use in estimating signals in noise, which will be derived and used in Chapter 7 (image processing).

EXAMPLE 5.2-4 *(SHP Wiener filter design)*

Using the Levenberg–Marquardt optimization method, the following SHP Wiener filter was designed in [10]. The FRF used 65 coefficients, approximately equally spread over the numerator and denominator and involving five input and output rows. Figures 5.20 and 5.21 show the ideal and designed magnitude response, and Figures 5.22 and 5.23 show the respective phase responses.

We see that all of the main features of the ideal magnitude and phase response of the Wiener filter are achieved by the fully recursive design. This is because the FRF impulse response is not constrained to have support on a wedge of the general causal half-space like the conventional NSHP recursive filter. The measured magnitude mean-square error (MSE) was 0.0028 and the phase MSE was 0.042. As seen in [10], the conventional NSHP recursive filter was not able to achieve this high degree of approximation. (See Chapter 7 for more on 2-D Wiener filters and their applications.)

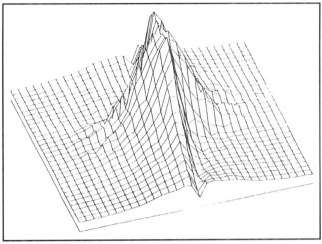

FIGURE 5.20   *Ideal SHP Wiener filter magnitude (origin in middle)*

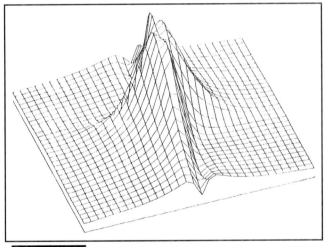

FIGURE 5.21   *SHP FRF-designed magnitude (origin in middle)*

5.3 SUBBAND/WAVELET FILTER DESIGN

Most spatial subband/wavelet or SWT filter design uses the separable product approach, so here we just discuss some 1-D design methods for the component filters of the 1-D analysis/synthesis system shown in Figure 5.24. Johnston [8] designed pairs of even-length quadrature mirror filters (QMFs) with linear phase using a computational optimization method. Aliasing was canceled out via the method of Esteban and Galand [5], which makes the synthesis choice $G_0(\omega) \triangleq H_1(\omega - \pi)$

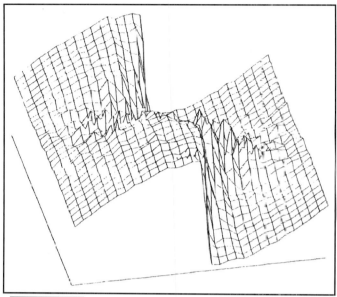

FIGURE 5.22   Ideal SHP Wiener phase response (origin in middle)

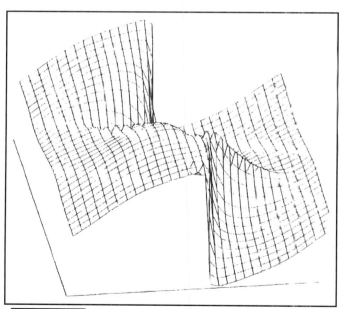

FIGURE 5.23   SHP FRF Wiener-designed phase response (origin in middle)

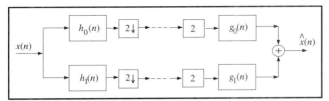

FIGURE 5.24   *1-D diagram of analysis/synthesis SWT/ISWT system*

and $G_1(\omega) \triangleq -H_0(\omega - \pi)$, and the lowpass–highpass property of the analysis filterbank was assured via $H_1(\omega) = H_0(\omega - \pi)$. So, the design can concentrate on the transfer function

$$T(\omega) \triangleq \frac{1}{2}\left[H_0^2(\omega) - H_0^2(\omega - \pi)\right].$$

In the perfect reconstruction case, $T(\omega) = 1$, and so $\hat{x}(n) = x(n)$. In the case of Johnston QMFs, perfect reconstruction is not possible, but we can achieve a very good approximation $\hat{x}(n) \simeq x(n)$, as will be seen below. For a given stopband spec ω_s and energy tolerance ε, he minimized

$$\int_0^{\pi/2}\left[|T(\omega)|^2 - 1\right]^2 d\omega \quad \text{subject to constraint} \quad \int_{\omega_s}^{\pi/2}|H_0(\omega)|^2 d\omega \leqslant \varepsilon,$$

using the Hooke and Jeeves nonlinear optimization algorithm [7]. The design took advantage of linear phase, by working with the real-valued quantity $\widetilde{H}_0$ defined by the relation

$$\widetilde{H}_0(\omega) \triangleq e^{+j\omega(N-1)/2}H_0(\omega).$$

He used a dense frequency grid to approximate the integrals and used the Lagrangian method to bring in the stopband constraint, resulting in minimization of total function E:

$$E \triangleq E_r + \lambda E_s,$$

with

$$E_r \triangleq 2 \sum_{\omega_i=0}^{\pi/2}\left[\widetilde{H}_0^2(\omega_i) + \widetilde{H}_0^2(\pi - \omega_i) - 1\right]^2,$$

$$E_s \triangleq \sum_{\omega_i=\omega_s}^{\pi/2}\widetilde{H}_0^2(\omega_i). \tag{5.3-1}$$

He thus generated a family of even-length, linear-phase FIR filters parameterized by their length N, stopband ω_s, and stopband tolerance ε. Results were reported in

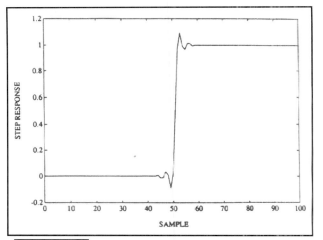

FIGURE 5.25   *Step response of Johnston 16C linear-phase QMF*

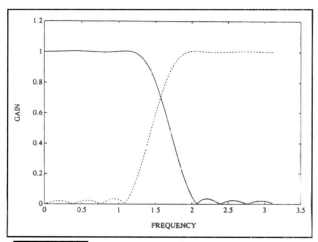

FIGURE 5.26   *Magnitude frequency response of Johnston 16C filter*

terms of transition bandwidth, passband ripple, and stopband attenuation. Johnston originally designed his filters for audio compression. Note that since we trade off E_r versus E_s, we do not get *perfect reconstruction* with these QMF designs, still the approximation in the total transfer function $T(\omega)$ can be very accurate, i.e., within 0.02 dB, which is quite accurate enough so that errors are not visible in typical image processing displays.

Note that the design equation, (5.3-1), is only consistent with $T(\omega) = 1/2$, so an overall multiplication by 2 would be necessary when using these Johnston filters.

EXAMPLE 5.3-1 (*Johnston's filter 16C*)

Here we look at the step response and frequency response of a 16-tap QMF designed by Johnston [8] using the above nonlinear optimization method. For the 16C filter, the value of $\lambda = 2$ achieved stopband attenuation of 30 dB. The transition bandwidth was 0.10 radians and the achieved overall transmission was flat to within 0.07 dB. Another similar filter, the 16B, achieved 0.02 dB transmission flatness. The step response in Figure 5.25 shows about a 5–10% amount of overshoot, or ringing. The frequency response in Figure 5.26 shows good lowpass and highpass analysis characteristics.

Other filters, due Simoncelli and Adelson [14], were designed using a similar optimization approach, but using a min-max or l^∞ norm and incorporating a modified $1/|\omega|$ weighting function with bias toward low frequency to better match the human visual system. Their designed filters are useful for both odd and even lengths N. Notable is their 9-tap filter, which has quite good performance for image coding.

5.3.1 WAVELET (BIORTHOGONAL) FILTER DESIGN METHOD

The 1-D subband filter pairs designed by wavelet analysis generally relate to what is called maximally flat design in the signal processing literature [18] and the filters are then used for separable 2-D subband analysis and synthesis as discussed in Section 5.3. The term *biorthogonal* is used in the wavelet literature to denote the case where the analysis filter set $\{h_0, h_1\}$ is different from the synthesis filter set $\{g_0, g_1\}$. Generally, the extra freedom in the design of biorthogonal filters results in a more accurate design for the lowpass filters combined with perfect reconstruction. On the other hand, departures from orthogonality generally have a negative effect on coding efficiency. So the best biorthogonal wavelet filters for image coding are usually nearly orthogonal.

Rewriting the perfect reconstruction SWT equations from Section 4.4 of Chapter 4 in terms of Z-transforms, we have the transfer function

$$H_0(z)G_0(z) + H_1(z)G_1(z) = 2, \qquad (5.3\text{-}2)$$

given the aliasing cancellation condition

$$H_0(-z)G_0(z) + H_1(-z)G_1(z) = 0. \qquad (5.3\text{-}3)$$

In the papers by Antonini *et al.* and Cohen *et al.* [2,3] on biorthogonal subband/wavelet filter design, a perfect reconstruction solution is sought under the constraint that the analysis and synthesis lowpass filters h_0 and g_0 be symmetric and FIR, i.e., linear phase with phase zero. They cancel out aliasing (5.3-3) via the choices [13]

$$H_1(z) = zG_0(-z) \quad \text{and} \quad G_1(z) = z^{-1}H_0(-z) \tag{5.3-4}$$

The transfer function (5.3-2) then becomes

$$H_0(z)G_0(z) + H_0(-z)G_0(-z) = 2,$$

which, when evaluated on the unit circle $z = e^{j\omega}$, gives the overall system frequency response

$$H_0(\omega)G_0(\omega) + H_0(\omega - \pi)G_0(\omega - \pi) = 2, \tag{5.3-5}$$

owing to the use of real and symmetric lowpass filters h_0 and g_0.

It now remains to design h_0 and g_0 to achieve this perfect reconstruction (5.3-2) or (5.3-5). For this they use a spectral factorization method and introduce a prescribed parameterized form for the product $H_0(\omega)G_0(\omega)$ that is known to satisfy (5.3-5) exactly. Since it is desired that both the analysis and reconstruction filters have a high degree of flatness, calling for zeros of the derivative at $\omega = 0$ and π, the following equation for the product H_0G_0 was proposed in Cohen *et al.* [3],

$$H_0(\omega)G_0(\omega) = \cos^{2K}(\omega/2)\left[\sum_{p=0}^{L-1} \binom{L-1+p}{p} \sin^{2p}(\omega/2)\right], \tag{5.3-6}$$

where $\cos(\omega/2)$ provides zeros at $\omega = \pi$ and the terms $\sin(\omega/2)$ provide zeros at $\omega = 0$. Then actual filter solutions can be obtained with varying degrees of spectral flatness and various lowpass characteristics, all guaranteed to give perfect reconstruction, by factoring the function on the right-hand side of (5.3-6) for various choices of K and L.

The polynomial on the right-hand side of (5.3-6)—call it $|M_0(\omega)|^2$—is half of a *power complementary* pair defined by the relation

$$\left|M_0(\omega)\right|^2 + \left|M_0(\omega - \pi)\right|^2 = 2,$$

which had been considered earlier by Smith and Barnwell [15] for the design of perfect reconstruction orthogonal subband/wavelet filters. A particularly nice treatment of this wavelet design method is presented in Taubman and Marcellin [17].

EXAMPLE 5.3-2 [9/7 Cohen–Doubechies–Feauveau (CDF) filter]

In Antonini *et al.* [2], the authors consider several example solutions. The one that works the best in their image coding example corresponds to factors H_0 and G_0 in (5.3-6), which make h_0 and g_0 have nearly equal lengths. A solution

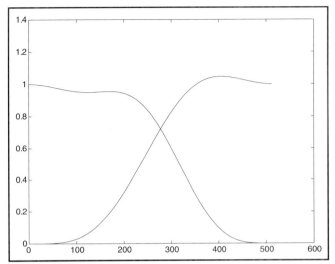

FIGURE 5.27   *Frequency response of CDF 9/7 analysis filters (512 pt. DFT)*

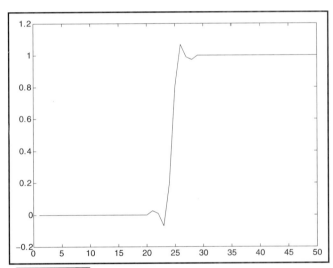

FIGURE 5.28   *Step response of CDF 9/7 analysis lowpass filter*

corresponding to $K = 4$ and $L = 4$ resulted in the following 9-tap/7-tap (or simply 9/7) filter pair:

n	0	±1	±2	±3	±4
$2^{-1/2}h_0(n)$	0.602949	0.266864	−0.078223	−0.016864	0.026749
$2^{-1/2}g_0(n)$	0.557543	0.295636	−0.028772	−0.045636	0.0

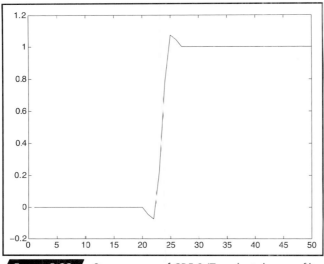

FIGURE 5.29   *Step response of CDF 9/7 synthesis lowpass filter*

Figure 5.27 shows the lowpass and highpass analysis system frequency response plotted from a 512-point FFT. The corresponding analysis lowpass filter step response is shown in Figure 5.28. The synthesis lowpass filter step response is given in Figure 5.29. This synthesis step response seems quite similar to that of the Johnston filter in Figure 5.25, with a little less ringing. This filter pair has become the most widely used method of SWT analysis and synthesis (ISWT) and is chosen as the default filter for the ISO standard JPEG 2000 [13,17]. It is well regarded for its generally high coding performance and reduced amount of ringing when the upper subband is lost for reasons due to coding noise or ambient noise. This filter set is generally denoted as "CDF 9/7," for the authors in [3], or often as just "Daubechies 9/7."

In comparing the QMFs with the wavelet-designed ones, note that the main difference is that the QMFs have a better passband response with a sharper cutoff, which can reduce aliasing error in the lowpass band. The wavelet-designed filters tend to have poorer passband response but less ringing in the synthesis filter step response, and hence less visual artifacts.

EXAMPLE 5.3-3 *(anti-alias filtering)*
Here we take a 4CIF[4] (704 × 576) monochrome image and subsample it to CIF (352 × 288) using two different anti-alias filters. On the left in Figure 5.30 is the result using the CDF 9/7 subband/wavelet filter. The image on the right was obtained using the MPEG4 recommended lowpass filter prior to the dec-

---

4. CIF, Common intermediate format.

FIGURE 5.30   *Result of using two different anti-alias filters for downsampling*

imator. The MPEG4 lowpass recommendation has the form

n	0	± 1	± 2	± 3	± 4	± 5	± 6
$64h(n)$	24	19	5	-3	-4	0	2

We can see that the image on the right in Figure 5.30 is softer, but also that the one on the left has significant spatial aliasing that shows up on the faces of some of the buildings. It should be noted that this image frame, from the MPEG test video *City*, has an unusual amount of high-frequency data. Most 4CIF images would not alias this badly with the CDF 9/7 lowpass filter.

5.4 CONCLUSIONS

In this chapter we have introduced 2-D or spatial filter design. We studied FIR filters with emphasis on the window design method, but also considered the 1-D to 2-D transformation method and the application of POCS to spatial filter design. In the case of IIR or recursive filters, we studied some simple spatial-domain design methods and then briefly overviewed computer-aided design of spatial IIR filters. We introduced fully recursive filters and saw an example of using an FRF to give a good approximation to the frequency response of an ideal spatial Wiener filter (see Chapter 7). Finally, we discussed the design problem for subband/wavelet filters as used in a separable SWT.

5.5 PROBLEMS

1   A certain real-valued filter $h(n_1, n_2)$ has support on $[-L, +L] \times [-L, +L]$ and has a Fourier transform $H(\omega_1, \omega_2)$ that is real-valued also. What kind of symmetry must h have in the n_1, n_2 plane?

2   Let the real-valued impulse response have quadrantal symmetry:

$$h(n_1, n_2) = h(-n_1, n_2) = h(n_1, -n_2) = h(-n_1, -n_2).$$

What symmetries exist in the Fourier transform? Is it real? How can this be used to ease the computational complexity of filter design?

3   A certain *ideal* lowpass filter with cutoff frequency $\omega_c = \frac{\pi}{2}$ has impulse response

$$h_d(n_1, n_2) = \frac{1}{\sqrt{n_1^2 + n_2^2}} J_1\left(\frac{\pi}{2}\sqrt{n_1^2 + n_2^2}\right).$$

(a)   What is the passband gain of this filter? (Hint: $\lim_{x \to 0} \frac{J_1(x)}{x} = \frac{1}{2}$.)

(b)   We want to design an $N \times N$-point, linear-phase, FIR filter $h(n_1, n_2)$ with *first-quadrant* support using the ideal function $h_d(n_1, n_2)$ given above. We choose a *separable* 2-D Kaiser window with parameter $\beta = 2$. The continuous-time 1-D Kaiser window is given as

$$w_\alpha(t) = \begin{cases} I_0(\beta\sqrt{1 - (t/\tau)^2})/I_0(\beta), & |t| < \tau \\ 0, & \text{else.} \end{cases}$$

Write an expression for the filter coefficients $h(n_1, n_2)$: $n_1 = 0, \ldots, N - 1$, $n_2 = 0, \ldots, N - 1$.

(c)   Suppose we use a 512×512-point row–column FFT to approximately implement the above designed filter h for a 512×512 pixel image:

    i   Evaluate the number of complex multiplies used as a function of M for an $M \times M$ image, assuming the component 1-D FFT uses the Cooley–Tukey approach. (The 1-D Cooley–Tukey FFT algorithm needs $\frac{1}{2}M\log_2 M$ complex multiplies when M is a power of 2 approach.) Please specialize your result to $M = 512$.

    ii   What portion of the output will be the correct (*linear convolution*) result? Specify which pixels will be correct.

4   We want to design an $N \times N$-point, *linear-phase*, FIR filter $h(n_1, n_2)$ with *first-quadrant support* using the ideal function $h_i(n_1, n_2)$ given as

$$h_i(n_1, n_2) = \frac{1}{4} \frac{J_1\left(\frac{\pi}{2}\sqrt{n_1^2 + n_2^2}\right)}{\sqrt{n_1^2 + n_2^2}}.$$

We choose a *separable* 2-D Kaiser window with parameter $\alpha = 2$. The continuous-time 1-D Kaiser window is given as

$$w_\alpha(t) = \begin{cases} I_0(\beta\sqrt{1 - (t/\tau)^2})/I_0(\beta), & |t| < \tau \\ 0, & \text{else.} \end{cases}$$

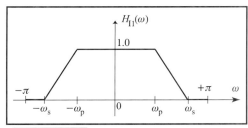

FIGURE 5.31   *Alternative ideal filter response*

    (a)   Write down the filter coefficients of $h(n_1, n_2)$: $n_1 = 0, \ldots, N-1, n_2 = 0, \ldots, N-1$.

    (b)   Use MATLAB to find the magnitude of the frequency response of the designed filter. Plot the result for $N = 11$.

5   Let $0 < \omega_p < \pi$, and define the 1-D ideal filter response

$$H_I(\omega) \triangleq \begin{cases} 1, & |\omega| \leqslant \frac{1}{2}(\omega_s + \omega_p) \\ 0, & \text{else} \end{cases} \quad \text{on } [-\pi, +\pi].$$

Then define $H_{I1}(\omega) \triangleq W_1(\omega) \otimes H_I(\omega)$, where $\otimes$ denotes periodic convolution, and where

$$W_1(\omega) \triangleq \begin{cases} \frac{1}{\omega_s - \omega_p}, & |\omega| \leqslant \frac{1}{2}(\omega_s - \omega_p) \\ 0, & \text{else} \end{cases} \quad \text{on } [-\pi, +\pi], \text{ with } \pi > \omega_s > \omega_p,$$

then $H_{I1}(\omega)$ is as shown in Figure 5.31.

    (a)   Find the impulse response $h_{I1}(n)$. Note that it is of infinite support.

    (b)   Define and sketch the 2-D ideal filter response $H_{I1}(\omega_1, \omega_2) \triangleq H_{I1}(\omega_1)H_{I1}(\omega_2)$ and give an expression for its impulse response $h_{I1}(n_1, n_2)$.

    (c)   Find the so-called l^2 or minimum least-squares error approximation to the ideal spatial filter you found in part b, with support $[-N, +N]^2$.

6   Modify the MATLAB program McDesFilt.m (available on the book CD-ROM) to design a near-circular highpass filter with stopband edge $\omega_s = 0.1\pi$ and $\omega_p = 0.3\pi$, and apply the filter to the *Lena* image (note: requires MATLAB with image processing toolbox). Is the output what you expected? Try scaling and shifting the output y to make it fit in the output display window [0, 255]. Hint: The output scaling $y = 128 + 2*y$ works well.

7   In deriving the McClellan transformation method, we need to find the $a'(n)$ coefficients in (5.1-2) in terms of the $a(n)$ coefficients in (5.1-1). In this problem we work out the case for $M = 4$.

(a) Use $T_n(x)$ for $n = 0, 4$ to evaluate $\cos n\omega$ in terms of powers of $\cos^k \omega$ for $k = 0, 4$.

(b) Substitute these expressions into (5.1-1) and determine $a'(n)$ in terms of $a(n)$ for $n = 0, 4$.

(c) Solve the equations in part b for $a(n)$ in terms of $a'(n)$.

8 Let the filter h with first-quadrant impulse response support have system function

$$H(z_1, z_2) = \frac{b_{00}}{1 + a_{10}z_1^{-1} + a_{01}z_2^{-1} + a_{11}z_1^{-1}z_2^{-1}}.$$

(a) Find the coefficients $\{b_{00}; a_{10}, a_{01}, a_{11}\}$ such that the impulse response h agrees with the four prescribed values:

$$\begin{aligned} h(0, 0) &= 1.0, & h(1, 0) &= 0.8, \\ h(0, 1) &= 0.7, & h(1, 1) &= 0.6. \end{aligned}$$

(b) Determine whether the resulting filter is stable or not. (Hint: Try the root mapping method.)

9 In 2-D filter design we often resort to a general optimization program such as Fletcher–Powell to perform the design. Such programs often ask for analytic forms for the partial derivatives of the specified error function with respect to each of the filter coefficients.

Consider a simple recursive filter with frequency response

$$H(\omega_1, \omega_2) = \frac{a}{1 - be^{-j\omega_1} - ce^{-j\omega_2}},$$

where the three filter parameters a, b, and c are real variables.

(a) Find the three partial derivatives of H with respect to each of a, b, and c, for fixed ω_1, ω_2.

(b) Find the corresponding partial derivatives of the conjugate function H^*.

(c) Choose a weighted mean-square error function as

$$\mathcal{E} = \frac{1}{(2\pi)^2} \int_{-\pi}^{+\pi} \int_{-\pi}^{+\pi} W(\omega) |I(\omega) - H(\omega)|^2 \, d\omega,$$

for given ideal frequency-response function I and positive weighting function W. Find the partial derivatives $\partial \mathcal{E}/\partial a$, $\partial \mathcal{E}/\partial b$, and $\partial \mathcal{E}/\partial c$, making use of your results in parts a and b. Hint: Rewrite $|A|^2 = AA^*$ first. Express your answers in terms of W, I, and H.

10 Consider a zero-phase or two-pass filter design with NSHP recursive filters

$$H(\omega_1, \omega_2) \triangleq H_{\oplus+}(\omega_1, \omega_2)H_{\ominus-}(\omega_1, \omega_2) = |H(\omega_1, \omega_2)|^2,$$

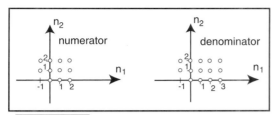

FIGURE 5.32   *Numerator and denominator coefficient-support region indicated by open circles*

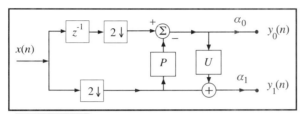

FIGURE 5.33   *Illustration of the lifted SWT*

where $H_{\ominus-}(\omega_1, \omega_2) \triangleq H^*_{\oplus+}(\omega_1, \omega_2)$. Equivalently, in the spatial domain,

$$h(n_1, n_2) = h_{\oplus+}(n_1, n_2) * h_{\ominus-}(n_1, n_2).$$

Let the NSHP filter $H_{\oplus+}(\omega_1, \omega_2)$ have real coefficients with numerator support and denominator support as indicated in Figure 5.32.

(a)   Find the spatial support of $h_{\oplus+}(n_1, n_2)$ assuming that this filter is stable in the $+n_1, +n_2$ direction.

(b)   State the spatial support of $h_{\ominus-}(n_1, n_2)$.

(c)   What is the support of $h(n_1, n_2)$?

11   Show that the biorthogonal subband/wavelet constraint (5.3-4) achieves alias cancellation in (5.3-3) and that the transfer function (5.3-2) becomes

$$H_0(z)G_0(z) + H_0(-z)G_0(-z) = 2,$$

which, when evaluated on the unit circle $z = e^{j\omega}$, equals

$$H_0(\omega)G_0(\omega) + H_0(\omega - \pi)G_0(\omega - \pi) = 2,$$

owing to the use of real and symmetric lowpass filters h_0 and g_0.

12   The *lifted* SWT is defined starting from the *Lazy* SWT, which just separates the input sequence x into even and odd terms. In the lifted SWT, this is followed by *prediction* and *update* steps as specified by the operators P and U, respectively, as shown in Figure 5.33. The output multipliers α_0 and α_1 may be needed for proper scaling.

(a) Find the corresponding prediction and update operators for the SWT that uses the Haar filter set

$$h_0(n) = \delta(n) + \delta(n - 1),$$
$$h_1(n) = \delta(n) - \delta(n - 1).$$

(b) A key property of lifting is that the operators P and U are not constrained at all. Show the ISWT for the lifted transform in Figure 5.33 by starting from the right-hand side of the figure and first undoing the update step and then undoing the prediction step. Note that this is possible even for nonlinear operators P and U.

(c) Can you do the same for the LeGall–Tabatabai (LGT) 5/3 analysis filter set?

$$h_0(n) = \left\{ -\frac{1}{8}, \frac{1}{4}, \frac{3}{4}, \frac{1}{4}, -\frac{1}{8} \right\},$$
$$h_1(n) = \left\{ -\frac{1}{2}, 1, -\frac{1}{2} \right\}.$$

Reference to [3] may be necessary.

REFERENCES

[1] A. Abo-Taleb and M. M. Fahmy, "Design of FIR Two-Dimensional Digital Filters by Successive Projections," *IEEE Trans. Circ. Syst.* CAS-31, 801–805, Sept. 1984.

[2] M. Antonini, M. Barlaud, P. Mathieu, and I. Daubechies, "Image Coding Using Wavelet Transform," *IEEE Trans. Image Process.* 1, 205–220, April 1992.

[3] A. Cohen, I. Daubechies, and J.-C. Feauveau, "Biorthogonal Bases of Compactly Supported Wavelets," *Commun. Pure Appl. Math.* XLV, 485–560, 1992.

[4] D. E. Dudgeon and R. M. Mersereau, *Multidimensional Digital Signal Processing*, Prentice-Hall, Englewood Cliffs, NJ, 1983.

[5] D. Esteban and C. Galand, "Application of Quadrature Mirror Systems to Split Band Voice Coding Schemes," *Proc. ICASSP*, 191–195, May 1977.

[6] D. B. Harris and R. M. Mersereau, "A Comparison of Algorithms for Minimax Design of Two-Dimensional Linear Phase FIR Digital Filters," *IEEE Trans. Accoust., Speech, Signal Process* ASSP-25, 492–500, Dec. 1977.

[7] R. Hooke and T. A. Jeeves, "Direct Search Solution of Numerical and Statistical Problems," *J. ACM* 8, 212–229, April 1961.

[8] J. D. Johnston, "A Filter Family Designed for Use in Quadrature Mirror Filter Banks," *Proc. IEEE Int. Conf. Accoust. Speech, Signal Process. (ICASSP)*, 291–294, Denver, CO, 1980.

[9] J. F. Kaiser, "Nonrecursive Digital Filter Design Using I_0 − sinh Window Function," *Proc. IEEE Symp. Circ. Syst.*, 20–23, 1974.

[10] J.-H. Lee and J. W. Woods, "Design and Implementation of Two-Dimensional Fully Recursive Digital Filters," *IEEE Trans. Accoust., Speech, Signal Process*, **ASSP-34**, 178–191, Feb. 1986.

[11] J. Lim, *Two-Dimensional Signal and Image Processing*, Prentice-Hall, Englewood Cliffs, NJ, 1990.

[12] J. H. McClellan and D. S. K. Chan, "A 2-D FIR Filter Structure Derived from the Chebyshev Recursion," *IEEE Trans. Circ. Syst.* **CAS-24**, 372–378, July 1977.

[13] M. Rabbani and R. Joshi, "An Overview of the JPEG 2000 Still Image Compression Standard," *Signal Process.: Image Commun. J.* **17**, 3–48, 2002.

[14] E. P. Simoncelli and E. H. Adelson, "Subband Transforms," in *Subband Image Coding*, (J. W. Woods, ed.), Kluwer Academic Publ., Dordrecht, NL, 1991. pp. 143–192.

[15] M. J. T. Smith and T. P. Barnwell III, "Exact Reconstruction for Tree-structured Subband Coders," *IEEE Trans. Accoust., Speech, Signal Process* **ASSP-34**, 431–441, June 1986.

[16] H. Stark and Y. Yang, *Vector Space Projections*, John Wiley, New York, NY, 1998.

[17] D. S. Taubman and M. W. Marcellin, *JPEG2000: Image Compression, Fundamentals, Standards, and Practice*, Kluwer Academic Publ., Norwell, MA, 2002.

[18] P. P. Vaidyanathan, *Multirate Systems and Filter Banks*, Prentice-Hall, Englewood Cliffs, NJ, 1993.

INTRODUCTORY IMAGE
PROCESSING

6

This chapter covers some basics of image processing that are helpful in the applications in later chapters. We will discuss light, sensors, and the human visual system. We will cover spatial and temporal properties of human vision, and present a basic spatiotemporal frequency response of the eye–brain system. This information, interesting in its own right, will be useful for the design and analysis of image and video signal processing systems that produce images and video to be seen by human observers. We also give some description of elementary image processing filters.

6.1 LIGHT AND LUMINANCE

We use the notation $c(x, y, t, \lambda)$ for radiant flux or light intensity as a function of position and time at the spatial wavelength λ. Now the human visual system (HVS) does not perceive this radiant flux directly. A *relative luminous efficiency* $v(\lambda)$ has been defined to relate our perceived *luminance* (brightness) to the incoming radiant flux

$$l(x, y, t) \triangleq \mathcal{K} \int_0^\infty c(x, y, t, \lambda) v(\lambda) \, d\lambda,$$

where the constant $\mathcal{K} = 685$ lumens/watt and the luminance l is expressed in lumens/square meter [11]. We can think of the relative luminous efficiency as a kind of filter response of the human eye. As such, it would be expected to vary from person to person, thus a so-called *standard observer* was defined by the international committee CIE[1] in 1929, as sketched in Figure 6.1, thereby defining a *standard luminance l*. Note that luminance is a quantity averaged over wavelength,[2] and as such, is called a monochrome or gray-level measure of the light intensity. Looking at Figure 6.1, normalized to 1, we see that the response of human vision peaks around 555 nanometers (nm), which is *green* light, and has a range from about 420 to 670 nm, with 450 nm called *blue* and 650 nm called *red*. The low end of visibility is *violet*, and the range below the visible is called *ultraviolet*. At the high end we have *red* light, and beyond this end of the visible region is called *infrared* light.

We can also define similar response functions for image sensors, and suppressing the position dependency x, y, t, we have

$$l \triangleq \mathcal{K}' \int_0^\infty c(\lambda) s_{\text{BW}}(\lambda) \, d\lambda,$$

---

1. Commission Internationale de l'Eclairage (International Commission on Illumination).

2. *Wavelength and spatial frequency*: Image and video sensors record energy (power × time) on a scale of micrometers or more (for typical 1-cm sensor chips) and so are not sensitive to the oscillations of the electromagnetic field at the nanometer level. Our spatial frequency is variation of this average power on the larger scale. So there is no interaction between the wavelength of the light and the spatial frequency of the image, at least for present-day sensors.

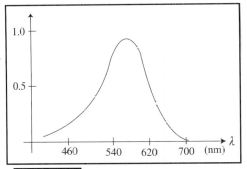

FIGURE 6.1 *Sketch of CIE standard observer relative luminous efficiency*

where s_{BW} is the black–white (gray-level) sensitivity of an imaging device. Examples of image sensors are charge-coupled devices, various types of image pickup tubes, and more recently CMOS devices. By inserting optical wavelength filters in front of these devices, we can get different responses from various color regions or channels. It is common in present-day cameras to employ optical filters to create three color channels, red, green, and blue, to match what is known about the color perception system in human vision, i.e., the eye–brain system. Similar response functions can characterize image displays.

The human eye is composed of an *iris*, a *lens*, and an imaging surface, or *retina*. The purpose of the iris is to let in the right amount of light, to avoid either saturation or underillumination. The lens focuses this light on the retina, which is composed of individual *cones* and *rods*. The most sensitive part of the retina is small in size and called the *fovea*. This is where the cones are densely located, and of three color efficiencies, responding principally to red, green, and blue. Labeling these luminous efficiencies as $s_R(\lambda), s_G(\lambda)$, and $s_B(\lambda)$, we see the result of psychovisual experiments, sketched in Figure 6.2, with the green and blue response being fairly distinct, and the red response maybe better called yellow-green response, overlapping with the green.

We can then characterize on the average the human cone response to an incident flux distribution $c(\lambda)$ as

$$R = \mathcal{K} \int_0^\infty c(\lambda) s_R(\lambda)\, d\lambda, \tag{6.1-1}$$

$$G = \mathcal{K} \int_0^\infty c(\lambda) s_G(\lambda)\, d\lambda, \tag{6.1-2}$$

and

$$B = \mathcal{K} \int_0^\infty c(\lambda) s_B(\lambda)\, d\lambda. \tag{6.1-3}$$

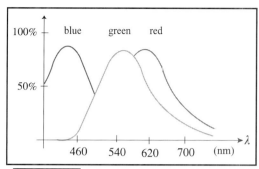

FIGURE 6.2   *Sketch of average sensitivity functions of three types of color receptors (cones) in the human eye*

The rods are spread over a very large area of the retina, but at a considerably lower spatial density than the cones. These rods are of one type, and so can only provide a monochrome response. They are very sensitive to light, though, and provide our main visual perception at dim light levels, such as at night or in a very darkened room. This is why we do not perceive color in night vision.

The historical development of color image sensors has been an attempt to mimic the three-channel nature of the cones in the human eye. This is true of both color film and color electronic sensors. For example, in the early 1950s a standards committee in the USA, the National Television Systems Committee (NTSC) [6,8,11], defined an RGB color system to be used for domestic television. The R, G, and B signals were defined in terms of the incoming radiant flux and a set of color sensitivity functions. An efficient transformation was then introduced to reduce the redundancy in these three signals,

$$\begin{bmatrix} Y \\ I \\ Q \end{bmatrix} = \begin{bmatrix} 0.299 & 0.587 & 0.114 \\ 0.596 & -0.274 & -0.322 \\ 0.211 & -0.523 & 0.312 \end{bmatrix} \begin{bmatrix} R \\ G \\ B \end{bmatrix},$$

and to provide compatibility with the then-existent monochrome TV standard. The Y signal is called the *luminance component*,[3] and can be seen as a weighted average of the three color channels, with more weight given to the visually dominant green channel. The other two components I and Q are color difference channels commonly referred to as *components*. The digital television standard International Telecommunications Union (ITU) 601 has changed this somewhat, resulting in the same luminance component Y, but the chrominance components were changed to $C_R = 0.713(R - Y)$ and $C_B = 0.564(B - Y)$, and are sometimes

---

3. In practice this color space transformation is usually performed after a nonlinear operation called "gamma correction," to be discussed in Section 6.5.1. In this typical case, we really do not have luminance. So the Y component is then called *luma* instead [5].

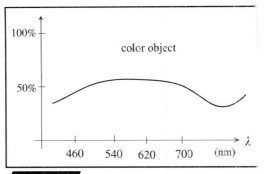

FIGURE 6.3   *Wavelength spectrum of a color object*

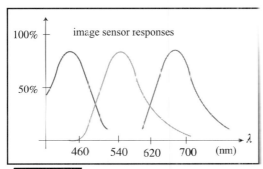

FIGURE 6.4   *Possible color sensor response functions*

referred to as the *color difference* components. The overall ITU 601 transformation is then

$$\begin{bmatrix} Y \\ C_R \\ C_B \end{bmatrix} = \begin{bmatrix} 0.299 & 0.587 & 0.114 \\ 0.500 & -0.418 & -0.082 \\ -0.169 & -0.331 & 0.500 \end{bmatrix} \begin{bmatrix} R \\ G \\ B \end{bmatrix}.$$

Effectively, we thus define a three-dimensional *color* (vector) *space*.

EXAMPLE 6.1-1 (*color image sensed and displayed*)

Consider a solid color object that is photographed (sensed) and then displayed, with color spectra of Figure 6.3. When being sensed, it is illuminated by a white light (i.e., a flat color spectrum and recorded with the image sensor's sensitivity functions, shown in Figure 6.4. The resulting sensed R, G, and B values are the integrals of the point-by-point product of these two curves [5]. Assuming no signal processing of these values, then the display device adds in its sensitivity curves based on its red, green, and blue light sources,

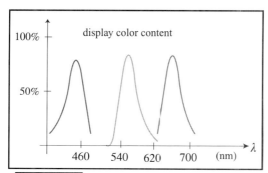

FIGURE 6.5   *Possible display color primaries*

with wavelength curves shown in Figure 6.5, and creates the displayed image.[4]

Since we just have a solid color here, the main question is whether the displayed color is "the same" as the original object. The displayed output will be a weighted combination of the three curves in Figure 6.5, which clearly cannot equal the object spectral response shown in Figure 6.3. However, from the viewpoint of human visual response, we just want the displayed color to be perceived as the same, called *metameric* [5]. For this we need only for the integrals in (6.1-1), (6.1-2), and (6.1-3) to be the same, because then, up to an approximation due to this linear model, the cones will output the same signal as when the color object was directly perceived. Clearly, without careful choice of these curves and scaling of their respective strengths, the perceived color will not be the same.

An approximate solution to this problem is to *white balance* the system. In this process, a pure white object is imaged first and then the sensor output R, G, and B values are scaled so as to produce a white perception in the HVS of the (standard) viewer. The actual signal captured by the camera may not be white because of possible (probable) nonwhite illumination, but the scaling is to make the output image be perceived as white when viewed in its environment.

6.2 STILL IMAGE VISUAL PROPERTIES

Here we look at several visual phenomena that are important in image and video perception. First we present Weber's law, which defines contrast and introduces

---

4. We do not intend to imply that these sensor and display response functions are common or typical. They are merely to introduce the concept involved in color reproduction. More typical curves for various display devices are given in [5].

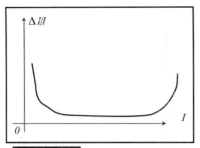

FIGURE 6.6   *Average just-noticeable intensity fraction versus background intensity*

the concept of *just-noticeable difference* (JND). We then present the contrast sensitivity function, also called the visual magnitude transfer function (MTF), as effectively the human visual frequency response. We introduce the concept of spatial adaptation of contrast sensitivity, and then look at temporal effects.

6.2.1 WEBER'S LAW

Psychovisual researchers early on found that the eye–brain response to a uniform step in intensity is not the same over the full range of human perception. In fact, they found that more nearly the just-noticeable percent is nearly constant over a wide range. This is known now as Weber's Law. Writing I for the incident intensity (or luminance) and ΔI for the just noticeable change, we have

$$\Delta I / I \approx \text{constant,}$$

with the constant value in the range $[0.01, 0.03]$, and this value holds constant for at least three decades in $\log I$, as sketched in Figure 6.6. We note that Weber's Law says we are more sensitive to light intensity changes in low light levels than in strong ones.

A new quantity called *contrast* was then defined via

$$\Delta C = \Delta I / I,$$

which, in the limit of small Δ, becomes the differential equation $dC/dI = I^{-1}$, which integrates to

$$C = \ln(I/I_0). \tag{6.2-1}$$

Sometimes we refer to signals I as being in the *intensity domain,* and the nonlinearly transformed signals C as being in the *contrast* or *density domain*, the latter name coming from a connection with photographic film (see Section 6.4.2). We note that the contrast domain is most likely the better choice for quantizing a value with uniform step size. This view should be tempered with an understanding of how the psychovisual data were taken to verify Weber's Law.

FIGURE 6.7   *Stimula that can be used to test Weber's Law*

FIGURE 6.8   *Equal increments at three different brightness values: 50, 100, 200, on range* [0, 255]

With reference to Figure 6.7, human subjects were presented with a disk of incremented intensity $I + \Delta I$ in a uniform background of intensity I, and were asked if they could notice the presence of the disk or not. These results are then averaged over many observers, whereby the threshold effect seen in Weber's Law was found. The actual JND threshold is typically set at the 50% point in these distributions. Figure 6.8 illustrates the concept of such a JND test, showing five small disks of increasing contrast ($+2, +4, +6, +8, +10$) on three different backgrounds, with contrasts 50, 100, and 200 on the 8-bit range [0, 255].

6.2.2 CONTRAST SENSITIVITY FUNCTION

Another set of psychovisual experiments has determined what has come to be regarded as the spatial frequency response of the HVS. In these experiments, a uniform plane wave is presented to viewers at a given distance, and the angular period of the image focused on their retina is calculated. The question is at what

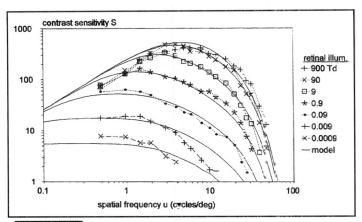

FIGURE 6.9 *Contrast sensitivity measurements of van Ness and Bouman. (Copyright JOSA, 1967, reproduced with permission.)*

intensity does the plane wave (sometimes called an optical grating) first become visible. The researcher then records this value as a function of angular spatial frequency, expressed in units of cycles/degree. These values are then averaged over many observers to come up with a set of threshold values for a so-called *standard observer*. The reciprocal of this function is then taken as the human visual frequency response and called the *contrast sensitivity function* (CSF) as seen plotted in Figure 6.9 (reprinted from Barten [1]). Of course we must use uniform background of a prescribed value, on account of Weber's Law. Otherwise the threshold would change. Also, for very low spatial frequencies, the threshold value should be given by Weber's observation, since he used a plain or *flat* signal. The CSF is based on the assumption that the above-threshold sensitivity of the HVS is the same as the threshold sensitivity. In fact, this may not be the case, but it is the current working assumption. Another objection too could be that the threshold may not be the same with two stimuli present, i.e., the sum of two different plane waves. Nevertheless, with the so-called *linear hypothesis*, we can weight a given disturbance presented to the HVS with a function that is the reciprocal of these threshold values, and effectively normalize them. If the overall intensity is then gradually reduced, all the gratings will become invisible (not noticeable to about 50% of human observers with normal acuity) at the same point. In that sense, then, the CSF is the frequency response of the HVS.

An often referenced formula that well approximates this curve was obtained by Mannos and Sakrison [12], and is expressed in terms of the radial frequency $\omega_r \triangleq \sqrt{\omega_1^2 + \omega_2^2}$ in cycles/degree,

$$H(\omega_r) = A\left(\alpha + \frac{\omega_r}{\omega_0}\right)\exp-\left(\frac{\omega_r}{\omega_0}\right)^\beta, \quad \omega_r \geq 0.$$

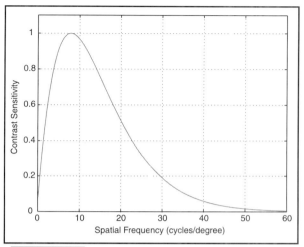

FIGURE 6.10   *Plot of the Mannos and Sakrison function*

with $A = 2.6$, $\alpha = 0.0192$, $\omega_0 = 8.772$, and $\beta = 1.1$. In this formula, the peak occurs at $\omega_r = 8$ (cycles/degree) and the function is normalized to maximum value 1. A linear amplitude plot of this function is given in Figure 6.10. Note that sensitivity is quite low at zero spatial frequency. So some kind of spatial structure or texture is almost essential if a feature is to get noticed by our human visual system.

6.2.3 LOCAL CONTRAST ADAPTATION

In Figure 6.11 we see a pair of small squares, one on a light background and the other on a dark background. While the small squares appear to have different gray-levels, in fact the gray-levels are the same. This effect occurs because the human visual system adapts to surrounding brightness levels when it interprets the brightness of an object.

There is also a local contrast adaptation wherein the JND moves upward as the background brightness moves away from the average contrast of the object. Such a test can be performed via the approach sketched in Figure 6.12, where we note that the small central square is split, and slightly darker on the left. For most people, this effect is more evident from the display on the left, where the contrast with the local background is only slight. However, on the right, the large local contrast with the local background makes this effect harder to perceive. Effectively, this means that the JND varies somewhat with local contrast, or said another way, there is some kind of local masking effect.

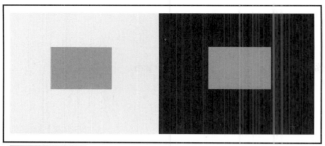

FIGURE 6.11   *Illustration of local adaptation property of the human visual system*

FIGURE 6.12   *Illustration of dependence of JND on local background brightness*

6.3 TIME-VARIANT HUMAN VISUAL SYSTEM PROPERTIES

The same kind of psychovisual experiments described in the preceding section can be employed to determine the visibility of time-variant features. By employing a spatial plane-wave grating and then modulating it sinusoidally over time, the contrast sensitivity of the human visual system can be measured in a spatiotemporal sense. Plots of experiments by Kelly [9,10], as reprinted in Barten [1], are shown in Figures 6.13 and 6.14. Figure 6.13 shows slices of the spatiotemporal CSF plotted versus spatial frequency, with the slices at various temporal frequencies, over the range 2 to 23 Hz. We see that maximum response occurs around 8 cycles/degree at 2 Hz. Also, we see no bandpass characteristic at the higher temporal frequencies. Figure 6.14 shows slices of the CSF as measured by Kelly, 1971, where there are two slices at spatial frequencies of 0 and 3 cycles/degree (cpd), plotted versus temporal frequency. Again we see the bandpass characteristic at the very low spatial frequency, while there is a lowpass characteristic at the higher spatial frequencies. Note that the highest contrast sensitivity is at temporal frequency 0 and spatial frequency 3 cpd.

A 3-D perspective log plot of the spatiotemporal CSF by Lambrecht and Kunt [14] is shown in Figure 6.15. Note that these 3-D CSFs are not separable functions. Note also that there is an order of magnitude or more difference between

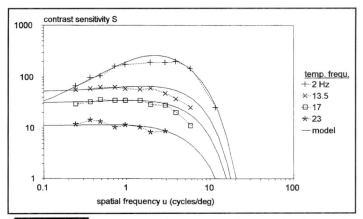

FIGURE 6.13   *Spatiotemporal CSF. (Copyright JOSA, 1979, reproduced with permission.)*

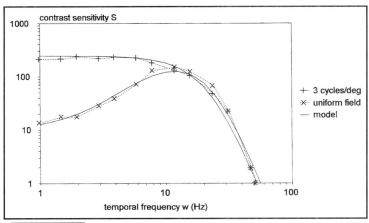

FIGURE 6.14   *Temporal CSF with spatial parameter. (Copyright JOSA, 1971, reproduced with permission.)*

the sensitivity at DC or origin and that at the peak of maybe about 10 Hz and 6 cpd.

A summary comment is that the HVS seems to be sensitive to change, either spatially or temporally, and that is what tends to get noticed. Lesser variations tend to be ignored. These results can be used to perceptually weight various error criteria for image and video signal processing, to produce a result more visually pleasing. An excellent source for further study is the thesis monograph of Barten [1]. Human visual perception is also covered in the review Chapter 8.2 by Pappas *et al.* in *Handbook of Image and Video Processing*, 2nd edn. [2].

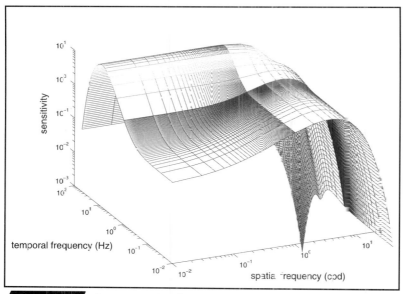

FIGURE 6.15 *Perspective plot of a spatiotemporal CSF (normalized to 10). (Reproduced with permission from Lambrecht and Kunt (Image Communication, 1998).)*

6.4 IMAGE SENSORS

6.4.1 ELECTRONIC

Electronic image sensors today are mainly solid state and of two types: *charge-coupled device* (CCD) and *complementary metal oxide semiconductor* (CMOS). At present, CCD sensors tend to occupy the nitch of higher quality and price, while CMOS sensors generally have somewhat lower quality but much lower cost, because they benefit more directly from extensive silicon memory technology. These sensors are present in virtually all current video cameras as well as all digital image cameras. They are generally characterized by the number of pixels: today running up to a total of 8M pixels for widely available still-image cameras, and very high-end professional video cameras. While image sensors can read out their frames rather slowly, e.g., 1–3 frames per second (fps), video sensors must output their image frames at a faster rate, i.e., 24–100 fps. Also, still-image sensors tend to have deeper cell wells to accommodate a wider dynamic range or bit depth, i.e., 12–14 bits per pixel (bpp) rather than the 8–10 bpp common with video sensors.

In some cameras there is only one sensor for the three primary colors. In this case some kind of color checkerboard pattern is applied to the front end of the array to sense color. In professional video cameras there is a prism together with three color filters, each with its own sensor, one channel for each of *R*, *G*, and *B*. Recently a multilayer CMOS sensor [4] has been developed that stacks *R*, *G*, and *B* pixels vertically and uses the wavelength filtering properties of silicon

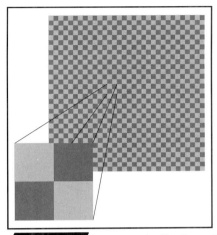

FIGURE 6.16   2 × 2 pixel cell in Bayer CFA (see color insert)

to avoid the use of the relatively heavy and cumbersome prism in a high-quality video camera. Both of these last two have an advantage over the color mosaic sensor, because three-color information is available at every pixel location, rather than only on a checkerboard pattern, common in today's digital still cameras. Of course, one can just put in more pixels to partially make up for the lost resolution.

The most used color mosaic filter is the Bayer[5] *color filter array* (CFA) [7] shown in Figure 6.16, where the basic 2 × 2 cell is repeated to tile the sensor surface, effectively subsampling each of the three color primaries R, G, and B. We see that it favors green as does the human visual system and that it subsamples red and blue by the overall factor 1:4. Some internal processing in the digital camera tries to make up for the resulting aliasing, which can be ameliorated by a slight defocusing of the camera lens. The in-camera postprocessing does a generally good job, but, still, for certain images with much high-frequency color information that repeats on a grid of some sort, aliasing errors can be visible [7]. We also note that the sensor does not really sample, rather it counts up photons inside a square or rectangular pixel. As such, it can better be modeled as a box filter followed by an ideal sampler. This gives some small degree of highpass filtering, which tends to suppress spatial alias energy by the separable product with the familiar $\sin x/x$ function.

The quality of an image sensor is also defined by what its light sensitivity is, and how much light it takes to saturate it. A large dynamic range for electronic image sensors is difficult to achieve without a large cell area to collect the light, hence forcing a trade-off of pixel count versus silicon area used. Another variable is the exposure time. Generally, higher dynamic range can be achieved by length-

---

5. Bryce Bayer of Eastman Kodak Company, U.S. Patent No. 3,971,065, *Color Imaging Array*, 1976.

ening the exposure time, but that can lead to either blurring or too low a frame rate for use in a video sensor.

These electronic sensors tend to provide an output that is linear in intensity, having simply counted the photons coming in. Nevertheless, it is common for camera makers to put in a nonlinearity to compensate for the nonlinearity of the previously common CRT display devices. This is the so-called *gamma compensation* function. We talk more about gamma when discussing image displays in Section 6.5.

The sensor noise of CCD and CMOS tends to be photon counting related. First there is a background radiation giving a noise floor, typically assumed to be Poisson distributed. Then the signal itself is simply modeled as a Poisson random variable, with mean equal to the average input intensity, say λ, at that pixel cell. A key characteristic of the Poisson distribution is that the variance is equal to the mean, so that the rms signal-to-noise ratio (SNR) becomes $\lambda/\sqrt{\lambda} = \sqrt{\lambda}$. This means the SNR is better at high light levels than at low levels, and the noise level itself is intensity dependent. For even modest counts, the Poisson distribution can be approximated as Gaussian [13], with mean λ and standard deviation $\sqrt{\lambda}$.

6.4.2 FILM

The highest resolution and dynamic range image capture today is still film based as of this writing (mid 2005). Video is captured on 70 mm film in IMAX cameras at resolutions in excess of 8K × 6K pixels. The spatial resolution of conventional 35 mm movie film is generally thought to be in a range from 3K to 4K pixels across, depending on whether we measure a camera negative or a so-called *distribution print*, which may be several generations removed from the camera. Also, film has traditionally excelled in dynamic range, i.e., the ability to simultaneously resolve small differences in gray-level in both very bright and very dark regions. Because of Weber's Law, the human eye is more sensitive to changes in the dark regions, where film generally has had an advantage over electronic sensors.

Digital images often were initially captured on film in an effort to achieve the highest quality. The film is then scanned by a device that shines white light through it and records the so-called *density* of the film. There are various names for such a device, e.g., scanner, microdensitometer, and telecine. Film is generally characterized by a D–$\log E$ curve, where D is the density and E is the exposure, defined as the time integral of intensity over the exposure interval. A typical D–$\log E$ curve is sketched out in Figure 6.17. We see that there is a broad "linear" range bordered by "saturation" in the bright region and "fog" in the dark region. The linear part of the D–$\log E$ curve gives the dynamic range of the film, often expressed in terms of f-stops. Note that density in the linear region of this curve is quite analogous to contrast as we have defined it in (6.2-1), and so is much more nearly perceptually uniform than is intensity. Digitization of film is often done in

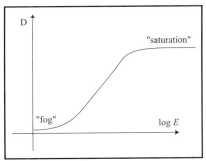

FIGURE 6.17   *Sketch of D–log E curve of film*

the "density domain," giving us effectively a record of contrast. Of course, since film is completely continuous, there is no spatial aliasing to worry about.

Film is composed of grains that have a probability of being "exposed" when struck by an incoming photon. As such, film also counts photons, and the "fog" area in Figure 6.17 corresponds to grains exposed by the background radiation. The saturation area in the D–$\log E$ curve corresponds to the situation where most all the grains are already exposed, so a longer exposure time or greater average incoming light intensity will not expose many more grains, and, finally, the density of the film will not increase.

6.5 IMAGE AND VIDEO DISPLAY

Images, once acquired and processed, can be displayed on many devices, both analog and digital. In the established technology area we have photographic prints and slides, both positive and negative. In the classic video area, we have film projectors and cathode ray tubes (CRT) for electronic video, such as television. Modern technology involves electronic projectors such as liquid crystal digital (LCD), liquid crystal on silicon (LCOS), and digital light processing (DLP). There are also flat-panel plasma displays made using miniature tubes, as well as the ubiquitous smaller LCD panels used in both stationary and portable computers. Each of these display categories has issues that affect its use for a particular purpose—the display resolution, frame rate, gamma, dynamic range, noise level, color temperature, etc. It can be said that no one electronic display technology is the best choice for all applications, so that the best display for a given application requires careful choice of alternatives available.

LCD displays are common for computer monitors and image projectors and offer good spatial resolution and color rendition. Their black-level response[6] has

---

6. Black level is the brightness on the screen when the incoming pixel value is 0. The lower, the better, especially in a darkened room, e.g., a theater.

been a problem, not matching that of the CRT. Also, LCDs have had a problem with transient response, which tends to leave a trail on the screen from moving objects. The LCD technology does not seem to scale well to large flat-panel displays. LCOS is a reflective technology similar to LCD, but with apparently better color and transient response. Very high resolution values have recently been achieved by LCOS in a 4K × 2K digital cinema projector. DLP chips contain micromirrors at each pixel location. The micromirror is tilted under control of the digital pixel value to control the amount of light thrown on the screen for that pixel. It offers both excellent black-levels and transient response in a properly designed projector.

Plasma panels are very bright and the technology can scale to large sizes, but there can be visible noise in the dark regions due to the pulsewidth modulation system used in their gray-level display method. In a bright room though, this is not usually a problem.

As of this writing (mid 2005), the cathode ray tube (CRT) is the only device that directly displays interlaced video, without deinterlacing it first.[7] Since deinterlacing is not perfect, this is an advantage for an interlaced source such as common SD video and high-definition 1080i. Also, CRTs currently have the highest spatial resolution, good temporal resolution (transient response), and the best black levels. Unfortunately, they tend to be large and heavy.

6.5.1 GAMMA

Until recently, the standard of digital image display was the CRT. These devices have a nonlinear display characteristic, usually parameterized by γ as follows,

$$I = v_{\text{in}}^{\gamma},$$

where v_{in} is the CRT input voltage, I is the light intensity output, and the gamma value γ usually ranges from 1.8 to 2.5. Because of this nonlinearity, and because of the widespread use of CRTs in the past, video camera manufacturers have routinely implemented a *gamma correction* into their cameras for nominal value $\gamma = 2.2$, thus precompensating the camera output voltage as

$$v_{\text{out}} = v_{\text{in}}^{1/\gamma}.$$

Since v_{in} is proportional to I for a CCD or CMOS sensor, this near square-root function behaves somewhat similar to the log used in defining contrast, and so the usual camera output is more nearly contrast than intensity, with the benefit of a more uniform perceptual space.

7. Deinterlacing is covered in Chapter 10.

This filter has been extensively used for image smoothing to reduce both high-frequency content and noise. It can also be applied recursively to create the so-called *Gaussian pyramid* [3,11].

6.6.3 PREWITT OPERATOR

The Prewitt operator is an unscaled, first-order approximation to the gradient of a supposed underlying continuous-space function $x(t_1, t_2)$ computed as

$$\frac{\partial x}{\partial t_1} \approx \frac{x(t_1 + \Delta t_1, t_2) - x(t_1 - \Delta t_1, t_2)}{2 \Delta t_1},$$

but given by a 3×3 FIR impulse response $h(n_1, n_2)$ with mask

$$\begin{bmatrix} -1 & 0 & 1 \\ -1 & 0 & 1 \\ -1 & 0 & 1 \end{bmatrix},$$

again centered on the origin to eliminate delay. Note that this Prewitt operator averages three first-order approximations to the horizontal gradient, from the three closest scan lines of the image to reduce noise effects. The corresponding approximation of the vertical gradient

$$\frac{\partial x}{\partial t_2} \approx \frac{x(t_1, t_2 + \Delta t_2) - x(t_1, t_2 - \Delta t_2)}{2 \Delta t_2},$$

is given by the 3×3 FIR impulse response $h(n_1, n_2)$ with mask

$$\begin{bmatrix} -1 & -1 & -1 \\ 0 & 0 & 0 \\ 1 & 1 & 1 \end{bmatrix},$$

again centered on the origin to eliminate delay, and again with the n_2 axis directed downward and the n_1 axis directed across, as above.

6.6.4 SOBEL OPERATOR

The Sobel operator is a slight variation on the Prewitt operator, with horizontal and vertical masks given as

$$\begin{bmatrix} -1 & 0 & 1 \\ -2 & 0 & 2 \\ -1 & 0 & 1 \end{bmatrix} \quad \text{and} \quad \begin{bmatrix} -1 & -2 & -1 \\ 0 & 0 & 0 \\ 1 & 2 & 1 \end{bmatrix},$$

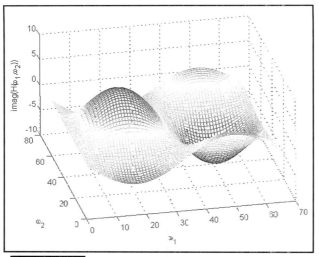

FIGURE 6.19   *Imaginary part of horizontal Sobel operator frequency response*

with the central row being weighted up to achieve some emphasis on the current row or column.

Both the Sobel and Prewitt operators are used widely in image analysis [8] to help locate edges in images. Location of edges in images can be a first step in image understanding and object segmentation, where the output of these "edge detectors" must be followed by some kind of regularization, i.e., smoothing, thinning, and gap filling [6]. A plot of the imaginary part of the frequency response of the Sobel operator, centered on zero, is shown in Figure 6.19, with a contour plot in Figure 6.20. Please note that the plots look like a scaled version of ω_1, but only for low frequencies. In particular, the side lobes at high ω_2 indicate that this "derivative" filter is only accurate on lowpass data, or alternatively should only be used in combination with a suitable lowpass filter to reduce these side lobes.

6.6.5 LAPLACIAN FILTER

In image processing, the name *Laplacian filter* often refers to the simple 3×3 FIR filter

$$\begin{bmatrix} 0 & -1 & 0 \\ -1 & 4 & -1 \\ 0 & -1 & 0 \end{bmatrix},$$

used as a first-order approximation to the Laplacian of an assumed underlying continuous-space function $x(t_1, t_2)$:

$$\nabla^2 x(t_1, t_2) = \frac{\partial^2 x(t_1, t_2)}{\partial^2 t_1} + \frac{\partial^2 x(t_1, t_2)}{\partial^2 t_2}.$$

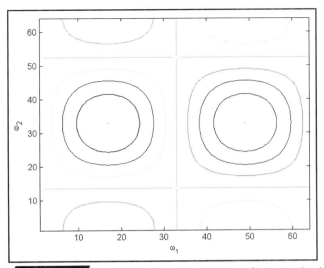

FIGURE 6.20   *Contour plot of imaginary part of horizontal Sobel operator*

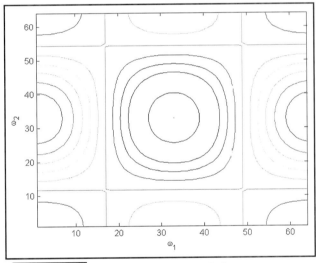

FIGURE 6.21   *Contour plot of magnitude frequency response of Laplace filter*

The magnitude of the frequency response of this Laplace filter is shown in Figure 6.21, where again we note the reasonable approximation at low frequencies only.

The zero-crossing property of the Laplacian can be used for edge location. Often these derivative filters are applied to a smoothed function to avoid problems with image noise amplification [6]. Another Laplacian approximation is available using the Burt and Adelson Gaussian filter [3].

6.7 CONCLUSIONS

The basic properties of images, sensors, displays, and the human visual system studied in this chapter are the essential topics that provide the link between multi-dimensional signals and images and video. Most applications to image and video signal processing make use of one or more of these properties. In this chapter we also reviewed some simple image processing filters that are commonly used in practice, and noted their shortcomings from the frequency-response viewpoint. Design techniques from Chapter 5 can usefully be applied here to improve on some of these simple image processing filters.

6.8 PROBLEMS

1   One issue with the YC_RC_B color space comes from the constraint of positivity for the corresponding RGB values.

    (a)   Using 10-bit values, the range of RGB data is $[0, 1023]$. What is the resulting range for each of Y, C_R, and C_B?

    (b)   Does every point in the range you found in part a correspond to a valid RGB value, i.e., a value in the range $[0, 1023]$?

2   From the Mannos and Sakrison human visual system (HVS) response function of Fig. 6.10, how many pixels should there be vertically on a 100-inch vertical screen, when viewed at a distance of 3H, i.e., 300 inche ? at 6H? Assume we want the HVS response to be down to 0.01 at the Nyquist frequency.

3   Assume the measured just-noticeable difference (JND) is $\Delta I = 1$ at $I = 50$. Hence

$$\Delta I/I = 0.02.$$

Consider a range of intensities $I = I_0 = 1$ to $I = 100$.

    (a)   If we quantize I with a fixed step-size, what should this step-size be so that the quantization will not be noticeable? How many steps will there be to cover the full range of intensities with this fixed step-size? How many bits will this require in a fixed-length indexing of the quantizer outputs?

    (b)   If we instead quantize contrast

$$C = \ln(I/I_0),$$

with a fixed step-size quantizer, what should ΔC be? How many steps are needed? How many bits for fixed-length coding of the quantizer output?

4   Referring to the three-channel theory of color vision (6.1-1)–(6.1-3), assume that three partially overlapping human visual response curves (luminous efficiencies) $s_R(\lambda)$, $s_G(\lambda)$, and $s_B(\lambda)$ are given. Let there be a camera with red, green, and blue sensors with response functions $s_{cR}(\lambda)$, $s_{cG}(\lambda)$, and $s_{cB}(\lambda)$, where the subscript c indicates "camera." Let the results captured by the camera be displayed with three very narrow wavelength [here modeled as monochromatic $s(\lambda) = \delta(\lambda)$] beams, additively superimposed, and centered at the wavelength peak of each of the camera luminous efficiencies. Assuming white incident light, i.e., uniform intensity at all wavelengths, what conditions are necessary so that we perceive the same R, G, and B sensation as if we viewed the scene directly? Neglect any nonlinear effects.

5   In "white balancing" a camera, we capture the image of a white card in the ambient light, yielding three R, G, and B values. Now, we might like these three values to be equal; however, because of the ambient light not being known, they may not be. How should we modify these R, G, and B values to "balance" the camera in this light? If the ambient light source changes, should we white balance the camera again? Why?

6   Consider the "sampling" that a CCD or CMOS image sensor does on the input Fourier transform $X(\Omega_1, \Omega_2)$. Assume the sensor is of infinite size and that its pixel size is uniform and square with size $T \times T$. Assume that, for each pixel, the incoming light intensity (monochromatic) is integrated over the square cell and that the sample value becomes this integral for each pixel (cell).

(a)   Express the Fourier transform of the resulting sample values; call it $X_{\mathrm{CCD}}(\omega_1, \omega_2)$ in terms of the continuous Fourier transform $X(\Omega_1, \Omega_2)$.

(b)   Assuming spatial aliasing is not a problem, find the resulting discrete-space transform $X(\omega_1, \omega_2)$.

7   The box filter was introduced in Section 6.6 as a simple FIR filter that finds much use in image processing practice. Here we consider a square $L \times L$ box filter and implement it recursively to reduce the required number of additions.

(a)   Consider an odd-length 1-D box filter with $L = 2M + 1$ points and unscaled output

$$Ly(n) = \sum_{k=n-M}^{n+M} x(n),$$

and show that this sum can be realized recursively by

$$Ly(n) = Ly(n-1) + x(n+M) - x(n-M).$$

How many adds and multiplies per point are required for this 1-D filter?

(b)   Find a 2-D method to realize the square $L \times L$ box filter for odd L. How much *intermediate storage* is required by your method? (Intermediate storage is the temporary storage needed for the processing. It does not include any storage that may be needed to store either the input or output arrays.)

8 For the horizontal derivative 3×3 FIR approximation called the Sobel operator, consider calculating samples of its frequency response with a 2-D DFT program. Assume the DFT is of size $N \times N$ for some large value of N.

(a) Where can you place the Sobel operator coefficients in the square $[0, N-1]^2$ if you are only interested in samples of the magnitude response?

(b) Where must you place the Sobel operator coefficients in the square $[0, N-1]^2$ if you are interested in samples of the imaginary part of the Fourier transform?

9 Image processing texts recommend that the Sobel operator (filter),

$$
\left\{
\begin{array}{ccc}
-1 & 0 & -1 \\
-2 & 0 & -2 \\
-1 & 0 & -1
\end{array}
\right\},
$$

should only be used on a smoothed image. So, consider the simple 3×3 smoothing filter,

$$
\left\{
\begin{array}{ccc}
1 & 1 & 1 \\
1 & 1 & 1 \\
1 & 1 & 1
\end{array}
\right\},
$$

with output taken at the center point. Then let us smooth first and then apply the Sobel operator to the smoothed output.

(a) What is the size of the resulting combined FIR filter?

(b) What is the combined filter's impulse response?

(c) Find the frequency response of the combined filter.

(d) Is it more efficient to apply these two 3×3 filters in series, or to apply the combined filter? Why?

REFERENCES

[1] P. G. J. Barten, *Contrast Sensitivity of the Human Eye and Its Effects on Image Quality*, Ph.D. thesis, Technical Univ. of Eindhoven, The Netherlands, 1999.

[2] A. C. Bovik, ed., *Handbook of Image and Video Processing*, 2nd edn., Elsevier Academic Press, Burlington, MA, 2005.

[3] P. J. Burt and E. H. Adelson, "The Laplacian Pyramid as a Compact Image Code," *IEEE Trans. Commun.*, COM-31, 532–540, April 1983.

[4] Foveon, Inc., Santa Clara, CA. Web site: www.foveon.com.

[5] E. J. Giorgianni and T. E. Madden, *Digital Color Management: Encoding Solutions*, Addison-Wesley, Reading, MA, 1998.

[6] R. C. Gonzalez and R. E. Woods, *Digital Image Processing*, 2nd edn., Prentice-Hall, Upper Saddle River, NJ, 2002.

[7] B. K. Gunturk, J. Glotzbach, Y. Altunbasak, R. W. Schafer, and R. M. Mersereau, "Demosaicking: Color Filter Array Interpolation," *IEEE Signal Process. Magazine*, **22**, 44–54, Jan. 2005.

[8] A. K. Jain, *Fundamentals of Digital Image Processing*, Prentice-Hall, Englewood Cliffs, NJ, 1989.

[9] D. H. Kelly, "Theory of Flicker and Transient Responses, II. Counterphase Gratings," *J. Optic. Soc. Am. (JOSA)*, **61**, 632–640, 1971.

[10] D. H. Kelly, "Motion and Vision, II. Stabilized Spatio-Temporal Threshold Surface," *J. Optic. Soc. Am. (JOSA)*, **69**, 1340–1349, 1979.

[11] J. Lim, *Two-Dimensional Signal and Image Processing*, Prentice-Hall, Englewood Cliffs, NJ, 1990.

[12] J. L. Mannos and D. J. Sakrison, "The Effects of a Visual Error Criteria on the Encoding of Images," *IEEE Trans. Inf. Theory*, **IT-20**, 525–536, July 1974.

[13] H. Stark and J. W. Woods, *Probability and Random Processes with Applications to Signal Processing*, 3rd edn., Prentice-Hall, Upper Saddle River, NJ, 2002.

[14] C. J. van den Branden Lambrecht and M. Kunt, "Characterization of Human Visual Sensitivity for Video Imaging Applications," *Signal Process.*, **67**, 255–269, June 1998.

[15] F. L. van Ness and M. A. Bouman, "Spatial Modulation Transfer Function of the Human Eye," *J. Optic. Soc. Am. (JOSA)*, **57**, 401–406, 1967.

IMAGE ESTIMATION AND RESTORATION

7

Here we apply the linear systems and basic image processing knowledge of the previous chapters to modeling, developing solutions, and experimental testing of algorithms for two dominant problems in digital image processing: *image estimation* and *image restoration*. By image estimation we mean the case where a clean image has been contaminated with noise, usually through sensing, transmission, or storage. We will treat the independent and additive noise case. The second problem, image restoration, means that in addition to the noise, there is some blurring due to motion or lack of focus. We attempt to "restore" the image in this case. Of course, the restoration will only be approximate.

We first develop the theory of linear estimation in two dimensions, and then present some example applications for monochrome image processing problems. We then look at some space-variant and nonlinear estimators. This is followed by a section on image and/or blur model parameter identification (estimation) and combined image restoration. Finally, we make some brief comments on extensions of these 2-D estimation algorithms to work with color images.

7.1 2-D RANDOM FIELDS

A random sequence in two dimensions is called a *random field*. Images that display clearly random characteristics include sky with clouds, textures of various types, and sensor noise. It may perhaps be surprising that we also model general images that are unknown as random fields with an eye to their estimation, restoration, and, in a later chapter, transmission and storage. But this is a necessary first step in designing systems that have an optimality across a class of images sharing common statistics. The theory of 2-D random fields builds upon random process and probability theory. Many references exist, including [20]. We start out with some basic definitions.

DEFINITION 7.1-1 (*random field*)
A 2-D random field $x(n_1, n_2)$ is a mapping from a probability *sample space* Ω to the class of two-dimensional sequences. As such, for each outcome $\zeta \in \Omega$, we have a deterministic sequence, and for each location (n_1, n_2) we have a random variable.

For a random sequence $x(n_1, n_2)$ we define the following low-order moments:

DEFINITION 7.1-2 (*mean function*)
The mean function of a random sequence $x(n_1, n_2)$ is denoted

$$\mu_x(n_1, n_2) \triangleq E\{x(n_1, n_2)\}.$$

Correlation function:

$$R_x(n_1, n_2; m_1, m_2) \triangleq E\{x(n_1, n_2)x^*(m_1, m_2)\}.$$

Covariance function:

$$K_x(n_1, n_2; m_1, m_2) \triangleq E\{x_c(n_1, n_2)x_c^*(m_1, m_2)\},$$

where $x_c(n_1, n_2) \triangleq x(n_1, n_2) - \mu_x(n_1, n_2)$ and is called the *centered version* of x.

We immediately have the following relation between these first- and second-order moment functions,

$$K_x(n_1, n_2; m_1, m_2) = R_x(n_1, n_2; m_1, m_2) - \mu_x(n_1, n_2)\mu_x^*(m_1, m_2).$$

We also define the *variance function*

$$\sigma_x^2(n_1, n_2) \triangleq K_x(n_1, n_2; n_1, n_2)$$

$$= \text{var}\{x(n_1, n_2)\} = E\{|x_c(n_1, n_2)|^2\}.$$

and note

$$\sigma_x^2(n_1, n_2) \geqslant 0.$$

When all the statistics of a random field do not change with position (n_1, n_2), we say that the random field is *homogeneous*, analogously to the stationarity property from random sequences [20]. This homogeneity assumption is often relied upon in order to be able to estimate the needed statistical quantities from a given realization or sample field or image. Often the set of images used to estimate these statistics is called the *training set.*

DEFINITION 7.1-3 (*homogeneous random field*)
A random field is homogeneous when the Nth-order joint probability density function (pdf) is invariant with respect to the N relevant positions, and this holds for all positive integers N, i.e , for all N and locations $\mathbf{n}_i$, we have

$$f_x(x(\mathbf{n}_1), x(\mathbf{n}_2), \dots, x(\mathbf{n}_N)) = f_x(x(\mathbf{n}_1 + \mathbf{m}), x(\mathbf{n}_2 + \mathbf{m}), \dots, x(\mathbf{n}_N + \mathbf{m})),$$

independent of the shift vector $\mathbf{m}$.[1]

Usually we do not have such a complete description of a random field as is afforded by the complete set of joint pdfs of all orders, and so must resort to partial descriptions, often in terms of the low-order moment functions. This situation calls for the following classification, which is much weaker than (strict) homogeneity.

---

1. Be careful: while the notation $f_x(x(\mathbf{n}_1), x(\mathbf{n}_2), \dots, x(\mathbf{n}_N))$ seems friendly enough, it is really shorthand for $f_x(x(\mathbf{n}_1), x(\mathbf{n}_2), \dots, x(\mathbf{n}_N); \mathbf{n}_1, \mathbf{n}_2, \dots, \mathbf{n}_N)$, where the locations are given explicitly. This extended notation *must* be used when we evaluate f_x at specific choices for the $x(\mathbf{n}_i)$.

DEFINITION 7.1-4 (*wide-sense homogeneous*)

A 2-D random field is wide-sense homogeneous (wide-sense stationary) if

1 $\mu_x(n_1, n_2) = \mu_x(0, 0) = \text{constant}.$
2 $R_x(n_1 + m_1, n_2 + m_2; n_1, n_2) = R_x(m_1, m_2; 0, 0)$ independent of n_1 and n_2.

In the homogeneous case, for notational simplicity we define the constant mean $\mu_x \triangleq \mu_x(0, 0)$ and the two-parameter correlation function $R_x(m_1, m_2) \triangleq R_x(m_1, m_2; 0, 0)$ and covariance function $K_x(m_1, m_2) \triangleq K_x(m_1, m_2; 0, 0)$.

EXAMPLE 7.1-1 (*IID Gaussian noise*)

Let $w(n_1, n_2)$ be independent and identically distributed (IID) Gaussian noise with mean function $\mu_w(n_1, n_2) = \mu_w$, and standard deviation $\sigma_w > 0$. Clearly we have wide-sense homogeneity here. The two-parameter correlation function is given as

$$R_w(m_1, m_2) = \sigma_w^2 \delta(m_1, m_2) + \mu_w^2,$$

and covariance function

$$K_w(m_1, m_2) = \sigma_w^2 \delta(m_1, m_2).$$

The first-order pdf becomes

$$f_x\big(x(\mathbf{n})\big) = \frac{1}{\sqrt{2\pi}\,\sigma_w} \exp -\frac{(x(\mathbf{n}) - \mu_w)^2}{2\sigma_w^2}$$
$$\sim N\big(\mu_w, \sigma_w^2\big).$$

The IID noise field is called *white noise* when its mean is zero.

7.1.1 FILTERING A 2-D RANDOM FIELD

Let $G(\omega_1, \omega_2)$ be the frequency response of a spatial filter, with impulse response $g(n_1, n_2)$. Consider the convolution of the filter impulse response with an IID noise field $w(n_1, n_2)$ whose mean is μ_w. Calling the output random field $x(n_1, n_2)$, we have

$$x(n_1, n_2) = g(n_1, n_2) * w(n_1, n_2)$$
$$= \sum_{k_1, k_2} g(k_1, k_2) w(n_1 - k_1, n_2 - k_2).$$

We can then compute the mean and covariance function of x as follows:

$$\mu_x(n_1, n_2) = E\{x(n_1, n_2)\}$$

$$= E\left\{ \sum_{k_1, k_2} g(k_1, k_2) w(n_1 - k_1, n_2 - k_2) \right\}$$

$$= \sum_{k_1, k_2} g(k_1, k_2) E\{w(n_1 - k_1, n_2 - k_2)\}$$

$$= \sum_{k_1, k_2} g(k_1, k_2) \mu_w(n_1 - k_1, n_2 - k_2)$$

$$= \sum_{k_1, k_2} g(k_1, k_2) \mu_w$$

$$= G(0, 0) \mu_w.$$

The covariance function of the generated random field $x(n_1, n_2)$ is calculated as

$$K_x(n_1, n_2; m_1, m_2)$$

$$= E\{x_c(n_1, n_2) x_c^*(m_1, m_2)\}$$

$$= \sum_{k_1, k_2} \sum_{l_1, l_2} g(k_1, k_2) g^*(l_1, l_2) E\{w_c(n_1 - k_1, n_2 - k_2) w_c^*(m_1 - l_1, m_2 - l_2)\}$$

$$= \sum_{k_1, k_2} \sum_{l_1, l_2} g(k_1, k_2) g^*(l_1, l_2) K_w(n_1 - k_1, n_2 - k_2; m_1 - l_1, m_2 - l_2)$$

$$= g(n_1, n_2) * K_w(n_1, n_2; m_1, m_2) * g^*(m_1, m_2)$$

$$= g(n_1, n_2) * \sigma_w^2 \delta(n_1 - m_1, n_2 - m_2) * g^*(m_1, m_2). \qquad (7.1\text{-}1)$$

Such a "coloring" of the IID noise is often useful in generating models for real correlated noise. A warning on our notation in (7.1-1): The first $*$ denotes 2-D convolution on the (n_1, n_2) variables, while the second $*$ denotes 2-D convolution on the (m_1, m_2) variables.

Generalizing a bit, we can write the output of a filter $H(\omega_1, \omega_2)$ with general colored noise input $x(n_1, n_2)$ as

$$y(n_1, n_2) = h(n_1, n_2) * x(n_1, n_2)$$

$$= \sum_{k_1, k_2} h(k_1, k_2) x(n_1 - k_1, n_2 - k_2).$$

This could be a filtering of the random field model we just created. Calculating the mean function of the output y, we find

$$\mu_y(n_1, n_2) = E\{y(n_1, n_2)\}$$

$$= E\left\{ \sum_{k_1,k_2} h(k_1, k_2) x(n_1 - k_1, n_2 - k_2) \right\}$$

$$= \sum_{k_1,k_2} h(k_1, k_2) E\{x(n_1 - k_1, n_2 - k_2)\}$$

$$= \sum_{k_1,k_2} h(k_1, k_2) \mu_x(n_1 - k_1, n_2 - k_2)$$

$$= h(n_1, n_2) * \mu_x(n_1, n_2).$$

And for the correlation function,

$$R_y(n_1, n_2; m_1, m_2)$$

$$= E\{y(n_1, n_2) y^*(m_1, m_2)\}$$

$$= \sum_{k_1,k_2} \sum_{l_1,l_2} h(k_1, k_2) h^*(l_1, l_2) E\{x(n_1 - k_1, n_2 - k_2) x^*(m_1 - l_1, m_2 - l_2)\}$$

$$= \sum_{k_1,k_2} \sum_{l_1,l_2} h(k_1, k_2) h^*(l_1, l_2) R_x(n_1 - k_1, n_2 - k_2; m_1 - l_1, m_2 - l_2)$$

$$= h(n_1, n_2) * R_x(n_1, n_2; m_1, m_2) * h^*(m_1, m_2),$$

and covariance function,

$$K_y(n_1, n_2; m_1, m_2)$$

$$= E\{y_c(n_1, n_2) y_c^*(m_1, m_2)\}$$

$$= \sum_{k_1,k_2} \sum_{l_1,l_2} h(k_1, k_2) h^*(l_1, l_2) E\{x_c(n_1 - k_1, n_2 - k_2) x_c^*(m_1 - l_1, m_2 - l_2)\}$$

$$= \sum_{k_1,k_2} \sum_{l_1,l_2} h(k_1, k_2) h^*(l_1, l_2) K_x(n_1 - k_1, n_2 - k_2; m_1 - l_1, m_2 - l_2)$$

$$= h(n_1, n_2) * K_x(n_1, n_2; m_1, m_2) * h^*(m_1, m_2).$$

If we specialize to the homogeneous case, we obtain

$$\mu_y(n_1, n_2) = \sum_{k_1,k_2} h(k_1, k_2) \mu_x$$

or, more simply,

$$\mu_y = H(0,0)\mu_x.$$

The correlation function is given as

$$R_y(m_1, m_2) = \sum_{k_1, k_2} \sum_{l_1, l_2} h(k_1, k_2) h^*(l_1, l_2) R_x(m_1 - k_1 - l_1, m_2 - k_2 + l_2)$$

$$= \sum_{k_1, k_2} \sum_{l_1, l_2} h(k_1, k_2) R_x(m_1 - k_1 + l_1, m_2 - k_2 + l_2) h^*(l_1, l_2)$$

$$= h(m_1, m_2) * R_x(m_1, m_2) * h^*(-m_1, -m_2)$$

$$= \big(h(m_1, m_2) * h^*(m_1, m_2)\big) * R_x(m_1, m_2). \tag{7.1-2}$$

Similarly, for the covariance function,

$$K_y(m_1, m_2) = \big(h(m_1, m_2) * h^*(m_1, m_2)\big) * K_x(m_1, m_2).$$

Taking Fourier transforms, we can move to the *power spectral density* (PSD) domain

$$S_x(\omega_1, \omega_2) \triangleq \sum_{m_1, m_2} R_x(m_1, m_2) \exp -j2\pi(m_1\omega_1 - m_2\omega_2)$$

and obtain the Fourier transform of (7.1-2) as:

$$S_y(\omega_1, \omega_2) = H(\omega_1, \omega_2) S_x(\omega_1, \omega_2) H^*(\omega_1, \omega_2)$$

$$= |H(\omega_1, \omega_2)|^2 S_x(\omega_1, \omega_2). \tag{7.1-3}$$

EXAMPLE 7.1-2 (*power spectrum of images*)
One equation is often used to represent the power spectral density of a typical image [12]

$$S_x(\omega_1, \omega_2) = \frac{K}{\big(1 + (\omega_1^2 + \omega_2^2)/\omega_3^2\big)^{3/2}} \quad \text{for } |\omega_1|, |\omega_2| \leqslant \pi,$$

and was used in [10] with $K = \pi/42.19$ to model a video conferencing image frame. The resulting log plot from MATLAB is given in Figure 7.1 (ref. SpectImage.m). A 64×64-point DFT was used, and zero frequency is in the middle of the plot, at $k_1 = k_2 = 32$. Note that the image appears to be very lowpass in nature, even on this decibel or log scale. Note that this PSD is not rational and so does not correspond to a finite 2-D difference equation model.

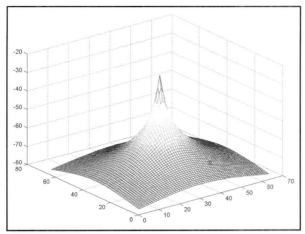

Figure 7.1　Log or dB plot of example image spectra

7.1.2 Autoregressive Random Signal Models

A particularly useful model for random signals and noises is the autoregressive (AR) model. Mathematically this is a 2-D difference equation driven by white noise. As such, it generates a correlated zero-mean field, which can be colored by the choice of its coefficients

$$x(n_1, n_2) = \sum_{(k_1,k_2)\in\mathcal{R}_a-(0,0)} a_{k_1,k_2} x(n_1 - k_1, n_2 - k_2) + w(n_1, n_2).$$

We can find the coefficients a_{k_1,k_2} by solving the 2-D linear prediction problem

$$\hat{x}(n_1, n_2) = \sum_{(k_1,k_2)\in\mathcal{R}_a-(0,0)} a_{k_1,k_2} x(n_1 - k_1, n_2 - k_2), \qquad (7.1\text{-}4)$$

which can be done using the *orthogonality principle* of estimation theory [20].

Theorem 7.1-1 (*optimal linear prediction*)
The $(1, 0)$-step linear prediction coefficients in (7.1-4) that minimize the mean-square prediction error,

$$E\{|x(n_1, n_2) - \hat{x}(n_1, n_2)|^2\},$$

can be determined by the orthogonality principle, which states the error $e \triangleq \hat{x} - x$ must be *orthogonal* to the data used in the linear prediction, i.e.,

$$e(n_1, n_2) \perp x(n_1 - k_1, n_2 - k_2) \quad \text{for } (k_1, k_2) \in \mathcal{R}_a - (0, 0).$$

For two random variables, orthogonal means that their correlation is zero, so we have $E\{e(n_1, n_2)x^*(n_1 - k_1, n_2 - k_2)\} = 0$ for $(k_1, k_2) \in \mathcal{R}_a - (0, 0)$. To actually find the coefficients, we start with

$$E\{e(n_1, n_2)x^*(n_1 - k_1, n_2 - k_2)\} = 0,$$

and substitute $e = \hat{x} - x$, to obtain

$$E\{\hat{x}(n_1, n_2)x^*(n_1 - k_1, n_2 - k_2)\} = E\{x(n_1, n_2)x^*(n_1 - k_1, n_2 - k_2)\},$$

or

$$E\left\{\sum_{l_1, l_2} a_{l_1, l_2} x(n_1 - l_1, n_2 - l_2)x^*(n_1 - k_1, n_2 - k_2)\right\}$$

$$= E\{x(n_1, n_2)x^*(n_1 - k_1, n_2 - k_2)\},$$

$$\sum_{l_1, l_2} a_{l_1, l_2} R_x(n_1 - l_1, n_2 - l_2; n_1 - k_1, n_2 - k_2)$$

$$= R_x(n_1, n_2; n_1 - k_1, n_2 - k_2) \quad \text{for } (k_1, k_2) \in \mathcal{R}_a - (0, 0),$$

which is an ordinary set of $\mathcal{N}$ linear equations in $\mathcal{N}$ unknown coefficients, where $\mathcal{N}$ is the number of points in the set $\mathcal{R}_a - (0, 0)$. This assumes we know the correlation function R_x for the indicated range. Once these coefficients are found, the mean-square power (variance) of $e(n_1, n_2)$ can be determined.

In the homogeneous case, these equations simplify to

$$\sum_{l_1, l_2} a_{l_1, l_2} R_x(k_1 - l_1, k_2 - l_2) = R_x(k_1, k_2) \quad \text{for } (k_1, k_2) \in \mathcal{R}_a - (0, 0).$$

Defining a coefficient vector $\mathbf{a}$ of dimension $\mathcal{N}$, these equations can easily be put into matrix form. The resulting correlation matrix $\mathbf{R}$ will be invertible if the correlation function R_x is positive definite [20]. The mean-square prediction error is then easily obtained as

$$E\{|e(n_1, n_2)|^2\} = E\{e(n_1, n_2)(\hat{x} - x)^*(n_1, n_2)\}$$

$$= -E\{e(n_1, n_2)x^*(n_1, n_2)\}$$

$$= -E\{(\hat{x} - x)(n_1, n_2)x^*(n_1, n_2)\}$$

$$= E\{|x|^2(n_1, n_2)\} - E\{\hat{x}(n_1, n_2)x(n_1, n_2)^*\}$$

$$= \sigma_x^2(n_1, n_2) - \sum_{l_1, l_2} a_{l_1, l_2} R_x(n_1 - l_1, n_2 - l_2; n_1, n_2),$$

which, specialized to the homogeneous case, becomes

$$\sigma_e^2 = \sigma_x^2 - \sum_{l_1, l_2} a_{l_1, l_2} R_x(-l_1, -l_2).$$

EXAMPLE 7.1-3 [(1×1)-*order quarter plane (QP) predictor*]
Let the predictor coefficient support of 2-D linear predictor be given as

$$\mathcal{R}_a - (0,0) = \big\{(1,0), (0,1), (1,1)\big\}.$$

Assume the random field x is homogeneous with zero mean and correlation (covariance) function given as $R_x(m_1, m_2)$. Then the so-called Normal equations are given in matrix form as

$$\begin{bmatrix} R_x(0,0) & R_x(1,-1) & R_x(0,-1) \\ R_x(-1,1) & R_x(0,0) & R_x(-1,0) \\ R_x(0,1) & R_x(1,0) & R_x(0,0) \end{bmatrix} \begin{bmatrix} a_{1,0} \\ a_{0,1} \\ a_{1,1} \end{bmatrix} = \begin{bmatrix} R_x(1,0) \\ R_x(0,1) \\ R_x(1,1) \end{bmatrix}.$$

7.2 ESTIMATION FOR RANDOM FIELDS

Here we consider the problem of estimating a random field x from observations of another random field y. We assume that x and y are zero mean, and write the estimate as

$$\hat{x}(n_1, n_2) = \sum_{(k_1, k_2) \in \mathcal{R}_h} h_{k_1, k_2} y(n_1 - k_1, n_2 - k_2).^2$$

Using the orthogonality principle, with $e = \hat{x} - x$, we have

$$\hat{x}(n_1, n_2) - x(n_1, n_2) \perp \big\{ y(n_1 - k_1, n_2 - k_2) \text{ for } (k_1, k_2) \in \mathcal{R}_h \big\},$$

which becomes

$$E\big\{\hat{x}(n_1, n_2) y^*(n_1 - k_1, n_2 - k_2)\big\} = E\big\{x(n_1, n_2) y^*(n_1 - k_1, n_2 - k_2)\big\},$$

---

2. If x and y had nonzero means, μ_x and μ_y, respectively, then the appropriate estimate would have the form $\hat{x} = \mu_x + h * (y - \mu_y)$.

or

$$E\left\{\sum_{l_1,l_2} h_{l_1,l_2} y(n_1 - l_1, n_2 - l_2) y^*(n_1 - k_1, n_2 - k_2)\right\}$$

$$= E\{x(n_1, n_2) y^*(n_1 - k_1, n_2 - k_2)\},$$

$$\sum_{l_1,l_2} h_{l_1,l_2} R_{yy}(n_1 - l_1, n_2 - l_2; n_1 - k_1, n_2 - k_2)$$

$$= R_{xy}(n_1, n_2; n_1 - k_1, n_2 - k_2) \quad \text{for } (k_1, k_2) \in \mathcal{R}_h.$$

Specialized to the homogeneous case, we have

$$\sum_{(l_1,l_2)\in\mathcal{R}_h} h_{l_1,l_2} R_{yy}(k_1 - l_1, k_2 - l_2) = R_{xy}(k_1, k_2) \quad \text{for } (k_1, k_2) \in \mathcal{R}_h. \qquad (7.2\text{-}1)$$

7.2.1 INFINITE OBSERVATION DOMAIN

In the case where the observation region $\mathcal{R}_h$ is infinite, (7.2-1) becomes a convolution. For example, if $\mathcal{R}_h = (-\infty, +\infty)^2$, we get

$$h(n_1, n_2) * R_{yy}(n_1, n_2) = R_{xy}(n_1, n_2), \quad \text{where } -\infty < n_1, n_2 < +\infty.$$

We can express this convolution in the frequency domain,

$$H(\omega_1, \omega_2) S_{yy}(\omega_1, \omega_2) = S_{xy}(\omega_1, \omega_2),$$

or

$$H(\omega_1, \omega_2) = \frac{S_{xy}(\omega_1, \omega_2)}{S_{yy}(\omega_1, \omega_2)},$$

at those frequencies where $S_{yy} > 0$,[3] which is the general equation for the 2-D noncausal (unrealizable) *Wiener filter*. We consider a few special cases.

1 $y = x + n$ with $x \perp n$, i.e., x and n are orthogonal,

$$S_{xy}(\omega_1, \omega_2) = S_{xx}(\omega_1, \omega_2),$$

and

$$S_{yy}(\omega_1, \omega_2) = S_{xx}(\omega_1, \omega_2) + S_{nn}(\omega_1, \omega_2)$$

---

3. At those frequencies where $S_{yy} = 0$, the exact value of H does not matter.

so that the Wiener filter is

$$H(\omega_1, \omega_2) = \frac{S_{xx}(\omega_1, \omega_2)}{S_{xx}(\omega_1, \omega_2) + S_{nn}(\omega_1, \omega_2)},$$

the so-called *estimation* case. Notice that the filter is approximately 1 at those frequencies where the SNR is very high, with SNR defined as $S_{xx}(\omega_1, \omega_2)/S_{nn}(\omega_1, \omega_2)$. The filter then attenuates other lesser values.

2 $y = g * x + n$ with $x \perp n$, then

$$S_{xy}(\omega_1, \omega_2) = S_{xx}(\omega_1, \omega_2)G^*(\omega_1, \omega_2),$$

and

$$S_{yy}(\omega_1, \omega_2) = |G(\omega_1, \omega_2)|^2 S_{xx}(\omega_1, \omega_2) + S_{nn}(\omega_1, \omega_2),$$

so that the Wiener filter is

$$H(\omega_1, \omega_2) = \frac{G^*(\omega_1, \omega_2)S_{xx}(\omega_1, \omega_2)}{|G(\omega_1, \omega_2)|^2 S_{xx}(\omega_1, \omega_2) + S_{nn}(\omega_1, \omega_2)},$$

the so-called *restoration* case. Notice that the filter looks like an inverse filter at those frequencies where the SNR is very high. The filter tapers off and provides less gain at other frequency values.

3 $y = u + n$ with $u \perp n$, and we want to estimate $x \triangleq b * u$. In this case the Wiener filter is just b convolved with the estimate in case 1,

$$H(\omega_1, \omega_2) = \frac{B(\omega_1, \omega_2)S_{uu}(\omega_1, \omega_2)}{S_{uu}(\omega_1, \omega_2) + S_{nn}(\omega_1, \omega_2)}.$$

4 $y(n_1, n_2) = x(n_1 - 1, n_2)$, the linear prediction case, as treated earlier.

A good example of case 3 would be estimation of the derivative of a signal.

Wiener Filter — Alternative Derivation

A direct derivation of the Wiener filter through the concept of a 2-D *whitening filter* is possible. We first give a brief summary of 2-D spectral factorization, whose factor will yield the needed whitening filter. Then we re-derive the noncausal Wiener filter and go on to find the causal Wiener filter.

Theorem 7.2-1 (*spectral factorization*)

Given a homogeneous random field $x(n_1, n_2)$ with power spectral density $S_{xx}(\omega_1, \omega_2) > 0$ on $[-\pi, +\pi]^2$, there exists a *spectral factorization*

$$S_{xx}(\omega_1, \omega_2) = \sigma^2 B_{\oplus+}(\omega_1, \omega_2)B_{\ominus-}(\omega_1, \omega_2),$$

with $\sigma > 0$, and $B_{\oplus+}(z_1, z_2)$ is stable and causal with a stable causal inverse.

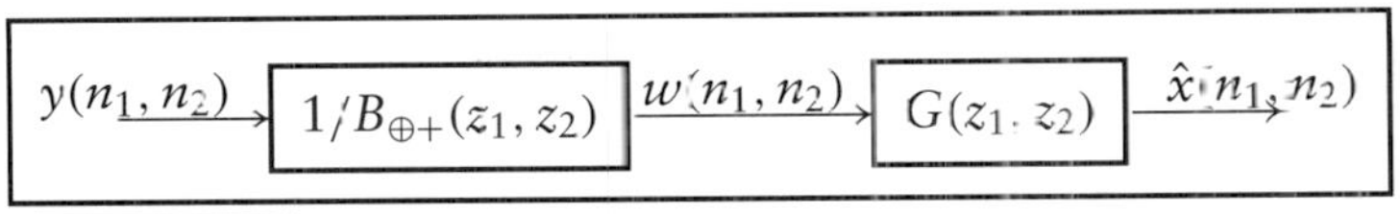

FIGURE 7.2   *Whitening filter realization of 2-D Wiener filter*

In terms of Z-transforms, we have $B_{\ominus-}(z_1, z_2) = B_{\oplus+}(z_1^{-1}, z_2^{-1})$, where we assume real coefficients in the spectral factors. Unfortunately, even when the PSD $S_x(\omega_1, \omega_2)$ is rational, the spectral factors $B_{\oplus+}$ and $B_{\ominus-}$ will generally be infinite order, and can be used as ideal functions in a 2-D filter design approximation. Alternatively, the factors can be related to the linear prediction filters of the last section, and approximation can be obtained in that way too. More on spectral factorization is contained in [7,18].

Consider observations consisting of the homogeneous random field $y(n_1, n_2)$ and assume the causal spectral factor of $S_{yy}(\omega_1, \omega_2)$ is given as $B_{\oplus+}(z_1, z_2)$. Since this factor has a stable and causal inverse, we can use $B_{\oplus+}^{-1}(z_1, z_2)$ to whiten the spectra of y and obtain a whitened output $w(n_1, n_2)$, with variance σ_w^2, often called the 2-D *innovations sequence*. We can then just as well base our estimate on w and filter it with $G(z_1, z_2)$ to obtain the estimate $\hat{x}(n_1, n_2)$, as shown in Figure 7.2.

We can define the estimation error at the output of G as

$$e(n_1, n_2) \triangleq x(n_1, n_2) - \sum_{l_1, l_2} g(l_1, l_2) w(n_1 - l_1, n_2 - l_2), \qquad (7.2\text{-}2)$$

and then express the mean-square estimation error in the real-valued case, as

$$E\{e^2(n_1, n_2)\} = E\left\{ \left[x(n_1, n_2) - \sum_{l_1, l_2} g(l_1, l_2) w(n_1 - l_1, n_2 - l_2) \right]^2 \right\}$$

$$= R_{xx}(0, 0) - 2E\left\{ \sum_{l_1, l_2} g(l_1, l_2) x(n_1, n_2) w(n_1 - l_1, n_2 - l_2) \right\}$$

$$+ E\left\{ \left[\sum_{l_1, l_2} g(l_1, l_2) w(n_1 - l_1, n_2 - l_2) \right]^2 \right\}.$$

Next we use the whiteness of the innovations $w(n_1, n_2)$ to obtain

$$E\{e^2(n_1, n_2)\} = R_x(0, 0) - 2 \sum_{l_1, l_2} g(l_1, l_2) R_{xw}(l_1, l_2) + \sigma_w^2 \sum_{l_1, l_2} g^2(l_1, l_2)$$

$$= R_x(0, 0) + \sum_{l_1, l_2} \left[\sigma_w g(l_1, l_2) - R_{xw}(l_1, l_2)/\sigma_w \right]^2 - \frac{1}{\sigma_w^2} \sum_{l_1, l_2} R_{xw}^2(l_1, l_2),$$

$$(7.2\text{-}3)$$

where the last step comes from completing the square. At this point we observe that minimizing over choice of g is simple because it only affects the middle term, which clearly has its minimum at 0, easily obtained by setting

$$g(l_1, l_2) = \frac{1}{\sigma_w^2} R_{xw}(l_1, l_2).$$

The overall linear minimum mean-square error filter is obtained by convolving the whitening filter with g. In the Z-transform domain we have

$$G(z_1, z_2) = \frac{1}{\sigma_w^2} S_{xw}(z_1, z_2)$$

$$= \frac{1}{\sigma_w^2} \frac{S_{xy}(z_1, z_2)}{B_{\oplus+}(z_1^{-1}, z_2^{-1})}.$$

So

$$H(z_1, z_2) = \frac{1}{B_{\oplus+}(z_1, z_2)} G(z_1, z_2)$$

$$= \frac{S_{xy}(z_1, z_2)}{\sigma_w^2 B_{\oplus+}(z_1, z_2) B_{\oplus+}(z_1^{-1}, z_2^{-1})}$$

$$= \frac{S_{xy}(z_1, z_2)}{S_y(z_1, z_2)}.$$

NSHP CAUSAL WIENER FILTER

If we restrict the sum over l_1 and l_2 in (7.2-2) to the infinite NSHP causal region

$$\mathcal{R}_{\oplus+} = \{n_1 \geqslant 0, n_2 \geqslant 0\} \cup \{n_1 < 0, n_2 > 0\},$$

we have a causal Wiener filter. Equation (7.2-3) changes to

$$E\{e^2(n_1, n_2)\}$$

$$= R_{xx}(0, 0) + \sum_{(l_1, l_2) \in \mathcal{R}_{\oplus+}} [\sigma_w g(l_1, l_2) - R_{xw}(l_1, l_2)/\sigma_w]^2 - \frac{1}{\sigma_w^2} \sum_{l_1, l_2} R_{xw}^2(l_1, l_2),$$

which is minimized at

$$g(l_1, l_2) = \begin{cases} \frac{1}{\sigma_w^2} R_{xw}(l_1, l_2), & (l_1, l_2) \in \mathcal{R}_{\oplus+} \\ 0, & \text{else.} \end{cases}$$

In the Z-transform domain, we have

$$G(z_1, z_2) = \frac{1}{\sigma_w^2} \left[S_{xw}(z_1, z_2) \right]_{\oplus+},$$

where the subscript notation $\oplus+$ on the cross-power spectra S_{xw} means that we only include the part whose inverse transform has support on $\mathcal{R}_{\in-}$. We can write the full expression for the 2-D causal Wiener filter as

$$\begin{aligned}
G(z_1, z_2) &= \frac{1}{\sigma_u^2} S_{xw}(z_1, z_2) \\
&= \frac{1}{\sigma_u^2 B_{\oplus+}(z_1, z_2)} \left[S_{xu}(z_1, z_2) \right]_{\oplus+} \\
&= \frac{1}{\sigma_u^2 B_{\oplus+}(z_1, z_2)} \left[\frac{S_{xy}(z_1, z_2)}{E_{\ominus+}(z_1^{-1}, z_2^{-1})} \right]_{\ominus+}.
\end{aligned}$$

Again, this is an infinite-order ideal filter. For approximation, any of the filter design methods we have considered can be used. However, if the design is in the frequency domain, then both magnitude and phase error must be taken into account. An example was shown at the end of Chapter 5 in the design of a fully recursive Wiener filter.

7.3 2-D RECURSIVE ESTIMATION

First we review recursive estimation in one dimension, where we focus on the Kalman filter as the linear estimation solution for an AR signal model. After a brief review of the 1-D Kalman filter, we then extend this approach to two dimensions.

7.3.1 1-D KALMAN FILTER

We start with an AR signal model in the 1-D case

$$x(n) = \sum_{k=1}^{M} c_k x(n - k) + w(n),$$

where $n \geqslant 0$. The mean $\mu_w = 0$ and correlation $R_w(m) = \sigma_w^2 \delta(m)$ for the white model noise input w. The initial condition is *initial rest*: $x(-1) = x(-2) = \cdots = x(-M) = 0$. We do not observe the signal x directly. Instead we see the signal x convolved with an M-tap FIR filter h in the presence of an observation noise v

that is orthogonal to the model noise w. The so-called *observation equation* is then given as

$$y(n) = h(n) * x(n) + v(n) \quad \text{for } n \geqslant 0.$$

Re-expressing this model in vector form, we have, upon defining the M-dimensional signal column vectors

$$\mathbf{x}(n) \triangleq \big[x(n), x(n-1), \ldots, x(n-M+1)\big]^T, \qquad \mathbf{w}(n) \triangleq \big[w(n), 0, \ldots, 0\big]^T,$$

and *system matrix*

$$\mathbf{C} \triangleq \begin{bmatrix} c_1 & c_2 & c_3 & \cdots & c_M \\ 1 & 0 & 0 & \cdots & 0 \\ 0 & 1 & 0 & \ddots & \vdots \\ \vdots & \ddots & \ddots & \ddots & \ddots \\ 0 & \cdots & 0 & 1 & 0 \end{bmatrix}, \quad \text{and} \quad \mathbf{Q}_w \triangleq \begin{bmatrix} \sigma_w^2 & 0 & 0 & \cdots & 0 \\ 0 & 0 & 0 & \cdots & 0 \\ 0 & 0 & 0 & \cdots & 0 \\ \vdots & \vdots & \vdots & \vdots & \vdots \\ 0 & \cdots & 0 & 0 & 0 \end{bmatrix},$$

the signal *state equation*,

$$\mathbf{x}(n) = \mathbf{C}\mathbf{x}(n-1) + \mathbf{w}(n), \quad \text{for } n \geqslant 0, \quad \text{subject to} \quad \mathbf{x}(-1) = 0.$$

The advantage of this vector notation is that the Mth-order case can be treated with vector algebra just as simply as the first-order case. The scalar observation equation, based on a linear combination of the elements of the state-vector, can be written in vector form as

$$y(n) = \mathbf{h}^T \mathbf{x}(n) + v(n) \quad \text{for } n \geqslant 0,$$

by defining the column vector $\mathbf{h}$ as

$$\mathbf{h} \triangleq [h_0, h_1, \ldots, h_{M-1}]^T,$$

where $h_l = h(l)$, $l = 0, \ldots, M-1$. In these equations the observation noise v is a white noise, with $R_v(m) = \sigma_v^2 \delta(m)$, but orthogonal to model noise w, i.e., $v(n) \perp x(m)$ for all n and m.

Since we have only specified the second-order properties, i.e., mean and correlation functions, of these signals and noises, we do not have a complete stochastic characterization of this estimation problem. However, we can find the best estimator that is constrained to be linear from this information, as we did with the Wiener filter nonrecursive formulation in the last section. The result is the so-called *linear minimum mean-square error* (LMMSE) estimator.

Some comments are in order:

1 The signal model is Markov-p with $p = M$ when the signal generating noise w is Gaussian distributed.

2   When we have also that the observation noise is Gaussian, it turns out the LMMSE estimator is globally optimal in the unconstrained or nonlinear MMSE case also. This occurs because, in the joint Gaussian case, the optimal MMSE solution is linear in the data.

3   In applications, the support of the smoothing function h may not match the order of the state equation. In that case, either equation may be padded with zeros to come up with a match. Equivalently, we set $M = \max$[these two integers].

4   In many applications, the matrix C is time-variant. Dynamic physics-based models are used in the mechanical and aerospace industries to stochastically model vehicles and aircraft. The 1-D Kalman filter is widely used there.

From the study of estimation theory [20], we know that the MMSE solution to the problem is equivalent to finding the conditional mean

$$E\big\{\mathbf{x}(n)|y(n), y(n-1), \ldots, y(0)\big\} \triangleq \hat{\mathbf{x}}_a(n),$$

called the *filtering estimate*. As a by-product, we also obtain the so-called *one-step predictor estimate*

$$E\big\{\mathbf{x}(n)|y(n-1), y(n-2), \ldots, y(0)\big\} \triangleq \hat{\mathbf{x}}_b(n).$$

The solution for this estimation problem, when the signal and noise are jointly Gaussian distributed, is contained in many texts on 1-D statistical signal processing, e.g., [20].

1-D KALMAN FILTER EQUATIONS

Prediction

$$\hat{\mathbf{x}}_b(n) = \mathbf{C}\hat{\mathbf{x}}_a(n-1), \quad n \geqslant 0.$$

Update

$$\hat{\mathbf{x}}_a(n) = \hat{\mathbf{x}}_b(n) + \mathbf{g}(n)\big[y(n) - \mathbf{h}^T\hat{\mathbf{x}}_b(n)\big].$$

Here, the *Kalman gain* vector, $\mathbf{g}(n)$ scales the observation prediction error term $y(n) - \mathbf{h}^T\hat{\mathbf{x}}_b(n)$ prior to employing it as an additive update term to the state prediction vector $\hat{\mathbf{x}}_b(n)$. This gain vector $\mathbf{g}(n)$ is determined from the *error covariance equations* of the state vector $\mathbf{x}(n)$, which proceed via the nonlinear iterations:

Error covariance

$$\mathbf{P}_b(n) = \mathbf{C}\mathbf{P}_a(n-1)\mathbf{C}^T + \mathbf{Q}_w, \quad n > 0,$$
$$\mathbf{P}_a(n) = \big(\mathbf{I} - \mathbf{g}(n)\mathbf{h}^T\big)\mathbf{P}_b(n), \quad n \geqslant 0.$$

Here, $\mathbf{P}_b(n)$ and $\mathbf{P}_a(n)$ are the *error-covariance matrices*, before and after updating, respectively. Their definition is

$$\mathbf{P}_b(n) \triangleq E\left\{\left(\mathbf{x}(n) - \hat{\mathbf{x}}_b(n)\right)\left(\mathbf{x}(n) - \hat{\mathbf{x}}_b(n)\right)^T\right\},$$

and

$$\mathbf{P}_a(n) \triangleq E\left\{\left(\mathbf{x}(n) - \hat{\mathbf{x}}_a(n)\right)\left(\mathbf{x}(n) - \hat{\mathbf{x}}_a(n)\right)^T\right\}.$$

The gain vector then is given in terms of these as

Gain vector

$$\mathbf{g}(n) \triangleq \mathbf{P}_b(n)\left(\mathbf{h}^T\mathbf{P}_b(n)\mathbf{h} + \sigma_v^2\right)^{-1}, \quad n \geqslant 0.$$

We note from the AR signal model that $x(0) = w(0)$, or equivalently $\mathbf{x}(0) = (w(0), 0, \ldots, 0)^T$, from which follows the initial conditions $\hat{\mathbf{x}}_b(0) = \mathbf{0}$ and $\mathbf{P}_b(0) = \mathbf{Q}_w$ for the preceding filter equations. (For a complete derivation, the reader is referred to Section 11.4 in [20].)

Since the underlying signal model and observation model are scalar, these vector equations can be recast into equivalent scalar equations to offer more insight, which we do next.

SCALAR KALMAN FILTERING EQUATIONS

Predictor

$$\hat{x}_b^{(n)}(n) = \sum_{l=1}^{M} c_l \hat{x}_a^{(n-1)}(n - l), \quad \text{with } \hat{x}_b^{(n)}(m) = \hat{x}_a^{(n-1)}(m) \text{ for } m < n. \qquad (7.3\text{-}4)$$

Updates

$$\hat{x}_a^{(n)}(m) = \hat{x}_b^{(n)}(m) + g^{(n)}(n-m)\left[y(n) - \sum_{l=1}^{M} h_l \hat{x}_b^{(n)}(n - l)\right], \quad \text{for } n-(M-1) \leqslant m \leqslant n.$$
$$(7.3\text{-}5)$$

We see that there is first a prediction estimate $\hat{x}_b^{(n)}(n)$ based on the *past* observations $y(m), m < n$, but after this there are multiple updated estimates $\hat{x}_a^{(n)}(m)$ based on the current observation $y(n)$, beginning at time $n = m$ and proceeding up to *future* time $n = m + (M - 1)$. The estimate resulting from the first update is called the *Kalman filter* estimate, and the remaining ones are called *Kalman smoother* estimates. The former is a causal estimate, while the latter involve a fixed lag behind the observations $y(n)$. Note that the estimates must not get worse, and normally will get better, as they are successively updated. This follows from the corresponding optimality of this recursive estimation procedure for the MSE criteria. Thus we expect the fixed-lag smoother estimate of maximum lag $\hat{x}_a^{(n)}(n - (M - 1))$ will be the best one.

The scalar error-covariance equations become

Before update

$$R_b^{(n)}(n; m) = \sum_{l=1}^{M} c_l R_a^{(n)}(n - l; m), \quad \text{for } m < n$$

$$R_b^{(n)}(n; n) = \sum_{l=1}^{M} c_l R_a^{(n)}(n; n - l) + \sigma_w^2,$$

and

After update

$$R_a^{(n)}(m; l) = R_b^{(n)}(m; l) - g^{(r)}(n - m) R_b^{(n}(n; l), \quad n - (M - 1) \leqslant m, l \leqslant n.$$

The Kalman gain vector becomes in scalar form the *gain array*.

$$g^{(n)}(l) = R_b^{(n)}(n; n - l) / (R_b^{(n)}(n; n) + \sigma_l^2), \quad \text{for } 0 \leqslant l \leqslant M - 1.$$

7.3.2 2-D Kalman Filtering

If we consider a 2-D or spatial AR signal model

$$x(n_1, n_2) = \sum_{(k_1, k_2) \in \mathcal{R}_{\oplus+}} c_{k_1, k_2} x(n_1 - k_1, n_2 - k_2) + w(n_1, n_2). \tag{7.3-6}$$

and scalar observation model

$$y(n_1, n_2) = \sum_{k_1, k_2} h(k_1, k_2) x(n_1 - k_1, n_2 - k_2) + v(n_1, n_2). \tag{7.3-7}$$

over a finite observation region $\mathcal{O}$, we can process these observations in row-scanning order (also called a raster scan) to map them to an equivalent 1-D problem. The 2-D AR model then maps to a 1-D AR model, but with much larger state vector with large internal gaps. If the $\oplus+$ AR model is of order (M, M), and the observation region is a square of side N, i.e., $\mathcal{O} = [0, N - 1]^2$, then the equivalent 1-D AR model is of order $O(MN)$. This situation is illustrated in Figure 7.3. Looking at the scalar filtering equations (7.3-4) and (7.3-5), we see that the prediction term is still of low order $O(M^2)$, but the update term is of order $O(MN)$. The Kalman gain array is then also of order $O(MN)$, so that the nonlinear error-covariance equations must be run to calculate each of these update gain coefficients, and are hence also of order $O(MN)$. Since an image size is normally very large compared to the AR signal model orders, this situation is completely unsatisfactory. We thus conclude that the very efficient computational advantage of the Kalman filter in 1-D is a special case, and is lost in going to higher dimensions. Still there are useful approximations that can be employed based on the observation that, for most reasonable image signal models, the gain terms should be mainly confined to a small region surrounding the current observations. This process then results in the reduced update Kalman filter that has found good use in image processing when a suitable image signal model (7.3-6) and image blur model (7.3-7) are available.

The scalar filtering equations for the 2-D Kalman filter are as follows:

Predictor

$$\hat{x}_{\text{b}}^{(n_1,n_2)}(n_1,n_2) = \sum_{(l_1,l_2)\in\mathcal{R}_{\oplus+}} c_{l_1 l_2}\hat{x}_{\text{a}}^{(n_1-1,n_2)}(n_1-l_1,n_2-l_2),$$

$$\hat{x}_{\text{b}}^{(n_1,n_2)}(m_1,m_2) = \hat{x}_{\text{a}}^{(n_1-1,n_2)}(m_1,m_2), \quad \text{for } (m_1,m_2)\in\mathcal{S}_{\oplus+}(n_1,n_2)$$

Updates

$$\hat{x}_{\text{a}}^{(n_1,n_2)}(m_1,m_2) = \hat{x}_{\text{b}}^{(n_1,n_2)}(m_1,m_2) + g^{(n_1,n_2)}(n_1-m_1,n_2-m_2)$$

$$\times \left[y(n_1,n_2) - \sum_{(l_1,l_2)\in\mathcal{R}_{\oplus+}} b_{l_1,l_2}\hat{x}_{\text{b}}^{(n_1,n_2)}(n_1-l_1,n_2-l_2) \right],$$

$$\text{for } (m_1,m_2)\in\mathcal{S}_{\oplus+}(n_1,n_2),$$

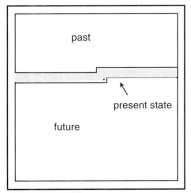

past

present state

future

FIGURE 7.3   *Illustration of the global state vector of a spatial Kalman filter*

where $\mathcal{S}_{\oplus+}(n_1, n_2)$ is the 2-D *global state region* that must be updated at scan location (n_1, n_2), as shown in the gray area in Figure 7.-. The corresponding error-covariance equations are

Before update

$$R_b^{(n_1, n_2)}(n_1, n_2; m_1, m_2)$$

$$= \sum_{(l_1, l_2) \in \mathcal{R}_{\oplus+}} c_{l_1 l_2} R_a^{(n_1, n_2)}(n_1 - l_1, n_2 - l_2; m_1, m_2), \quad \text{for } (m_1, m_2) \in \mathcal{S}_{\oplus+}(n_1, n_2),$$

$$R_b^{(n_1, n_2)}(n_1, n_2; n_1, n_2)$$

$$= \sum_{(l_1, l_2) \in \mathcal{R}_{\oplus+}} c_{l_1 l_2} R_a^{(n_1, n_2)}(n_1, n_2; n_1 - l_1, n_2 - l_2) + \sigma_w^2,$$

and

After update

$$R_a^{(n_1, n_2)}(m_1, m_2; l_1, l_2)$$

$$= R_b^{(n_1, n_2)}(m_1, m_2; l_1, l_2) - g^{(n_1, n_2)}(n_1 - m_1, n_2 - m_2) R_b^{(n_1, n_2)}(n_1, n_2; l_1, l_2),$$

$$\text{for } (m_1, m_2) \in \mathcal{S}_{\oplus+}(n_1, n_2).$$

The 2-D gain array is given as

$$g^{(n_1, n_2)}(l_1, l_2) = R_b^{(n_1, n_2)}(n_1, n_2; n_1 - l_1, n_2 - l_2) / (R_b^{(n_1, n_2)}(n_1, n_2; n_1, n_2) + \sigma_v^2).$$

$$(7.3\text{-}8)$$

7.3.3 REDUCED UPDATE KALMAN FILTER

As a simple approximation to the preceding spatial Kalman filtering equations, we see that the problem is with the update, as the prediction is already very efficient. So, we decide to update only nearby previously processed data points in our raster scan. We will definitely update those points needed directly in the upcoming predictions, but will omit update of points further away. Effectively, we define a local update region $\mathcal{U}_{\oplus+}(n_1, n_2)$ of order $O(M^2)$ with the property that

$$\mathcal{R}_{\oplus+}(n_1, n_2) \subset \mathcal{U}_{\oplus+}(n_1, n_2) \subset \mathcal{S}_{\oplus+}(n_1, n_2),$$

where

$$\mathcal{R}_{\oplus+}(n_1, n_2) = \{(n_1 - k_1, n_2 - k_2) \text{ with } (k_1, k_2) \in \mathcal{R}_{\oplus+}\}.$$

This is an approximation, since all the points in $S_{\oplus-}(n_1, n_2)$ will be used in some future prediction. However, if the current update is only significant for points in a local yet $O(M^2)$ neighborhood of (n_1, n_2), then omitting the updates of points further away should have negligible effect. The choice of how large to make the update region $\mathcal{U}_{\oplus+}(n_1, n_2)$ has been a matter of experimental determination.

The spatial recursive filtering equations for the *reduced update Kalman filter* (RUKF) then become the manageable set:

$$\hat{x}_{\mathrm{b}}^{(n_1, n_2)}(n_1, n_2) = \sum_{(l_1, l_2) \in \mathcal{R}_{\oplus+}} c_{l_1 l_2} \hat{x}_{\mathrm{a}}^{(n_1-1, n_2)}(n_1 - l_1, n_2 - l_2),$$

$$\hat{x}_{\mathrm{b}}^{(n_1, n_2)}(m_1, m_2) = \hat{x}_{\mathrm{a}}^{(n_1-1, n_2)}(m_1, m_2), \quad \text{for } (m_1, m_2) \in \mathcal{U}_{\oplus+}(n_1, n_2),$$

$$\hat{x}_{\mathrm{a}}^{(n_1, n_2)}(m_1, m_2) = \hat{x}_{\mathrm{b}}^{(n_1, n_2)}(m_1, m_2)$$
$$+ g^{(n_1, n_2)}(n_1 - m_1, n_2 - m_2)$$
$$\times \left[y(n_1, n_2) - \sum_{(l_1, l_2) \in \mathcal{R}_{\oplus+}} h_{l_1, l_2} \hat{x}_{\mathrm{b}}^{(n_1, n_2)}(n_1 - l_1, n_2 - l_2) \right],$$
$$\text{for } (m_1, m_2) \in \mathcal{U}_{\oplus+}(n_1, n_2),$$

with computation per data point of $O(M^2)$. In this way experience has shown that a good approximation to spatial Kalman filtering can often be obtained [24].

7.3.4 APPROXIMATE RUKF

While the RUKF results in a very efficient set of prediction and update equations, the error covariance equations are still very high order computationally. This is because the error covariance of each updated estimate must itself be updated with each nonupdated estimate, i.e., for $(m_1, m_2) \in \mathcal{U}_{\oplus+}(n_1, n_2)$ and, most importantly, all $(l_1, l_2) \in \mathcal{S}_{\oplus+}(n_1, n_2)$. To further address this problem, we approximate the RUKF by also omitting these error-covariance updates beyond a larger *covariance update region* $\mathcal{T}_{\oplus+}(n_1, n_2)$ that is still $O(M^2)$. Experimentally it is observed that the resulting *approximate RUKF* can be both very computationally efficient as well as very close to the optimal linear MMSE estimator [24]. However, the choice of the appropriate size for the update region $\mathcal{U}_{\oplus+}$ and the covariance update region $\mathcal{T}_{\oplus+}$ is problem dependent, and so some trial and error may be necessary to get near the best result. These various regions then satisfy the inclusion relations

$$\mathcal{R}_{\oplus+}(n_1, n_2) \subset \mathcal{U}_{\oplus+}(n_1, n_2) \subset \mathcal{T}_{\oplus+}(n_1, n_2) \subset \mathcal{S}_{\oplus+}(n_1, n_2),$$

where only the last region $\mathcal{S}_{\oplus+}$, the global state region, is $O(MN)$.

7.3.5 STEADY-STATE RUKF

For the preceding approximations to work, we generally need a stable AR signal model. In this case, experience has shown that the approximate reduced update filter converges to an LSI filter as $(n_1, n_2) \nearrow (\infty, \infty)$. For typical AR image models, as few as 5–10 rows and columns are often sufficient [24]. In that case, the gain array becomes independent of the observation position, and, to an excellent approximation, the error-covariance equations have no longer to be calculated. The resulting LSI filter, for either RUKF or approximate RUKF, then becomes

$$\hat{x}_{\mathrm{b}}^{(n_1,n_2)}(n_1, n_2) = \sum_{(l_1,l_2)\in\mathcal{R}_{\oplus+}} c_{l_1 l_2}\hat{x}_{\mathrm{a}}^{(n_1-1,n_2)}(n_1 - l_1, n_2 - l_2),$$

$$\hat{x}_{\mathrm{b}}^{(n_1,n_2)}(m_1, m_2) = \hat{x}_{\mathrm{a}}^{(n_1-1,n_2)}(m_1, m_2), \quad \text{for } (m_1, m_2) \in \mathcal{U}_{\oplus+}(n_1, n_2),$$

$$\hat{x}_{\mathrm{a}}^{(n_1,n_2)}(m_1, m_2) = \hat{x}_{\mathrm{b}}^{(n_1,n_2)}(m_1, m_2)$$

$$+ g(n_1 - m_1, n_2 - m_2)$$

$$\times \left[y(n_1, n_2) - \sum_{(l_1,l_2)\in\mathcal{R}_{\ominus+}} k_{l_1,l_2}\hat{x}_{\mathrm{b}}^{(n_1,n_2)}(n_1 - l_1, n_2 - l_2) \right],$$

$$\text{for } (m_1, m_2) \in \mathcal{U}_{\oplus+}(n_1, n_2).$$

7.3.6 LSI ESTIMATION AND RESTORATION EXAMPLES WITH RUKF

We present three examples of the application of the steady-state RUKF linear recursive estimator to image estimation and restoration problems. The first example is an estimation problem.

EXAMPLE 7.3-1 (*image estimation*)
Figure 7.4 shows an original image *Lena* (A), which is 256×256 and monochrome. On the right (B) is seen the same image plus white Gaussian noise added in the contrast domain (also called density domain) of Chapter 5. Figure 7.5 shows the steady-state RUKF estimate based on the noisy data at SNR $= 10$ dB in (B) of Figure 7.4. The SNR of the estimate is 14.9 dB, so that the *SNR improvement* (ISNR) is 4.9 dB. These results come from [13]. We can see the visible smoothing effect of this filter.

The next two examples consider the case where the blur function h is operative, i.e., image restoration. The first one is for a linear 1-D blur, which can simulate linear camera or object motion. If a camera moves uniformly in the horizontal direction for exactly M pixels during its exposure time, then an $M \times 1$ FIR blur can result,

$$h(n) = \frac{1}{M}\big[u(n) - u(n - M)\big].$$

Nonuniform motion will result in other calculable but still linear blur functions.

FIGURE 7.4   (A) 256 × 256 Lena—*original;* (B) Lena + *noise at* 10 *dB input SNR*

FIGURE 7.5   *RUKF estimate of* Lena *from* 10 *dB noisy data*

EXAMPLE 7.3-2 (*image restoration from 1-D blur*)

Figure 7.6 shows the 256 × 256 monochrome cameraman image blurred horizontally by a 10 × 1 FIR blur. The input blurred SNR (BSNR) = 40 dB. Figure 7.7 shows the result of a 3-gain restoration algorithm [13] making use of RUKF. The SNR improvement is 12.5 dB. Note that while there is considerable increase in sharpness, there is clearly some ringing evident. There is more on this example in Jeng and Woods [13]. This example uses the inhomogenous image model of Section 7.4.

FIGURE 7.6 *Cameraman blurred by horizontal FIR blur of length 10; BSNR = 40 dB*

FIGURE 7.7 *Inhomogeneous Gaussian using 3-gains and residual model*

The next example shows RUKF restoration performance for a simulated uniform area blur of size 7×7. Such area blurs are often used to simulate the camera's lack of perfect focus, or simply being out of focus. Also, certain types of sensor vibrations can cause the information obtained at a pixel to be close to a local average of incident light. In any case, a uniform blur is a challenging case, compared to more realistic tapered blurs that are more concentrated for the same blur support.

EXAMPLE 7.3-3 (*image restoration from area blur*)

This is an example of image restoration from a 7×7 uniform blur. The simulation was done in floating point, and white Gaussian noise was added to the blurred *cameraman* image to produce a BSNR = 40 dB. We investigate the effect of boundary conditions on this restoration RUKF estimate. We use the image model of Banham and Katsaggelos [2]. Figure 7.8 shows the restoration using a circulant blur, where we can see ringing coming from the frame boundaries where the circulant blur does not match the assumed linear blur of the RUKF, and Figure 7.9 shows the restoration using a linear blur model, where we notice much less ringing due to the assumed linear blur model. The update regions for both RUKF steady-state estimates were $(-12, 10, 12)$, meaning (−west, +east, +north), a 12×12 update region "northwest" of and including the present observation, plus 10 more columns "northeast." For Figure 7.8, the ISNR was 2.6 dB, or leaving out an image border of 25 pixels on all sides, 3.8 dB. For Figure 7.9, the corresponding ISNR = 4.4 dB, and leaving out the 25 pixel image border, 4.4 dB. In both cases the covariance update region was four columns wider than the update region. Interestingly, the corresponding improvement of the DFT-implemented Wiener filter [2] was reported at 3.9 dB. We should note that the RUKF results are not very good for small error-covariance update regions. For example, Banham and Katsaggelos [2] tried $(-6, 2, 6)$ for both the update and covariance up-

FIGURE 7.8   *RUKF restoration using circulant blur model from area blur*

FIGURE 7.9 *RUKF restoration from linear area blur model*

date region and found only 0.6 dB improvement. But then again, 7×7 is a very large uniform blur support.

REDUCED ORDER MODEL (ROM)

An alternative to the RUKF is the *reduced order model* (ROM) estimator. The ROM approach to Kalman filtering is well established [8], with applications in image processing [19]. In that method, approximation is made to the signal (image) model to constrain its global state to order $O(M^2)$, and then there is no need to reduce the update. The ROM Kalman filter (ROMKF) of Angwin and Kaufman [1] has been prized for its ready adaptation capabilities. A relation between ROMKF and RUKF has been explained in [16].

7.4 INHOMOGENEOUS GAUSSIAN ESTIMATION

Here we extend our AR image model to the inhomogeneous case by including a local mean and local variance [13]. Actually, these two must be estimated from the noisy data, but usually the noise remaining in them is rather small. So we write our *inhomogeneous Gaussian image model* as

$$x(n_1, n_2) \triangleq x_r(n_1, n_2) + \mu_x(n_1, n_2),$$

where $\mu_x(n_1, n_2)$ is the space-variant mean function of $x(n_1, n_2)$, and $x_r(n_1, n_2)$ is a *residual model* given as

$$x_r(n_1, n_2) = \sum_{(k_1,k_2)\in\mathcal{R}_c-(0,0)} c_{k_1,k_2} x_r(n_1 - k_{1,2} - k_2) + w_r(n_1, n_2), \qquad (7.4\text{-}1)$$

where, as usual, $w_r(n_1, n_2)$ is a zero-mean, white Gaussian noise with variance function $\sigma_{w_r}^2(n_1, n_2)$, a space-variant function. The *model noise* w_r describes the uncertainty in the residual image signal model.

Now, considering the estimation case first, the observation equation is

$$y(n_1, n_2) = x(n_1, n_2) + v(n_1, n_2),$$

which, upon subtraction of the mean $\mu_y(n_1, n_2)$, becomes

$$y_r(n_1, n_2) = x(n_1, n_2) + v(n_1, n_2), \qquad (7.4\text{-}2)$$

where $y_r(n_1, n_2) \triangleq y(n_1, n_2) - \mu_y(n_1, n_2)$, using the fact that the observation noise is assumed zero mean. We then apply the RUKF to the model and observation pair, (7.4-1) and (7.4-2). The RUKF will give us an estimate of $x_r(n_1, n_2)$ and the residual RUKF estimate of the signal then becomes

$$\hat{x}(n_1, n_2) = \hat{x}_r(n_1, n_2) + \mu_x(n_1, n_2).$$

In a practical application, we would have to estimate the space-variant mean $\mu_x(n_1, n_2)$, and this can be accomplished in a simple way, using a box filter (cf. Section 6.6.1) as

$$\hat{\mu}_x(n_1, n_2) \triangleq \frac{1}{(2M+1)^2} \sum_{(m_1,m_2)\in[-M,+M]^2} y(n_1 - m_1, n_2 - m_2),$$

and relying on the fact that $\mu_x = \mu_y$, since the observation noise v is zero mean. We can also estimate the local variance of the residual signal x_r through an estimate of the local variance of the residual observations y_r again using a box filter, as

$$\hat{\sigma}_{y_r}^2(n_1, n_2) \triangleq \frac{1}{(2M+1)^2} \sum_{(m_1,m_2)\in[-M,+M]^2} \left(y(n_1 - m_1, n_2 - m_2) - \hat{\mu}_x(n_1, n_2) \right)^2,$$

and

$$\hat{\sigma}_{x_r}^2(n_1, n_2) \triangleq \max\left\{ \hat{\sigma}_{y_r}^2(n_1, n_2) - \sigma_v^2, 0 \right\}.$$

The overall system diagram is shown in Figure 7.10, where we see a two-channel system, consisting of a low-frequency *local mean channel* and a high-frequency *residual channel*. The RUKF estimator of x_r can be adapted or modified

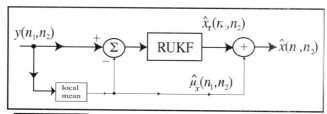

FIGURE 7.10   *System diagram for inhomogeneous Gaussian estimation with RUKF*

by the local spatial variance $\hat{\sigma}_{x_r}^2(n_1, n_2)$ on the residual channel, i.e., by the residual signal variance estimate.

EXAMPLE 7.4-1 (*simple inhomogeneous Gaussian estimation*)
Here we assume that the residual signal model is white Gaussian noise with zero mean and fixed variance $\sigma_{x_r}^2$. Since the residual observation noise is also white Gaussian, with variance σ_v^2, we can construct the simple Wiener filter for the residual observations as

$$h(n_1, n_2) = \frac{\sigma_{x_r}^2}{\sigma_{x_r}^2 + \sigma_v^2}\delta(n_1, n_2).$$

The overall filter then becomes

$$\hat{x}(n_1, n_2) = \frac{\sigma_{x_r}^2}{\sigma_{x_r}^2 + \sigma_v^2}\hat{y}_r(n_1, n_2) + \mu_x(n_1, n_2),$$

with approximate realization through box filtering as

$$\hat{x}(n_1, n_2) = \frac{\hat{\sigma}_{x_r}^2}{\hat{\sigma}_{x_r}^2 + \sigma_v^2}\hat{y}_r(n_1, n_2) + \hat{\mu}_x(n_1, n_2).$$

This filter has been known for some time in the image-processing literature, where it is referred to often as the simplest adaptive Wiener filter, first discovered by Wallis [22].

7.4.1 INHOMOGENEOUS ESTIMATION WITH RUKF

Now, for a given fixed residual model (7.4-1), there is a multiplying constant A that relates the model residual noise variance $\sigma_{w_r}^2$ to the residual signal variance $\sigma_{x_r}^2$. The residual model itself can be determined from a least-squares prediction on similar, but noise-free, residual image data. We can then construct a space-variant RUKF from this data, where the residual image model parameters are constant except for changes in the signal model noise variance. Instead of actually

Table 7.1. Obtained ISNRs

Input SNR	10 dB	3 dB
LSI RUKF	4.9	8.6
Simple Wiener (Wallis)	4.3	7.8
3-Gain residual RUKF	6.0	9.3
3-Gain normalized RUKF	5.2	9.1

running the space-variant RUKF error covariance equations, we can simplify the problem by quantizing the estimate $\hat{\sigma}_{x_r}^2$ into three representative values and then used steady-state RUKF for each of these three, just switching between them as the estimated model noise variance changes [13].

EXAMPLE 7.4-2 (*inhomogeneous RUKF estimate*)
Here we observe the *Lena* image in 10 dB white Gaussian noise, but we assume an inhomogeneous Gaussian model. The original image and input 10 dB noisy image are the same as in Figure 7.4. Four estimates are then shown in Figure 7.11: (A) is an LSI estimate, (B) is the simple Wallis filter estimate, (C) is the residual RUKF estimate, and (D) is a normalized RUKF estimate that is defined in [13]. We can notice that all three adaptive or inhomogeneous estimates look better than the LSI RUKF estimate, even though the Wallis estimate has the lowest ISNR.

The actual SNR improvements in these estimates areas shown in Table 7.1. An example of inhomogenous RUKF restoration was given in Example 7.3-2.

7.5 ESTIMATION IN THE SUBBAND/WAVELET DOMAIN

Generalizing the above two-channel system, consider that first a subband/wavelet decomposition is performed on the input image. Then we would have one low-frequency channel, similar to the local mean channel, and many intermediate and high-frequency subband channels. As we have already described, we can, in a suboptimal fashion, perform estimates of the signal content on each subband separately. Finally, we create an estimate in the spatial domain by performing the corresponding inverse SWT. A basic contribution in this area was by Donoho [6], who applied noise-thresholding to the subband/wavelet coefficients. Later Zhang *et al.* [25] introduced simple adaptive Wiener filtering for use in *wavelet image denoising*.

In these works, it has been found useful to employ the so-called *overcomplete* subband/wavelet transform (OCSWT), which results when the decimators are omitted in order to preserve shift-invariance of the transform. Effectively there are four phases computed at the first splitting stage and then this progresses geometrically down the subband/wavelet tree. In the reconstruction phase, an average

FIGURE 7.11   *Various inhomogeneous Gaussian estimates:* $\begin{bmatrix} a & b \\ c & d \end{bmatrix}$ *(A) LS'; (B) Wallis filter; (C) residual RUKF; (D) normalized RUKF*

over these phases must be done to arrive at the overall estimate in the image domain. Both hard and soft thresholds have been considered, the basic idea being that small coefficients are most likely due to the assumed white observation noise. If the wavelet coefficient, aka subband value, is greater than a well-determined threshold, it is likely to be "signal" rather than "noise." The estimation of the threshold is done based on the assumed noise statistics. While only constant thresholds are considered in [6], a spatially adaptive thresholding is considered in [4].

EXAMPLE 7.5-1 (*SWT denoising*)

Here we look at estimates obtained by thresholding in the SWT domain. We start with the 256×256 monochrome *Lena* image and then add Gaussian

FIGURE 7.12 *Estimate using hard threshold in SWT domain*

white noise to make the SNR = 22 dB. Then we input this noisy image into a five-stage SWT using an 8-tap orthogonal wavelet filter due to Daubechies. Calling the SWT domain image $y(n_1, n_2)$, we write the thresholding operation as

$$\hat{y} = \begin{cases} 0, & |y| < t \\ y, & |y| \geqslant t, \end{cases}$$

where the noise-threshold value was taken as $t = 40$ on the 8-bit image scale $[0, 256]$. Using this *hard threshold*, we obtain the image in Figure 7.12. For a soft threshold, given as

$$\hat{y} = \begin{cases} \text{sgn}(y)(|y| - t), & |y| < t \\ y, & |y| \geqslant t, \end{cases}$$

the result is shown in Figure 7.13. Using the OCSWT and hard threshold, we obtain Figure 7.14. Finally, the result of soft threshold in OCSWT domain is shown in Figure 7.15. For the SWT, the choice of this threshold $t = 40$ resulted in output peak SNR (PSNR) = 25 dB in the case of hard thresholding, while the soft threshold gave 26.1 dB. For the OCSWT, we obtained 27.5 dB for the hard threshold and 27 dB for the soft threshold, resulting in ISNRs of 5.5 and 5.0 dB, respectively. In the case of OCSWT, there is no inverse transform because it is overcomplete, so a least-squares inverse was used.

FIGURE 7.13 *Estimate using soft threshold in the SWT domain*

FIGURE 7.14 *Hard threshold $t = 40$ in OCSWT domain, ISNR = 5.5 dB*

Besides these simple noise-thresholding operations, we can perform the Wiener or Kalman filter in the subband domain. An example of this appears at the end of Section 7.7 on image and blur model identification.

FIGURE 7.15   *Estimate using soft threshold in OCSWT domain, ISNR = 5.0 dB*

7.6 BAYESIAN AND MAP ESTIMATION

All of the estimates considered thus far have been Bayesian, meaning that they optimized over not only the *a posteriori* observations but also jointly over the *a priori* image model in order to obtain an estimate that was close to minimum mean-square error. We mainly used Gaussian models that resulted in linear, and sometimes space-variant linear, estimates. In this subsection, we consider a nonlinear Bayesian image model that must rely on a global iterative recursion for its solution. We use a so-called Gibbs model for a conditionally Gaussian random field, where the conditioning is on a lower level (unobserved) line field that specifies the location of edge discontinuities in the upper level (observed) Gaussian field. The line field is used to model edge-like lineal features in the image. It permits the image model to have edges, across which there is low correlation, as well as smooth regions of high correlation. The resulting estimates then tend to retain image edges that would otherwise be smoothed over by an LSI filter. While this method can result in much higher quality estimation and restoration, it is by far the most computationally demanding of the estimators presented thus far. The needed Gauss Markov and compound Gauss Markov models are introduced next.

7.6.1 GAUSS MARKOV IMAGE MODELS

We start with the following noncausal, homogeneous Gaussian image model,

$$x(n_1, n_2) = \sum_{(k_1, k_2) \in \mathcal{R}_c} c(k_1, k_2) x(n_1 - k_1, n_2 - k_2) + u(n_1, n_2), \qquad (7.6\text{-}1)$$

where the coefficient support $\mathcal{R}_c$ is a noncausal neighborhood region centered on 0, but excluding 0, and the image model noise u is Gaussian, zero-mean, and with correlation given as

$$R_u(m_1, m_2) = \begin{cases} \sigma_u^2, & (m_1, m_2) = 0 \\ -c(m_1 - k_1, m_2 - k_2)\sigma_u^2, & (m_1 - k_1, m_2 - k_2) \in \mathcal{R}_c \\ 0, & \text{elsewhere.} \end{cases} \qquad (7.6\text{-}2)$$

This image model is then Gaussian and Markov in the 2-D or spatial sense [23]. The image model coefficients provide a minimum mean square error (MMSE) interpolation $\hat{x}$ based on the neighbor values in $\mathcal{R}_c$,

$$\hat{x}(n_1, n_2) = \sum_{(k_1, k_2) \in \mathcal{R}_c} c_{k_1, k_2} x(n_1 - k_1, n_2 - k_2),$$

and the image model noise u is then the resulting interpolation error.

Note that u is not a white noise, but is colored and with a finite correlation support of small size $\mathcal{R}_c$, beyond which it is uncorrelated, and because it is Gaussian, also independent. This noncausal notion of Markov is somewhat different from the NSHP causal Markov we have seen earlier. The *noncausal Markov random field* is defined by

$$f_x\big(x(\mathbf{n})\,|\, \text{all other } x\big) = f_x\big(x(\mathbf{n})\,|\,x(\mathbf{n} - \mathbf{k}), \ \mathbf{k} \in \mathcal{R}_c\big),$$

where $\mathcal{R}_c$ is a small neighborhood region centered on, but not including, 0. If the random field is Markov in this noncausal or spatial sense, then the best estimate of $x(\mathbf{n})$ can be obtained using only those values of x that are neighbors in the sense of belonging to the set $\{x | x(\mathbf{n} - \mathbf{k}), \mathbf{k} \in \mathcal{R}_c\}$. A helpful diagram illustrating the noncausal Markov concept is Figure 7.16, which shows a central region $\mathcal{G}^+$ where the random field x is conditionally independent of its values on the outside region $\mathcal{G}^-$, given its values on a boundary region $\partial\mathcal{G}$ of the minimum width given by the neighborhood region $\mathcal{R}_c$. The "width" of $\mathcal{R}_c$ then gives the "order" of the Markov field. If we would want to compare this noncausal Markov concept to that of the NSHP causal Markov random fields considered earlier, we can think of the boundary region $\partial\mathcal{G}$ as being stretched out to infinity in such a way that we get the situation depicted in Figure 7.17a, where the boundary region $\partial\mathcal{G}$ then becomes just the global state vector support in the case of the NSHP Markov model

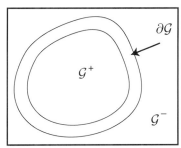

FIGURE 7.16   *Illustration of dependency regions for noncausal Markov field*

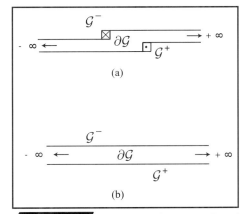

FIGURE 7.17   *Illustration of two causal Markov concepts*

for the 2-D Kalman filter. Another possibility is shown in Figure 7.17b, which denotes a vector concept of causality wherein scanning proceeds a full line at a time. Here, the *present* is the whole vector line $\mathbf{x}(n)$, with the n axis directed downwards. In all three concepts, the region $\mathcal{G}^-$ is called the *past*, boundary region $\partial\mathcal{G}$ is called the *present*, and region $\mathcal{G}^-$ is called the *future*. While the latter two are consistent with some form of causal or sequential processing in the estimator, the noncausal Markov, illustrated in Figure 7.16, is not, and thus requires iterative processing for its estimator solution.

Turning to the Fourier transform of the correlation of the image model noise u, we see that the PSD of the model noise random field u is

$$S_u(\omega_1, \omega_2) = \sigma_u^2 \left[1 - \sum_{(k_1, k_2) \in \mathcal{R}_c} c(k_1, k_2) \exp -j(k_1\omega_1 + k_2\omega_2) \right],$$

and that the PSD of the image random field x is then given as, via application of (7.1-3),

$$S_x(\omega_1, \omega_2) = \frac{\sigma_u^2}{[1 - \sum_{(k_1, k_2) \in \mathcal{R}_c} c(k_1, k_2) \exp -j(k_1\omega_1 - k_2\omega_2)]}. \tag{7.6-3}$$

In order to write the pdf of the Markov random field, we turn to the theory of *Gibbs distributions* as in Besag [3]. Based on this work, it can be shown that the unconditional joint pdf of a Markov random field x can be expressed as

$$f_x(\mathbf{X}) = K \exp -U_x(\mathbf{X}), \tag{7.6-4}$$

where the matrix $\mathbf{X}$ denotes x restricted to the finite region $\mathcal{X}$, and $U_x(\mathbf{X})$ is an *energy function* defined in terms of *potential functions* V as

$$U_x(\mathbf{X}) \triangleq \sum_{c_n \in C_n} V_{c_n}(\mathbf{X}), \tag{7.6-5}$$

and C_n denotes a *clique system* in $\mathcal{X}$, and K is a normalizing constant. Here a *clique* is a link between $x(\mathbf{n})$ and its immediate neighbors in the neighborhood region $\mathcal{R}_c$. An example of this concept is given next.

EXAMPLE 7.6-1 (*first-order clique system*)
Let the homogeneous random field x be noncausal Markov with neighborhood region $\mathcal{R}_c$ of the form

$$\mathcal{R}_c = \{(1, 0), (0, 1), (-1, 0), (0, -1)\},$$

which we call first-order noncausal Markov. Then the conditional pdf of $x(\mathbf{n}) = x(n_1, n_2)$ can be given as

$$f_x(x(\mathbf{n})| \text{ all other } x)$$
$$= f_x(x(\mathbf{n})|\{x(\mathbf{n} - (1, 0)), x(\mathbf{n} - (0, 1)), x(\mathbf{n} - (-1, 0)), x(\mathbf{n} - (0, -1))\}),$$

and the joint pdf of x over some finite region $\mathcal{X}$ can be written in terms of (7.6-5) with potentials V_{c_n} of the form

$$V_{c_n} = \frac{x^2(\mathbf{n})}{2\sigma_u^2} \quad \text{or} \quad V_{c_n} = -\frac{c(k)x(\mathbf{n})x(\mathbf{n} - \mathbf{k})}{2\sigma_u^2} \quad \text{for each } \mathbf{k} \in \mathcal{R}_c.$$

COMPOUND GAUSS MARKOV IMAGE MODEL

The partial difference equation model for the compound Gauss Markov (CGM) random field is given as

$$x(n_1, n_2) = \sum_{(k_1, k_2) \in \mathcal{R}} c^{l(n_1, n_2)}(k_1, k_2)x(n_1 - k_1, n_2 - k_2) + u^{l(n_1, n_2)}(n_1, n_2),$$

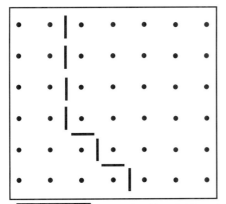

FIGURE 7.18   *Example of line field modeling edge in portion of fictitious image*

where $l(n_1, n_2)$ is a vector of four nearest neighbors from a line field $l(n_1, n_2)$, which originated in Geman and Geman [9]. This line field exists on an interpixel grid and takes on binary values to indicate whether a *bond* is inplace or broken between two pixels, both horizontal and vertical neighbors. An example of this concept is shown in Figure 7.18, which shows a portion of a fictitious image with an edge going downwards as modeled by a line field. Black line indicates broken bond ($l = 0$).

The model interpolation coefficients $c^{l(n_1, n_2)}(k_1, k_2)$ vary based on the values of the (in this case four) nearest neighbor line-field values as captured in the line field vector $l(n_1, n_2)$. The image model does not attempt to smooth in the direction of a broken bond. For the line field potentials V, we have used the model suggested in [9] and shown in Figure 7.19, where a broken bond is indicated by a line, an inplace bond by no line, and the pixel locations by large dots. Note that only one rotation of the indicated neighbor pattern is shown. We see that the potentials favor (with $V = 0.0$) bonds being inplace, with the next most favorable situation being a horizontal or vertical edge (with $V = 0.9$), and so on to less often occurring configurations.

Then the overall Gibbs probability mass function (pmf) for the line field can be written as

$$p(\mathbf{L}) = K_1 \exp -U_l(\mathbf{L}),$$

with

$$U_l(\mathbf{L}) \triangleq \sum_{c_l \in C_l} V_{c_l}(\mathbf{L}),$$

with $\mathbf{L}$ denoting a matrix of all line field values for an image. The overall joint mixed pdf/pmf for the CGM field over the image can then be written as

$$f(\mathbf{X}, \mathbf{L}) = K_2 \exp -\big(U_x(\mathbf{X}) + U_l(\mathbf{L})\big).$$

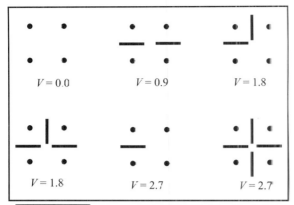

FIGURE 7.19 *Line field potential V values for indicated nearest neighbors (black lines indicate bond broken)*

We should note that **L** needs about twice as many points in it as **X**, since there is a link to the right and below each observed pixel, except for the last row and last column.

7.6.2 SIMULATED ANNEALING

In the simulated annealing method, a *temperature* parameter T is added to the above joint mixed pdf, so that it becomes

$$f(\mathbf{X}, \mathbf{L}) = K_2 \exp -\frac{1}{T}\left(U_x(\mathbf{X}) + U_l(\mathbf{L})\right).$$

At $T = 1$, we have the correct joint mixed pdf, but as we slowly lower $T \searrow 0$, the global maximum moves relatively higher than the local max max. and this property can be used to iteratively locate the *maximum a posteriori* (MAP) estimate

$$(\widehat{\mathbf{X}}, \widehat{\mathbf{L}}) \triangleq \arg\max_{\mathbf{X},\mathbf{L}} f(\mathbf{X}, \mathbf{L}|\mathbf{R}).$$

An iterative solution method, called *simulated annealing* (SA), developed for continuous valued images in [14], alternates between pixel update and line field update, as it completes passes through the received data **R**. At the end of each complete pass, the temperature T is reduced a small amount in an effort to eventually *freeze* the process at the joint MAP estimates $(\widehat{\mathbf{X}}, \widehat{\mathbf{L}})$. A key aspect of SA is determining the conditional pdfs for each pixel and line field location, given the observations **R** and all the other pixel and line field values. The following conditional distributions are derived in [14],

$$f_x(n_1, n_2) = K_3 \exp - \left[\frac{(x(n_1, n_2) - \sum_{(k_1,k_2)\in\mathcal{R}} c^{l(n_1,n_2)}(k_1, k_2)x(n_1 - k_1, n_2 - k_2))^2}{2T\sigma^2_{u,l(n_1,n_2)}} \right.$$

$$- \frac{\sum_{(k_1,k_2)\in\mathcal{R}_b}(x(n_1 + k_1, n_2 + k_2)}{2T\sigma^2_v}$$

$$\left. - \frac{\sum_{(m_1,m_2)\in\mathcal{R}_b} h(m_1, m_2)x(n_1 + k_1 - m_1, n_2 + k_2 - m_2))^2}{2T\sigma^2_v} \right]$$

and for each line field location (n_1, n_2)[4] between pixels (i_1, i_2) and (j_1, j_2), both vertically and horizontally, the conditional updating pmf

$$p_l(n_1, n_2) = K_4 \exp - \left[\frac{x^2(i_1, i_2)}{2T\sigma^2_{u,l(i_1,i_2)}} - \frac{c^{l(j_1,j_2)}(i_1 - j_1, i_2 - j_2)x(i_1, i_2)x(j_1, j_2)}{T\sigma_{u,l(i_1,i_2)}\sigma_{u,l(j_1,j_2)}} \right.$$

$$\left. + \frac{x^2(j_1, j_2)}{2T\sigma^2_{u,l(j_1,j_2)}} + \frac{1}{T} \sum_{(n_1,n_2)\in C_l} V_{c_l}(\mathbf{L}) \right],$$

where K_3 and K_4 are normalizing constants. These conditional distributions are *sampled* as we go through the sweeps, meaning that at each location **n**, the previous estimates of l and x are replaced by values drawn from these conditional distributions. In evaluating these conditional distributions, the latest available estimates of the other l or x values are always used. This procedure is referred to as a *Gibbs sampler* [9]. The number of iterations that must be done is a function of how fast the temperature decreases. Theoretical proofs of convergence [9,14] require a very slow logarithmic decrease of temperature with the sweep number n, i.e.,

$$T = C/\log(1 + n);$$

however, in practice a faster decrease is used. A typical number of sweeps necessary was experimentally found to be in the 100s.

EXAMPLE 7.6-2 (*simulated annealing for image estimation*)
In [14], we modeled the 256×256 *Lena* image by a compound Gauss Markov (CGM) model with order 1×1 and added white Gaussian noise to achieve an input SNR = 10 dB. For comparison purposes, we first show the Wiener filter result in Figure 7.20. There is substantial noise reduction but also visible

---

4. We note that the line field l is on the interpixel grid, with about twice the number of points as pixels in image x, so that x(**n**) is not the same location as l(**n**).

FIGURE 7.20 *Example of Wiener filtering for Gauss Markov model at input SNR = 10 dB*

FIGURE 7.21 *Simulated annealing estimate for CGM model at input SNR = 10 dB*

blurring. The simulated annealing result, after 200 iterations, is shown in Figure 7.21, where we see a much stronger noise reduction combined with an almost strict preservation of important visible edges.

FIGURE 7.22 *Input blurred image as BSNR = 40 dB*

FIGURE 7.23 *Blur restoration via Wiener filter*

EXAMPLE 7.6-3 (*simulated annealing for image restoration*)
In [14], we blurred the *Lena* image with a 5×5 uniform blur function and
then added a small amount of white Gaussian noise to achieve a BSNR =

FIGURE 7.24 *Blur restoration via simulated annealing*

40 dB, with the result shown in Figure 7.22. Then we restored this noisy and blurred image with a Wiener filter to produce the result shown in Figure 7.23, which is considerably sharper but contains some ringing artifacts. Finally, we processed the noisy blurred image with simulated annealing at 200 iterations to produce the image shown in Figure 7.24, which shows a sharper result with reduced image ringing.

An NSHP causal compound Gaussian Markov image model called *doubly stochastic Gaussian* (DSG) was also formulated [14] and a sequential noniterative solution method for an approximate MMSE was developed using a so-called *M-algorithm*. Experimental results are provided [14] and also show considerable improvement over the LSI Wiener and RUKF filters. The M-algorithms are named for their property of following M paths of chosen directional NSHP causal image models through the data. A relatively small number of paths sufficed for the examples in [14].

7.7 IMAGE IDENTIFICATION AND RESTORATION

Here we consider the additional practical need in most applications to estimate the signal model parameters, as well as the parameters of the observation, including the blur function in the restoration case. This can be a daunting problem, and does not admit of a solution in all cases. Fortunately, this *identification* or model parameter estimation problem can be solved in many practical cases.

Here we present a method of combined identification and restoration using the *expectation-maximization* (EM) algorithm.

7.7.1 EXPECTATION-MAXIMIZATION ALGORITHM APPROACH

We will use the 2-D vector, 4-D matrix notation of problem 17 in Chapter 4, and follow the development of Lagendijk *et al.* [17]. First we establish an AR signal model in 2-D vector form as

$$\mathbf{X} = \mathcal{C}\mathbf{X} + \mathbf{W}, \tag{7.7-1}$$

and then write the observations also in 2-D vector form as

$$\mathbf{Y} = \mathcal{D}\mathbf{X} + \mathbf{V}, \tag{7.7-2}$$

with the signal model noise $\mathbf{W}$ and the observation noise $\mathbf{V}$ both Gaussian, zero mean and independent of one another, with variances σ_w^2 and σ_v^2, respectively. We assume that the 2-D matrices $\mathcal{C}$ and $\mathcal{D}$ are parameterized by the filter coefficients $c(k_1, k_2)$ and $d(k_1, k_2)$, respectively, each being finite order and restricted to an appropriate region, that is to say, an NSHP support region of the model coefficients c, and typically a rectangular support region centered on the origin for the blur model coefficients d. All model parameters can be conveniently written together in the vector parameter

$$\Theta \triangleq \left\{ c(k_1, k_2), d(k_1, k_2), \sigma_w^2, \sigma_v^2 \right\}.$$

To ensure uniqueness, we assume that the blurring coefficients $d(k_1, k_2)$ are normalized to sum to one, i.e.,

$$\sum_{k_1, k_2} d(k_1, k_2) = 1, \tag{7.7-3}$$

which is often reasonable when working in the intensity domain, where this represents a conservation of power, as would be approximately true with a lens.

This parameter vector is unknown and must be estimated from the noisy data $\mathbf{Y}$ and we seek the maximum-likelihood (ML) estimate [20] of the unknown but nonrandom parameter Θ,

$$\widehat{\Theta}_{ML} \triangleq \arg \max_{\Theta} \left\{ \log f_{\mathbf{Y}}(\mathbf{Y}; \Theta) \right\}, \tag{7.7-4}$$

where $f_{\mathbf{Y}}$ is the pdf of the noisy observation vector $\mathbf{Y}$. Since $\mathbf{X}$ and $\mathbf{Y}$ are jointly Gaussian, we can write

$$f_{\mathbf{X}}(\mathbf{X}) = \sqrt{\frac{\det(\mathcal{I} - \mathcal{C})^2}{(2\pi)^{N^2} \det \mathcal{Q}_{\mathbf{W}}}} \exp -\frac{1}{2} \left\{ \mathbf{X}^T (\mathcal{I} - \mathcal{C})^T \mathcal{Q}_{\mathbf{W}}^{-1} (\mathcal{I} - \mathcal{C}) \mathbf{X} \right\}$$

and

$$f_{Y|X}(Y|X) = \frac{1}{\sqrt{(2\pi)^{N^2} \det Q_V}} \exp -\frac{1}{2} \{ (Y - \mathcal{D}X)^T \mathcal{C}_V^{-1} (Y - \mathcal{D}X) \}.$$

Now, combining (7.7-1) and (7.7-2), we have

$$Y = \mathcal{D}(\mathcal{I} - \mathcal{C})^{-1} W + V,$$

with covariance matrix

$$\mathcal{K}_{YY} = \mathcal{D}(\mathcal{I} - \mathcal{C})^{-1} Q_W (\mathcal{I} - \mathcal{C})^{-1} \mathcal{D}^T + Q_V,$$

so the needed ML estimate of the parameter vector Θ can be expressed as

$$\widehat{\Theta}_{ML} = \arg \max_{\Theta} \{ -\log(\det(\mathcal{K}_{YY})) - Y^T \mathcal{K}_{YY}^{-1} Y \}, \qquad (7.7\text{-}5)$$

which is unfortunately highly nonlinear and not amenable to closed-form solution. Further, there are usually several to many local maxima to worry about. Thus we take an alternative approach, and arrive at the *expectation maximization* (EM) algorithm, an iterative method that converges to the local optimal points of this equation.

The EM method (see Section 9.4 in [20]), talks about so-called *complete* and *incomplete* data. In our case the complete data is $\{X, Y\}$ and the incomplete data is just the observations $\{Y\}$. Given the complete data, we can easily solve for $\widehat{\Theta}_{ML}$; specifically, we would obtain $c(k_1, k_2)$ and σ_w^2 as the solution to a 2-D linear prediction problem, expressed as a 2-D normal equation (see Section 7.1). Then the parameters $d(k_1, k_2)$ and σ_v^2 would be obtained via

$$\hat{d}(k_1, k_2), \hat{\sigma}_v^2 = \arg \max f(Y|X) \sim N(\mathcal{D}X, Q_V),$$

which is easily solved via classical system theory. Note that this follows only because the pdf of the complete data separates, i.e.,

$$f(X, Y) = f(X)f(Y|X),$$

with the image model parameters only affecting the first factor, and the observation parameters only affecting the second factor. The EM algorithm effectively converts the highly nonlinear problem (7.7-5) into a sequence of problems that can be solved as simply as if we had the complete data. It is crucial in an EM problem to formulate it, if possible, so that the ML estimate is easy to solve, given the complete data. Fortunately, this is the case here.

E-M ALGORITHM

Start at $k = 0$ with an initial guess $\widehat{\Theta}^0$. Then alternate the following E-steps and M-steps till convergence:

$$\text{E-step } \mathcal{L}(\Theta; \widehat{\Theta}^{(k)}) \triangleq E[\log f(\mathbf{X}, \mathbf{Y}; \Theta) | Y; \widehat{\Theta}^{(k)}]$$

$$= \int \log f(\mathbf{X}, \mathbf{Y}; \Theta) f(\mathbf{X}|\mathbf{Y}; \widehat{\Theta}^{(k)}) \, d\mathbf{X},$$

$$\text{M-step } \widehat{\Theta}^{(k+1)} = \arg \max_{\Theta} \mathcal{L}(\Theta; \widehat{\Theta}^{(k)}).$$

It is proven in [5] that this algorithm will monotonically improve the likelihood of the estimate $\widehat{\Theta}^{(k)}$ and so result in a local optimum of the objective function given in (7.7-4).

We now proceed with the calculation of $\mathcal{L}(\Theta; \widehat{\Theta}^{(k)})$, by noting that

$$f(\mathbf{X}, \mathbf{Y}; \Theta) = f(\mathbf{Y}|\mathbf{X}) f(\mathbf{X}; \Theta),$$

and

$$f(\mathbf{X}|\mathbf{Y}; \widehat{\Theta}^{(k)}) = \frac{f(\mathbf{X}, \mathbf{Y}; \widehat{\Theta}^{(k)})}{f(\mathbf{Y}; \widehat{\Theta}^{(k)})}$$

$$= \frac{1}{\sqrt{(2\pi)^{N^2} \det \widehat{\mathcal{K}}_{\text{EE}}^{(k)}}} \exp -\frac{1}{2} \{(\mathbf{X} - \widehat{\mathbf{X}}^{(k)})^T (\widehat{\mathcal{K}}_{\text{EE}}^{(k)})^{-1} (\mathbf{X} - \widehat{\mathbf{X}}^{(k)})\},$$

where $\widehat{\mathbf{X}}^{(k)}$ and $\widehat{\mathcal{K}}_{\text{EE}}^{(k)}$ are, respectively, the conditional mean and conditional variance matrices of $\mathbf{X}$ at the kth iteration. Here we have

$$\widehat{\mathbf{X}}^{(k)} \triangleq E[\mathbf{X}|\mathbf{Y}; \widehat{\Theta}^{(k)}] = \widehat{\mathcal{K}}_{\text{EE}}^{(k)} \mathcal{D}^T \mathcal{Q}_{\text{V}}^{-1} \mathbf{Y},$$

with estimated error covariance

$$\widehat{\mathcal{K}}_{\text{EE}}^{(k)} \triangleq \text{cov}[\mathbf{X}|\mathbf{Y}; \widehat{\Theta}^{(k)}] = \{(\mathcal{I} - \mathcal{C})^T \mathcal{Q}_{\text{w}}^{-1} (\mathcal{I} - \mathcal{C})^T + \mathcal{D}^T \mathcal{Q}_{\text{v}}^{-1} \mathcal{D}\}^{-1}.$$

Thus the spatial Wiener filter designed with the parameters $\widehat{\Theta}^{(k)}$ will give us also the likelihood function $\mathcal{L}(\Theta; \widehat{\Theta}^{(k)})$ to be maximized in the M-step.

Turning to the M-step, we can express $\mathcal{L}(\Theta; \widehat{\Theta}^{(k)})$ as the sum of terms

$$\mathcal{L}(\Theta; \widehat{\Theta}^{(k)}) = C - N^2 \log(\sigma_{\text{w}}^2 \sigma_{\text{v}}^2) + \log \det(\mathcal{I} - \mathcal{C})^2$$

$$- \frac{1}{\sigma_{\text{v}}^2} \mathbf{Y}^T \mathbf{Y} + \frac{2}{\sigma_{\text{v}}^2} \text{tr}(D\widehat{\mathcal{K}}_{\text{XY}}) - \frac{1}{\sigma_{\text{v}}^2} \text{tr}(\mathcal{D}\widehat{\mathcal{K}}_{\text{XX}}^{(k)} \mathcal{D}^T)$$

$$- \frac{1}{\sigma_{\text{w}}^2} \text{tr}\{(\mathcal{I} - \mathcal{C}) \widehat{\mathcal{K}}_{\text{XX}}^{(k)} (\mathcal{I} - \mathcal{C})^T\}, \qquad (7.7\text{-}6)$$

where C is a constant and

$$\widehat{\mathcal{K}}_{XY}^{(k)} \triangleq E[\mathbf{XY}^T|\mathbf{Y}; \widehat{\Theta}^{(k)}] = \widehat{\mathbf{X}}^{(k)}\mathbf{Y}^T \qquad (7.7\text{-}7)$$

and

$$\widehat{\mathcal{K}}_{XX}^{(k)} \triangleq E[\mathbf{XX}^T|\mathbf{Y}; \widehat{\Theta}^{(k)}] = \widehat{\mathcal{K}}_{EE}^{(k)} + \widehat{\mathbf{X}}^{(k)}\widehat{\mathbf{X}}^{(k)T}. \qquad (7.7\text{-}8)$$

Notice that (7.7-6) separates into two parts, one of which depends on the signal model parameters and one of which only depends on the observation model parameters, the maximization of $\mathcal{L}(\Theta; \widehat{\Theta}^{(k)})$ can be separated into two distinct parts. The first part involving image model identification becomes

$$\arg \max_{c(k_1,k)_2,\sigma_w^2} \left\{ \log\det(\mathcal{I} - \mathcal{C})^2 - N^2\log\sigma_w^2 - \frac{1}{\sigma_w^2}\text{tr}\{(\mathcal{I} - \mathcal{C})\widehat{\mathcal{K}}_{XX}^{(k)}(\mathcal{I} - \mathcal{C})^T\} \right\},$$

where the first term can be deleted since an NSHP causal model must satisfy $\det(\mathcal{I} - \mathcal{C}) = 1$. The resulting simplified equation is quadratic in the image model coefficients $c(k_1, k_2)$ and is solved by the 2-D Normal equations. The second part of (7.7-6) becomes

$$\arg \max_{d(k_1,k_2),\sigma_v^2} \left\{ -N^2\log\sigma_v^2 - \frac{1}{\sigma_v^2}\mathbf{Y}^T\mathbf{Y} + \frac{2}{\sigma_v^2}\text{tr}(\mathcal{D}\widehat{\mathcal{K}}_{XY}) - \frac{1}{\sigma_v^2}\text{tr}(\mathcal{D}\widehat{\mathcal{K}}_{XX}^{(k)}\mathcal{D}^T) \right\},$$

which is quadratic in the blur coefficients $d(k_1, k_2)$ and thus easy to maximize for any σ_v^2. Then the estimate of σ_v^2 can be found.

If the preceding equations are circulant approximated, then due to the assumption of constant coefficients and noise variances, we can write scalar equations for each of these two identification problems. From (7.7-7) and (7.7-8) we get

$$\widehat{K}_{XX}^{(k)}(m_1, m_2) = \widehat{K}_{EE}^{(k)}(m_1, m_2) + \frac{1}{N^2}\sum_{k_1,k_2}\widehat{X}^{(k)}(k_1, k_2)\widehat{X}^{(k)}(k_1 - m_1, k_2 - m_2),$$

$$(7.7\text{-}9)$$

$$\widehat{K}_{XY}^{(k)}(m_1, m_2) = \frac{1}{N^2}\sum_{k_1,k_2}\widehat{X}^{(k)}(k_1, k_2)Y(k_1 - m_1, k_2 - m_2). \qquad (7.7\text{-}10)$$

where we have summed over all of the element-pairs with shift distance (m_1, m_2). Then, for the image model identification, we get the Normal equations

$$\widehat{K}_{XX}^{(k)}(m_1, m_2) = \sum_{\mathcal{S}_c}\hat{c}(k_1, k_2)\widehat{K}_{XX}^{(k)}(m_1 - k_1, m_2 - k_2), \quad \text{for all } (m_1, m_2) \in \mathcal{S}_c,$$

$$\hat{\sigma}_w^2 = \widehat{R}_{XX}^{(k)}(0, 0) - \sum_{k_1,k_2}\hat{c}(k_1, k_2)\widehat{K}_{XX}^{(k)}(k_1, k_2),$$

where $\mathcal{S}_c$ is the NSHP support of the AR signal model c.

For the blur model identification, the following equations can be derived in the homogeneous case [17],

$$\widehat{K}_{XY}^{(k)}(-m_1, -m_2) = \sum_{(k_1, k_2) \in \mathcal{S}_d} \hat{d}(k_1, k_2) \widehat{K}_{XX}^{(k)}(m_1 - k_1, m_2 - k_2),$$

$$\text{for all } (m_1, m_2) \in \mathcal{S}_d,$$

$$\hat{\sigma}_v^2 = \frac{1}{N^2} \sum_{(n_1, n_2) = (0,0)}^{(N-1, N-1)} Y^2(n_1, n_2) - \sum_{k_1, k_2} \hat{d}(k_1, k_2) \widehat{K}_{XY}^{(k)}(-k_1, -k_2),$$

where $\mathcal{S}_d$ is the support of the FIR blur impulse response d. The resulting d values can then be normalized via (7.7-3).

7.7.2 EM METHOD IN THE SUBBAND/WAVELET DOMAIN

In Kim and Woods [15], the preceding joint identification and restoration method was extended to work in the subband/wavelet domain. In this case we have separate image models in each of the subbands, where we can employ a local power level to modulate the signal power in the various subbands (equivalently, wavelet coefficients). This can result in much improved clarity in the restored image, as illustrated in the set of images shown in Figure 7.25. We note that the linear space-variant (LSV) restoration of the individual subbands works best from a visual standpoint. Its ISNR is 7.3 dB while that of fullband restoration achieved only 5.8 dB improvement.

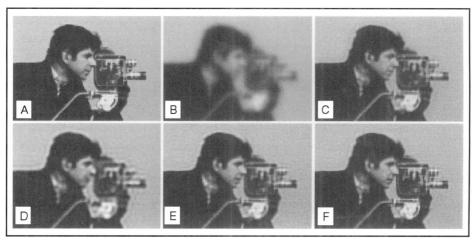

FIGURE 7.25 *Subband EM restoration of* cameraman *at BSNR* $= 40$ *dB,* $\begin{bmatrix} a & b & c \\ d & e & f \end{bmatrix}$: *(A) original; (B) blurred; (C) fullband restored; (D) LL subband restored; (E) subband (LSI); (F) subband (LSV)*

7.8 Color Image Processing

Since color images are simply multicomponent or vector images, usually of the three components, red, green, and blue, one could apply all of the methods of this chapter to each component separately. Alternatively, one can develop vector versions of the estimation methods for application to the vector images. Usually, though, color image processing is conducted in the YUV or YC_rC_b domain, by just processing the luma component Y. For example, in color image deblurring, often just the luma image is restored, while the chroma components are left alone, and reliance is given to the lower chroma bandwidth of the HVS and lesser sensitivity to error. Regarding blur function estimation, often this is determined from the luminance data alone, and then the same blur function is assumed for the chrominance channels. Such relatively simple color processing, though, has been criticized as leading to so-called *color shifts* that can sometimes lead to unnatural colors in the restored images. Vector restoration procedures have been especially developed to deal with this problem. A recent special issue of *IEEE Signal Processing Magazine* [21] deals extensively with the overall topic of processing color images.

7.9 Conclusions

This chapter has introduced the problem of estimation and restoration for spatial systems. We present both nonrecursive and recursive approaches for 2-D estimation. These methods extend the well-known 1-D discrete-time Wiener filter and predictor to the 2-D or spatial case. We considered recursive estimation, reviewed the 1-D Kalman filter, and then extended it to the 2-D case, where we saw that the state vector spread out to the width of the signal field, and so was not as efficient as the 1-D Kalman filter for the chosen AR model. We then presented a reduced update version of 2-D recursive linear estimation that has regained some of the lost properties. We also looked at space-variant extensions of these LSI models for adaptive Wiener and Kalman filtering and provided some examples. Then we considered estimation in the subband/wavelet domain as an extension of a simple adaptive inhomogenous method involving local mean and variance functions. We looked at so-called Bayesian methods using compound Gaussian Markov models and simulated annealing in their estimator solution. Finally, we introduced the problem of parameter estimation combined with image estimation and restoration and looked at how this EM technique can be carried to the subband/wavelet transformed observations.

7.10 Problems

1   Carry out the indicated convolutions in (7.1-1) to obtain the covariance function of white noise of variance σ_w^2, having been filtered with the im-

pulse response $g(n_1, n_2)$. Denoting $g_-^*(n_1, n_2) \triangleq g^*(-n_1, -n_2)$, express your answer in terms of σ_w^2, g, and g_-.

2   Extend the method in Example 7.1-3 to solve for the $(1, 1)$-order NSHP predictor, where the coefficient support of the 2-D linear predictor is given as

$$\mathcal{R}_a - (0, 0) \triangleq \{(1, 0), (-1, 1), (0, 1), (1, 1)\}.$$

Write the 4×4 matrix normal equation for this case in terms of a homogeneous random field correlation function $R_x(m_1, m_2)$.

3   Write the equations that specify a 5×5-tap FIR Wiener filter h with support on $[-2, +2]^2$ for the signal correlation function $R_x(m_1, m_2) = \sigma_x^2 \rho^{|m_1| + |m_2|}$, where σ_x^2 is a given positive value and ρ satisfies $|\rho| < 1$. Assume the observation equation is specified as

$$y(n_1, n_2) = 5x(n_1, n_2) + 3w(n_1, n_2),$$

where w is a white noise of variance σ_w^2 and $w \perp x$.

4   Reconsider problem 3, but now let the Wiener filter h have infinite support and actually write the solution in the frequency domain, i.e., find $H(\omega_1, \omega_2)$.

5   To determine a general causal Wiener filter, we need to first find the spectral factors of the noisy observations $S_y(\omega_1, \omega_2)$, in the sense of Theorem 7.2-1. Show that the power spectral density

$$S_y(\omega_1, \omega_2) = \frac{1}{1.65 + 1.6 \cos\omega_1 + 0.2 \cos\omega_2 + 0.16 \cos(\omega_1 - \omega_2)}$$

has stable and QP causal spectral factor

$$B_{++}(z_1, z_2) = \frac{1}{1 + 0.8z_1^{-1} + 0.1z_2^{-1}}.$$

What is $B_{--}(z_1, z_2)$ here? What is σ^2?

6   The 2-D Wiener filter has been designed using the orthogonality principle of linear estimation theory. Upon setting

$$\hat{x}(n_1, n_2) \triangleq \sum_{(k_1, k_2) \in \mathcal{R}_h} h_{k_1, k_2} y(n_1 - k_1, n_2 - k_2),$$

we found in the homogeneous, zero-mean case that h is determined as the solution to the Normal equations

$$\sum_{(l_1, l_2) \in \mathcal{R}_h} h_{l_1, l_2} R_{yy}(k_1 - l_1, k_2 - l_2) = R_{xy}(k_1, k_2) \quad \text{for } (k_1, k_2) \in \mathcal{R}_h.$$

By ordering the elements $(k_1, k_2) \in \mathcal{R}_j$ onto vectors, this equation can be put into matrix-vector form and then solved for vector $\mathbf{h}$ in terms of correlation matrix $\mathbf{R}_{yy}$ and cross-correlation vector $\mathbf{r}_{xy}$, as determined by the chosen element order. In this problem, our interest is in determining the resultant mean-square value $E[|e(n_1, n_2)|^2]$ of the estimation error $e(n_1, n_2) = \hat{x}(n_1, n_2) - x(n_1, n_2)$.

(a) First, using the orthogonality principle, show that

$$E[|e(n_1, n_2)|^2] = -E[x(n_1, n_2)e^*(n_1, n_2)]$$

$$= E[|x(n_1, n_2)|^2] - E[x(n_1, n_2)\hat{x}^*(n_1, n_2)]$$

$$= R_{xx}(0, 0) - \sum_{(k_1, k_2) \in \mathcal{R}_h} h^*_{k_1, k_2} R_{xy}(k_1, k_2).$$

(b) Then show that the solution to the Normal equations and the result of part a can be combined and written in matrix-vector form as

$$\sigma_e^2 = \sigma_x^2 - \mathbf{r}_{xy}^T \mathbf{R}_{yy}^{-1} \mathbf{r}_{xy}^*.$$

Note that $E[|e(n_1, n_2)|^2] = \sigma_e^2$ since we assume the means of x and y are zero.

7 In the 2-D Kalman filter, what is the motivation for omitting updates that are far from the observations? How does the reduced update Kalman filter (RUKF) differ from the approximate RUKF? For constant coefficient AR signal models and observation models, we experimentally find that a steady state is reached fairly rapidly as we process into the image. What role does model stability play in this?

8 Write out the steady-state RUKF equations for a 1×1-order $\oplus+$ model and update region

$$\mathcal{U}_{\oplus+}(n_1, n_2) = \{(n_1, n_2), (n_1 - 1, n_2), (n_1 - 2, n_2),$$

$$(n_1 + 2, n_2 - 1), (n_1 + 1, n_2 - 1), (n_1, n_2 - 1),$$

$$(n_1 - 1, n_2 - 1), (n_1 - 2, n_2 - 1)\}.$$

Write these equations in terms of model coefficients $\{c_{10}, c_{01}, c_{11}, c_{-1,1}\}$ and assumed steady-state gain values

$$\{g(0, 0), g(1, 0), g(2, 0), g(-2, 1), g(-1, 1), g(0, 1), g(1, 1), g(2, 1)\}.$$

9 With reference to problem 8, please list the various estimates provided by the above steady-state RUKF in terms of their increasing 2-D delay. For example, the first estimate is the prediction estimate $\hat{x}_b^{(n_1, n_2)}(n_1, n_2)$ with a delay of $(-1, 0)$, i.e., an advance of one pixel horizontally.

10 Continuing along the lines of problem 7 of Chapter 6, show how to efficiently calculate the local variance used in Section 7.4 by extending the box-filter concept. Assume the box is $2M + 1 \times 2M + 1$. How many multiplies per pixel? adds per pixel? How much intermediate storage is needed?

11 Show that the Markov random field satisfying the noncausal 2-D difference equation (7.6-1), with random input satisfying (7.6-2), has the PSD (7.6-3).

12 Derive the scalar equations (7.7-9) and (7.7-10) from the 2-D matrix equations (7.7-7) and (7.7-8). Make use of the fact that, in the homogeneous case,

$$\left(\widehat{\mathcal{K}}_{\mathbf{XX}}^{(k)}\right)_{n_1,n_2;n_1-k_1,n_2-k_2} = \left(\widehat{\mathcal{K}}_{\mathbf{XX}}^{(k)}\right)_{0,0;-k_1,-k_2},$$

for all (n_1, n_2).

REFERENCES

[1] D. L. Angwin and H. Kaufman, "Image Restoration Using Reduced Order Models," *Signal Process.*, **16**, 21–28, Jan. 1988.

[2] M. R. Banham and A. K. Katsaggelos, "Digital Image Restoration," *IEEE Signal Process. Magazine*, 24–41, March 1997 [see also "RUKF performance revisited" in *Signal Process. Forum* section of the November 1997 issue of the same magazine, pp. 12 and 14].

[3] J. Besag, "On the Statistical Analysis of Dirty Pictures," *J. Royal Statist. Soc. B*, **48**, 259–302, 1986.

[4] S. G. Chang, B. Yu, and M. Vetterli, "Spatially Adaptive Wavelet Thresholding with Context Modeling for Image Denoising," *IEEE Trans. Image Process.*, **9**, 1522–1531, Sept. 2000.

[5] A. P. Demster, N. M. Laird, and D. B. Rubin, "Maximum Likelihood from Incomplete Data via the EM Algorithm," *J. Royal Statist. Soc. B.*, **39**(1), 1–38, 1977.

[6] D. L. Donoho, "De-Noising by Soft-Thresholding," *IEEE Trans. Inf. Theory*, **41**, 613–627, May 1995.

[7] M. P. Ekstrom and J. W. Woods, "Two-Dimensional Spectral Factorization with Applications in Recursive Digital Filtering," *IEEE Trans. Acoust., Speech, Signal Process.*, **ASSP-24**, 115–128, April 1976.

[8] B. Friedland, "On the Properties of Reduced-Order Kalman Filters," *IEEE Trans. Auto. Control*, **34**, 321–324, March 1989.

[9] S. Geman and D. Geman, "Stochastic Relaxation, Gibbs Distributions, and the Bayesian Restoration of Images," *IEEE Trans. Pattern Anal. Machine Intel.*, **PAMI-6**, 721–741, Nov. 1984.

[10] B. Girod, "The Efficiency of Motion-Compensating Prediction for Hybrid Coding of Video Sequences," *IEEE J. Select. Areas Commun.*, **SAC-5**, 1140–1154, Aug. 1987.

[11] A. K. Jain, *Fundamentals of Digital Image Processing* Prentice-Hall, Englewood Cliffs, NJ, 1989.

[12] J. R. Jain and A. K. Jain, "Displacement Measurement and its Application in Interframe Image Coding," *IEEE Trans. Commun.*, COM-29, 1799–1804, Dec. 1981.

[13] F.-C. Jeng and J. W. Woods, "Inhomogeneous Gaussian Image Models for Estimation and Restoration," *IEEE Trans. Acoust., Speech, Signal Process.*, ASSP-36, 1305–1312, August 1988.

[14] F.-C. Jeng and J. W. Woods, "Compound Gauss–Markov Random Fields for Image Estimation," *IEEE Trans. Signal Process.*, 39, 633–697, March 1991.

[15] J. Kim and J. W. Woods, "Image Identification and Restoration in the Subband Domain," *IEEE Trans. Image Process.*, 3, 312–314, May 1994 [see also "Erratum," p. 873, Nov. 1994].

[16] J. Kim and J. W. Woods, "A New Interpretation of ROMKF," *IEEE Trans. Image Process.*, 6, 599–601, April 1997.

[17] R. L. Lagendijk, J. Biemond, and D. E. Boekee, "Identification and Restoration of Noisy Blurred Images using the Expectation-Maximization Algorithm," *IEEE Trans. Acoust., Speech, Signal Process.*, ASSP-38, 1180–1191, July 1990.

[18] J. Lim, *Two-Dimensional Signal and Image Processing*, Prentice-Hall, Englewood Cliffs, NJ, 1990.

[19] A. J. Patti, A. M. Tekalp, and M. I. Sezan, "A New Motion-Compensated Reduced-Order Model Kalman Filter for Space-Varying Restoration of Progressive and Interlaced Video," *IEEE Trans. Image Process.*, 7, 543–554, April 1998.

[20] H. Stark and J. W. Woods, *Probability and Random Processes with Applications to Signal Processing*, 3rd edn., Prentice-Hall, Englewood Cliffs, NJ, 2002.

[21] H. J. Trussell, E. Saber, and M. Vrhel, eds., "Color Image Processing Special Issue," *IEEE Signal Process. Magazine*, 22, Jan. 2005.

[22] R. Wallis, "An Approach to the Space Variant Restoration and Enhancement of Images," *Proc. Symp. Curr. Math. Problems in Image Science*, Naval Postgraduate School, Monterey, CA, Nov. 1976, pp. 107–111.

[23] J. W. Woods, "Two-dimensional Discrete Markovian Random Fields," *IEEE Trans. Inf. Theory*, IT-18, 232–240, March 1972.

[24] J. W. Woods and V. K. Ingle, "Kalman Filtering in Two Dimensions: Further Results," *IEEE Trans. Acoust., Speech, Signal Process.*, ASSP-29, 188–197, April 1981.

[25] H. Zhang, A. Nosratinia, and R. O. Wells, Jr., "Image Denoising via Wavelet-Domain Spatially Adaptive FIR Wiener Filtering," *Proc. ICASSP 2000*, Vancouver, BC, pp. 2179–2182, May 2000.

DIGITAL IMAGE COMPRESSION 8

Images are perhaps the most influential of media that we encounter on a daily basis. This chapter will apply 2-D statistical signal processing to their compression for both efficient transmission and storage. We consider a generic model consisting of first transformation, then quantization, and then entropy coding. We cover the popular DCT-based image coding as well as the newer SWT-based scalable image coders such as JPEG 2000.

8.1 INTRODUCTION

With reference to Figure 8.1, we see a universal diagram for the transmission or storage of digital images. On the extreme left we have the input image file. We consider the case where the input image is digital, with word format either floating point or fixed point. The theory, though, assumes that it is composed of real numbers, and, in fact, our statistical model for the input image is a continuous valued random field. From the viewpoint of the designer of an image communication system, the image is certainly unknown. Over the lifetime of the communications device, we see an ensemble of images that will be transmitted over it. This ensemble will have various statistical properties, such as histogram, joint histogram, mean function, correlation function, etc. So, it is reasonable to impose a statistical model on the *image source*.

The output of this image source is input to the *source coder*, whose purpose is to compress the image file by representing it as a finite string of binary digits (bits). If the input image were finite wordlength, then the source coder can be *lossless*, meaning that the input image can be reconstructed exactly from the compressed data. Similarly, a *lossy* coder is one for which only approximate reconstruction of the input image is possible. Lossy source coding is the main topic of this chapter. Next in the general communication diagram of Figure 8.1 is the *channel coder*, whose purpose is to adapt or strengthen the compressed bitstream to survive the digital channel. One way it does this is by appending error correction bits to the compressed data words that make up the source-coded bitstream. Examples of a digital channel are digital radio and television broadcast, digital subscriber line (DSL), telephone modem, optical communications, wireless networks, and digital storage channels as found in disk memory systems. At a certain level of abstraction we have the Internet and its protocol as a very common example of the digital channel. The remaining parts of Figure 8.1 show the decoding; first, the *channel decoder* attempts to recover the source-coded bitstream, then the *source decoder* tries to reconstruct the source image. Figure 8.2 shows a useful decomposition valid for many source coders, which consist of a *transformation*, followed by a *quantizer* and then an *entropy coder*.

The purpose of the transformation is to remove or suppress the redundant parts of the data, hopefully yielding components that are independent or at least

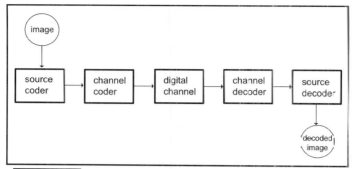

FIGURE 8.1 *Generic digital image communication system*

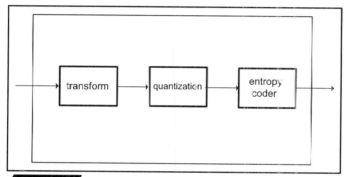

FIGURE 8.2 *Generic source-coding system diagram*

uncorrelated. Examples are *differential pulse-code modulation* (DPCM),[1] transforms based on frequency decompositions (DFT, DCT, SWT), and the optimal KLT. These transforms can be applied to the entire image or, as is commonly done, to relatively small blocks of pixel values. Usually these blocks do not overlap, but in some methods [e.g., *lapped orthogonal transform* (LOT) and SWT)] they do. The quantization block in Figure 8.2 is very important, because this is where the actual data compression occurs, by a shortening of the transformed data wordlength. The transform block is invertible, so that no data compression actually occurs there. A quantizer is a nonlinear, zero-memory device that chooses representative values for ranges of input data coming from the transform one-at-a-time, called *scalar quantization*, or several-at-a-time, called *vector quantization*. Most quantizers are scalar and common examples are uniform quantization (also called an A/D converter), nonuniform quantization, optimal uniform, optimal nonuniform, etc. The entropy coder block in Figure 8.2 has

---

1. In DPCM the transform and the quantizer are linked together in a feedback loop, so it is not right to say that the transform precedes the quantizer always. In DPCM the roles of the transform and the quantizer are intertwined.

the job of converting the representative values, outputs of the quantizer, to efficient variable-length codewords. If we omit this last block, then the quantizer output is converted into fixed-length code words, usually less efficient. Examples of variable-length codes are Huffman and arithmetic codes, which will be briefly described in Section 8.4.1.

8.2 TRANSFORMATION

The role of the transformation in data compression is to decorrelate and, more generally, to remove dependency between the data values, but without losing any information. For example, the transformation should be invertible, at least with infinite-wordlength arithmetic. For Gaussian random fields, rate Distortion Theory [6,7] states that the optimal transformation for this purpose is the KLT. The transform can be applied in small blocks or to the entire image. Common block transforms include the DFT, DCT, and an overlapped extension of DCT called LOT. The DCT is motivated by being close to the KLT for a one-dimensional first-order Gauss–Markov random sequence, when the correlation coefficient $\rho \lesssim 1.0$. The practical advantage of the DCT in these cases, however, is that it has a fast algorithm, unlike the general KLT. For images, the 2-D DCT is motivated by the Gaussian random field with correlation function given as separable product of the 1-D Markov correlation function, i.e.,

$$R_x(m_1, m_2) = \sigma_x^2 \rho_1^{|m_1|} \rho_2^{|m_2|}, \tag{8.2-1}$$

where ρ_1 and ρ_2 are the horizontal and the vertical correlation coefficients, respectively, which should both be close to one.

8.2.1 DCT

As we have seen in Chapter 4, for a data array of rectangular support $[0, N_1 - 1] \times [0, N_2 - 1]$, the 2-D DCT is given as

$$X_C(k_1, k_2) \triangleq \sum_{l_1=0}^{N_1-1} \sum_{l_2=0}^{N_2-1} 4x(n_1, n_2) \cos \frac{\pi k_1}{2N_1} (2n_1 + 1) \cos \frac{\pi k_2}{2N_2} (2n_2 + 1),$$

for $(k_1, k_2) \in [0, N_1 - 1] \times [0, N_2 - 1]$. Otherwise $X_C(k_1, k_2) \triangleq 0$, but in this chapter we only consider $(k_1, k_2) \in [0, N_1 - 1] \times [0, N_2 - 1]$. For the common block-DCT and an input image of size $L_1 \times L_2$, we decompose it into $N_1 \times N_2$ blocks and perform the DCT for each successive block. The total number of blocks then becomes $\lceil L_1/N_1 \rceil \times \lceil L_2/N_2 \rceil$, where the last block in each line and column may be a partial block. This transforms the image data into a sequence of DCT coefficients, which can be stored blockwise or can be stored with all like-frequency

coefficients together, effectively making a small "image" out of each DCT coefficient. Of these, the so-called DC image made up of the $X_C(0, 0)$ coefficients looks
like the original image, and in fact, is just an $(N_1 \times N_2) \downarrow$ subsampled version of
the original after $N_1 \times N_2$ box filtering.

Owing to the nearness of a typical natural image correlation function to
(8.2-1), the DCT coefficients in a given block are approximately uncorrelated.
We can expect the coefficients from different blocks to be approximately uncorrelated because of their distance from one another, if the blocksize is not too small.
Even for neighboring blocks, the argument of uncorrelatedness is quite valid for
the upper frequency or AC coefficients, and is only significantly violated by the
DC coefficients and a few low-frequency coefficients. This interblock coefficient-
correlation effect is almost ignored in standard coding algorithms such as JPEG,
where only the DC coefficients of blocks are singled out for further decorrela-
tion processing, usually consisting of a prediction followed by a coding of the
prediction residual.

A key property of a transform like DCT used in data compression is that it is
orthonormal, with its transformation matrix called *unitary*. This means that there
is a Parseval-like relation between the sum of the magnitude-square values in the
transform space, i.e., the coefficients, and the data in the image space. Since the
transform is linear, this energy balance must also hold for difference signals such
as the coding error. So if we are concerned with minimizing the sum magnitude-
squared error, we can just as well do this in the coefficient domain, where each
coefficient can be quantized (or coded more generally) *seperately*, using a *scalar
quantizer* (SQ). Specifically, we have

$$\sum_{n_1=0}^{N_1-1} \sum_{n_2=0}^{N_2-1} \left(x(n_1, n_2) - \hat{x}(n_1, n_2) \right)^2$$

$$= \frac{1}{4N_1 N_2} \sum_{k_1=0}^{N_1-1} \sum_{k_2=0}^{N_2-1} w(k_1) w(k_2) \left[X_C(k_1, k_2) - \hat{X}_C(k_1, k_2) \right]^2.$$

A generalization of the block-DCT transformation is the fast Lapped Orthogonal Transform (LOT) [16]. Here overlapping 16×16 LOT blocks are made up
by cascading neighboring 8×8 DCT blocks with an orthogonalizing transformation on their outputs. Actually, this is done in one dimension and then the
procedure is just copied to the 2-D case, via row–column separable processing.
An advantage of the LOT is that the objectionable blocking effect of DCT compression is reduced by the overlapping LOT input windows. Also, the LOT re-
spects the DSP theoretical method of 2-D sectioned convolution, by employing
block overlap in implementation of its block processing. Another way of providing overlapped blocks, and perhaps eliminating blocks altogether, is the SWT.

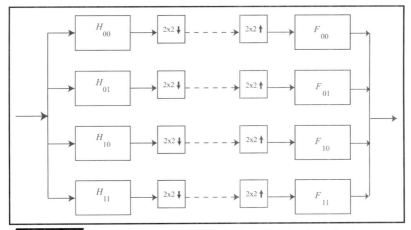

FIGURE 8.3   *General rectangular SWT/ISWT analysis/synthesis bank*

8.2.2 SWT

Subband/wavelet transformation (SWT) can be viewed as a generalization of block-based DCT. This is because there is a filter bank interpretation of the block-based DCT, wherein the filter impulse responses are the DCT basis functions, and the decimation is $(N_1 \times N_2) \downarrow$ downsampling. Usually in SWTs, the filter support is larger than the decimation ratio, which is normally $(2 \times 2) \downarrow$. So, SWT is seen as a generalization of DCT to overlapping blocks, with expected reductions or elimination of blocking artifacts. A general rectangular subband analysis/synthesis bank is shown in Figure 8.3.

Usually more stages of subband decomposition are conducted than just this one level. In the common dyadic (aka octave and wavelet) decomposition, each resulting LL subband is further split recursively, as illustrated in Figure 8.4, showing a three-level subband/wavelet decomposition. While this dyadic decomposition is almost universal, it is known to be suboptimal when a lot of high-spatial-frequency information is present (e.g., the well-known test image *Barbara*). When the subband splitting is optimized for a particular set of image statistics, the name *wavelet packet* [19] has emerged for the corresponding transformation.

In summary, SWT is seen to be a generalization of the block-DCT- and LOT-based transformations and as such, can only offer the possibility for better performance. Best performance in this transform class can then be obtained by optimizing over the subband/wavelet filters used and the nature and manner of the subband splittings. The goal of all such transformations is to decorrelate (i.e., remove linear dependencies in) the data prior to quantization.

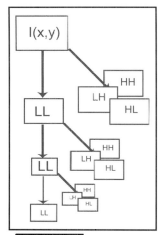

FIGURE 8.4   *Illustration of dyadic (octave, wavelet) subband decomposition to three levels*

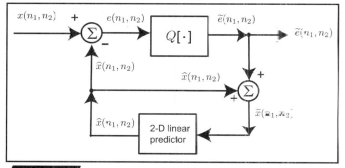

FIGURE 8.5   *Diagram of a 2-D DPCM coder*

8.2.3 DPCM

A two-dimensional extension of *differential pulse-code modulation* (DPCM) has been extensively used in image compression. We see with reference to Figure 8.5 that the main signal flow path, across the top of the figure, quantizes an error signal that is generated via a prediction $\hat{x}(n_1, n_2)$ that is fed back, and in turn, is generated from only the *past* values of the quantized error signal $\tilde{e}(n_1, n_2)$. Here the transformation and the quantization are intertwined, not matching our paradigm of transform coming before quantizer (see Figure 8.2). We will look at quantizers in some detail in the next section, but for now, just consider it a device that approximates the incoming prediction error $e(n_1, n_2)$ with a finite number of levels. The signal estimate $\hat{x}(n_1, n_2)$ is produced by 2-D linear $(1, 0)$-step prediction based on its input $\tilde{x}(n_1, n_2)$. Note that this feedback loop can also be run at the decoder, and, in fact, it *is* the decoder, with $\tilde{x}(n_1, n_2)$ as the decoder output.

Here we assume that the quantized error signal $\tilde{e}(n_1, n_2)$ is losslessly conveyed to the receiver and its source decoder, i.e., no channel errors occur.

One nice property of 2-D DPCM that directly extends from the 1-D case can be easily seen. Consider the overall error at any given data location (n_1, n_2):

$$x(n_1, n_2) - \tilde{x}(n_1, n_2) = \left[x(n_1, n_2) - \hat{x}(n_1, n_2) \right] - \left[\tilde{x}(n_1, n_2) - \hat{x}(n_1, n_2) \right]$$

$$= e(n_1, n_2) - \tilde{e}(n_1, n_2),$$

as can be seen from the two equations at the summing junctions in Figure 8.5. So if we design the quantizer Q to minimize the MSE between its input and output, we are also finding the quantizer that minimizes the MSE at the DPCM decoder output. This very useful property would be lost if we moved the quantizer Q to the right and outside of the feedback loop. This is because a prediction-error filter does not constitute an orthogonal transformation, and quantization errors would then accumulate at the decoder output.

There remains the problem of how to calculate the 2-D linear predictor coefficients. If we are concerned with high-quality (read high-bitrate) coding, then we can expect that $\tilde{x}(n_1, n_2) \approx x(n_1, n_2)$ to a sufficient degree of approximation to permit use of a linear $(1, 0)$-step prediction filter that was designed based on the quantization-noise-free input $x(n_1, n_2)$ itself. At low bitrates, some optimization can be done iteratively, based on starting with this filter, then generating the corresponding $\tilde{x}(n_1, n_2)$, and then generating a new linear $(1, 0)$-step prediction filter, and so on.[2] However, DPCM performance still deteriorates at low bitrates.

8.3 QUANTIZATION

The quantization operation is one of truncation or rounding. Figure 8.6 shows a quantizer Q, with input variable x and output variable $\hat{x} = Q(x)$, and Figure 8.7 shows a quantizer function $Q(x)$ plotted versus input x, which may be continuous valued or already quantized on a relatively fine scale to that of this quantizer.

The *scalar quantizer* (SQ) is specified in general by its number of output values or levels L, its *decision levels* d_i, and its *representation levels* r_i, also called output values. The defining equation for the quantizer function is then

$$Q(x) \triangleq \begin{cases} r_1, & d_0 < x \leqslant d_1 \\ \vdots & \vdots \\ r_i, & d_{i-1} < x \leqslant d_i \\ \vdots & \vdots \\ r_L, & d_{L-1} < x \leqslant d_L. \end{cases}$$

---

2. Starting this second time through, the filter is really a prediction estimator since its *design input* is $\tilde{x}$ and not x anymore.

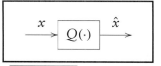

FIGURE 8.6 Scalar quantizer

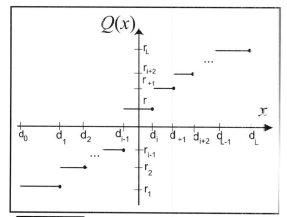

FIGURE 8.7 Quantizer characteristic that rounds down

Here we have taken the input range cells or *bins* as $(d_{i-1}, d_i]$, called round-down, but they could equally be $[d_{i-1}, d_i)$. With reference to Figure 8.7, the total range of the quantizer is $[r_1, r_L]$ and its total domain is $[d_0, d_L]$, where normally $d_0 = x_{\min}$ and $d_L = x_{\max}$. For example, in the Gaussian case, we would have $d_0 = x_{\min} = -\infty$, and $d_L = x_{\max} = +\infty$.

We define the quantizer error $e_Q(x) \triangleq Q(x) - x = \hat{x} - x$, and measure its distortion as

$$d(x, \hat{x}) \triangleq |e_Q|^2 \quad \text{or more generally} \quad |e_Q|^p, \quad \text{for positive integers } p.$$

The average quantizer square-error distortion D is given as

$$D = E\big[d^2(x, \hat{x})\big]$$

$$= \int_{-\infty}^{+\infty} d^2(x, \hat{x}) f_x(x)\, dx$$

$$= \sum_{i=1}^{L} \int_{d_{i-1}}^{d_i} |x - r_i|^2 f_x(x)\, dx,$$

where f_x is the pdf of the input random variable x.

8.3.1 UNIFORM QUANTIZATION

For $1 \leqslant i \leqslant L$, set $d_i - d_{i-1} = \Delta$, called the uniform quantizer *step size*. To complete specification of the uniform quantizer, we set the representation value as

$$r_i \triangleq \frac{1}{2}(d_i + d_{i-1}), \tag{8.3-1}$$

at the center of the input bin $(d_{i-1}, d_i]$, for $1 \leqslant i \leqslant L$.

EXAMPLE 8.3-1 (*9-level uniform quantizer for* $|x| \leqslant 1$)
Let $x_{\min} = -1$ and $x_{\max} = +1$. Then we set $d_0 = -1$ and $d_L = +1$ and take $L = 9$. If the step size is Δ, then the decision levels are given as $d_i = -1 + i\Delta$, for $i = 0, 9$, so that $d_9 = -1 + 9\Delta$. Now, this last value must equal $+1$, so we get $\Delta = 2/9$ [more generally we would have $\Delta = (x_{\max} - x_{\min})/L$]. Then from (8.3-1), we get

$$r_i = \frac{1}{2}(d_i + d_{i-1})$$

$$= -1 + \left(i - \frac{1}{2}\right)\Delta$$

$$= d_i - \frac{1}{2}\Delta.$$

Note that there is an output at zero. It corresponds to $i = 5$, with $d_5 = +1/9$. This input bin, centered on zero, is commonly called the *deadzone* of the quantizer.

If a uniform quantizer is symmetrically positioned with respect to zero, then when the number of levels L is an odd number, we have a *midtread quantizer*, characterized by an input bin centered on 0, also an output value. Similarly for such a uniform symmetric quantizer, when L is even, we have a *midrise quantizer*, characterized by a decision level at 0, and no zero output value.

8.3.2 OPTIMAL MSE QUANTIZATION

The average MSE of the quantizer output can be expressed as a function of the L output values and the $L - 1$ decision levels. We have, in the real-valued case,

$$D = \int_{-\infty}^{+\infty} d^2(x, \hat{x}) f_x(x) \, dx$$

$$= \sum_{i=1}^{L} \int_{d_{i-1}}^{d_i} (x - r_i)^2 f_x(x) \, dx$$

$$\triangleq g(r_1, r_2, \dots, r_L; d_1, d_2, \dots, d_{L-1}).$$

We can optimize this function by taking the partial derivatives with respect to r_i to obtain

$$\frac{\partial D}{\partial r_i} = \int_{d_{i-1}}^{d_i} \frac{\partial}{\partial r_i} (x - r_i)^2 f_x(x) \, dx$$

$$= \int_{d_{i-1}}^{d_i} 2(x - r_i) f_x(x) \, dx.$$

Setting these equations to zero we get a necessary condition for representation level r_i:

$$r_i = \frac{\int_{d_{i-1}}^{d_i} x f_x(x) \, dx}{\int_{d_{i-1}}^{d_i} f_x(x) \, dx} = E[x | d_{i-1} < x \leqslant d_i], \quad \text{for } 1 \leqslant i \leqslant L.$$

This condition is very reasonable; we should set the output (representation) value for the bin $(d_{i-1}, d_i]$, equal to the conditional mean of the random variable x, given that it is in this bin. Taking the partial derivatives with respect to the d_i, we obtain

$$\frac{\partial D}{\partial d_i} = \frac{\partial}{\partial d_i} \left\{ \int_{d_{i-1}}^{d_i} (x - r_i)^2 f_x(x) \, dx + \int_{d_i}^{d_{i+1}} (x - r_{i+1})^2 f_x(x) \, dx \right\}$$

$$= (d_i - r_i)^2 p_x(d_i) - (d_i - r_{i+1})^2 p_x(d_i).$$

Setting this equation to zero and assuming that $f_x(d_i) \neq 0$, we obtain the relation

$$d_i = \frac{1}{2}(r_i + r_{i+1}), \quad \text{for } 1 \leqslant i \leqslant L - 1,$$

remembering that d_0 and d_L are fixed. This equation gives the necessary condition that the optimal SQ decision points must be at the arithmetic average of the two neighboring representation values. This is somewhat less obvious than the first necessary condition, but can be justified as simply picking the output value nearest to input x, which is certainly necessary for optimality.

AN SQ DESIGN ALGORITHM

The following algorithm makes use of these necessary equations to iteratively arrive at the optimal MSE quantizer. It is experimentally found to converge. We assume the pdf has infinite support here.

1   Given L, and the probability density function $f_x(x)$, we set $d_0 = -\infty$, $d_L = +\infty$, and set index $i = 1$. We make a guess for r_1.

2   Use $r_i = \int_{d_{i-1}}^{d_i} x f_x(x)\, dx / \int_{d_{i-1}}^{d_i} f_x(x)\, dx$ to find d_i by integrating forward from d_{i-1} till a match is obtained.

3   Use $r_{i+1} = 2d_i - r_i$ to find r_{i+1}.

4   Set $i \leftarrow i + 1$ and go back to step 2, unless $i = L$.

5   At $i = L$, check

$$r_L \lessgtr \frac{\int_{d_{L-1}}^{\infty} x f_x(x)\, dx}{\int_{d_{L-1}}^{\infty} f_x(x)\, dx}.$$

If "$r_L >$," then reduce initial value r_1. Otherwise, increase r_1. Then return to step 2. Continue this till convergence.

The amount of increase/decrease in r_1 has to be empirically determined, but this relatively straightforward algorithm works well to find approximate values for the decision and representation levels. Note that numerical integration may be needed in step 2, depending on the pdf f_x. An alternative algorithm is the scalar version of the Linde, Buzo, and Gray (LBG) algorithm for vector quantization, to be shown next.

8.3.3 VECTOR QUANTIZATION

Information theory says that quantizing signal samples one at a time, i.e., scalar quantization, can never be optimal [3,6]. Certainly this is clear if the successive samples are correlated or otherwise dependent, and, in fact, that is the main reason for the transformation in our generic source compression system. But even in the case where the successive samples are independent, scalar quantization is still theoretically not the best way, although the expected gain would certainly be much less in this case. *Vector quantization* (VQ) addresses the nonindependent data problem by quantizing a group of data samples simultaneously [7]. As such, it is much more complicated than scalar quantization, with the amount of computation increasing exponentially in the vector size, with practical VQ sizes currently limited to 4×4 and below, although some experimental work has been done for 8×8 blocksize [28]. Another difference is that VQ is typically designed from a training set, while SQ is typically designed from a probability model. This difference gives a significant advantage to VQ in that it can use the "real" multidimensional pdf of the data, even in the absence of any viable theoretical multidimensional pdf model, other than joint Gaussian. As such, VQ coding results can be much better than those of SQ. In this regard, it should be noted that via its internal multidimensional pdf, the VQ effectively removes all dependencies in the data vector, and not just the so-called linear dependencies dealt with via a linear transformation. The down side, of course, is that 4×4 is a very small block. As a consequence, VQ is not often used alone without a transformation, but the two together can be quite powerful [8]. Also the overhead of transmitting the VQ codebook itself must be counted, and this can be very significant.

We later present a data-based design algorithm for VQ generally attributed to Linde, Buzo, and Gray (LBG), but we start out assuming the theoretical multi-dimensional pdf is known. We consider an $N \times N$ random vector $\mathbf{x}$ with joint pdf $f_{\mathbf{x}}(\mathbf{x})$ with support $\mathcal{X}$. Then we have a set of regions $\mathcal{C}_l$ in $\mathcal{X}$ that decompose this vector space in a mutually exclusive and collectively exhaustive way, i.e.,

$$\mathcal{X} = \bigcup_{l=1}^{L} \mathcal{C}_l \quad \text{and} \quad \mathcal{C}_l \cap \mathcal{C}_m = \varphi \quad \text{for all } l \neq m,$$

where ϕ is the null set.

On defining L *representation vectors* $\{\mathbf{r}_1, \mathbf{r}_2, \ldots, \mathbf{r}_L\}$, the vector quantizer then works as follows:

$$\hat{\mathbf{x}} = Q(\mathbf{x}) \triangleq \begin{cases} \mathbf{r}_1, & \mathbf{x} \in \mathcal{C}_1 \\ \vdots & \vdots \\ \mathbf{r}_L, & \mathbf{x} \in \mathcal{C}_L. \end{cases} \tag{8.3-2}$$

So, in VQ, the input-space *decision regions* $\mathcal{C}_l$ play the role of the input decision intervals (bins) in SQ. Unfortunately, these input regions are much harder to define than simple bins, and this makes the actual quantizing operation $Q(\mathbf{x})$ computationally difficult. In order to perform the mapping, (8.3-2), we have to find which region $\mathcal{C}_l$ contains $\mathbf{x}$.

As was the case for scalar quantization, we can write the VQ error $\mathbf{e}_Q \triangleq \hat{\mathbf{x}} - \mathbf{x}$, and express the total MSE as

$$D = E\left[(\hat{\mathbf{x}} - \mathbf{x})^T (\hat{\mathbf{x}} - \mathbf{x})\right]$$

$$= \int_{-\infty}^{+\infty} d^2(\mathbf{x}, \hat{\mathbf{x}}) f_{\mathbf{x}}(\mathbf{x}) \, d\mathbf{x}$$

$$= \sum_{l=1}^{L} \int_{\mathcal{C}_l} d^2(\mathbf{x}, \mathbf{r}_l) f_{\mathbf{x}}(\mathbf{x}) \, d\mathbf{x}.$$

One can show the following two necessary conditions for optimality:

- *Optimality Condition 1*: (useful for design of input regions $\mathcal{C}_l$ and for actually quantizing the data $\mathbf{x}$)

$$Q(\mathbf{x}) = \mathbf{r}_l \quad \text{iff} \quad d(\mathbf{x}, \mathbf{r}_l) \leqslant d(\mathbf{x}, \mathbf{r}_m) \quad \text{for all} \quad l \neq m.$$

- *Optimality Condition 2*: (useful for determining the representation vectors $\mathbf{r}_l$)

$$\mathbf{r}_l = E[\mathbf{x} \mid \mathbf{x} \in \mathcal{C}_l]. \tag{8.3-3}$$

The first optimality condition states that, for a given set of representation vectors $\{\mathbf{r}_1, \mathbf{r}_2, \ldots, \mathbf{r}_L\}$, the corresponding decision regions should decompose the input space in such a way as to cluster each input vector $\mathbf{x}$ to its nearest representation vector in the sense of Euclidean distance $d(\mathbf{x}, \mathbf{r}_l)$. The second optimality criteria says that if the input decision regions $\mathcal{C}_l$ are given, then the best choice for their representation vector is their conditional mean.

From the first optimality condition, we can write

$$\mathcal{C}_l = \big\{ \mathbf{x} \mid d(\mathbf{x}, \mathbf{r}_l) \leqslant d(\mathbf{x}, \mathbf{r}_m) \text{ for all } l \neq m \big\},$$

and such disjoint sets are called *Voronoi regions* [7]. So the large part of designing a VQ is to carve up the input space into these Voronoi regions, each determined by its property of being closer to its own representation vector $\mathbf{r}_l$ than to any other.

Now this is all well and good, but the fact remains that we have no suitable theoretical joint pdfs $f_x(\mathbf{x})$, other than multidimensional Gaussian, with which to calculate the conditional means in (8.3-3). However, if we are given access to a large set of input vectors $\{\mathbf{x}_1, \mathbf{x}_2, \ldots, \mathbf{x}_M\}$, all drawn from the same distribution, called the *training set*, the following data-based algorithm has been found to converge to a local minimum of the total square error over this training set [7], i.e., to provide a least-squares solution, with error metric

$$\sum_{i=1}^{M} d^2\big(\mathbf{x}_i, \mathbf{r}_l(\mathbf{x}_i)\big), \tag{8.3-4}$$

where $\mathbf{r}_l(\mathbf{x}_i)$ is just the VQ output for training vector $\mathbf{x}_i$.

8.3.4 LBG ALGORITHM [7]

1   Initially guess the L output representation vectors $\mathbf{r}_1, \mathbf{r}_2, \ldots, \mathbf{r}_L$.

2   Use optimality condition 1 to quantize each input vector in the training set, i.e., find the vector $\mathbf{r}_l$ that minimizes the distance $d(\mathbf{x}_i, \mathbf{r}_l)$. When this step finishes, we have partitioned the training set into the initial decision regions $\mathcal{C}_l$.

3 Use optimality condition 2 to update the representation vectors as

$$r_l = \frac{1}{M_i} \sum_{x_j \in C_l} x_j,$$

where $M_i \triangleq |\{x \mid x_j \in C_l\}|$, denoting the number of training vectors in C_l.

4 Go back to 2 and iterate till convergence.

A bit of consideration of the matter reveals that steps 2 and 3 can only improve the optimality, i.e., reduce the error (8.3-4), and never cause an increase in error. Hence the LBG algorithm converges to at least a local minimum of the total least-squares error. After design, the resulting VQ can be used on a separate set of input data called the *test set*, that is assumed to have the same or similar joint distribution.

EXAMPLE 8.3-2 (*vector quantization*)
Consider a simple example with $M = 5$ training vectors,

$$x_1 = \begin{bmatrix} 1 \\ 1 \end{bmatrix}, \qquad x_2 = \begin{bmatrix} 0 \\ 0 \end{bmatrix}, \qquad x_3 = \begin{bmatrix} 2 \\ 0 \end{bmatrix}, \qquad x_4 = \begin{bmatrix} -1 \\ 1 \end{bmatrix}, \qquad x_5 = \begin{bmatrix} 0 \\ -1 \end{bmatrix},$$

to be vector quantized to $L = 3$ output vectors, r_1, r_2, r_3, using the LBG algorithm. We start with the initial guess,

$$r_1 = \begin{bmatrix} 0 \\ 0 \end{bmatrix}, \qquad r_2 = \begin{bmatrix} 1 \\ 0 \end{bmatrix}, \qquad r_3 = \begin{bmatrix} -1 \\ 0 \end{bmatrix},$$

and proceed into iteration one of the LBG algorithm. At step two, we quantize (classify) the training data as follows:

$$x_1 \rightarrow r_2, \qquad x_2 \rightarrow r_1, \qquad x_3 \rightarrow r_2, \qquad x_4 \rightarrow r_3, \quad \text{and} \quad x_5 \rightarrow r_1.$$

Then at step three, we update the three representation vectors as

$$r_1 = \frac{1}{2}(x_2 + x_5) = \begin{bmatrix} 0 \\ -0.5 \end{bmatrix}, \qquad r_2 = \frac{1}{2}(x_1 + x_3) = \begin{bmatrix} 1.5 \\ 0.5 \end{bmatrix}$$

and

$$r_3 = x_4 = \begin{bmatrix} -1 \\ 1 \end{bmatrix}.$$

We then proceed into iteration two of the LBG algorithm. At step two, we get the mapping

$$x_1 \rightarrow r_2, \qquad x_2 \rightarrow r_1, \qquad x_3 \rightarrow r_2, \qquad x_4 \rightarrow r_3, \quad \text{and} \quad x_5 \rightarrow r_1,$$

which is unchanged from step two last time, so we are converged, and can omit another step three.

Vector quantization is not used in today's image coding standards, but this powerful method is asymptotically theoretically optimal and may re-emerge from the background with expected technology improvements in the future. Other than complexity, the most serious problem is the need to transmit the vector codebook, which can be quite large.

8.4 ENTROPY CODING

After the quantization, we must send the message set of quantizer outputs to the channel or channel coder. Without loss of generality, we can think in terms of binary digits, or bits, for this information.[3] In general, we can use a fixed-length or variable-length code for this purpose. Now, the mathematical theory of communication defines *information* [3,6] quantitatively for an independent, discrete-valued, stationary random source as

$$I(x_i) \triangleq \log_2 \frac{1}{p(x_i)},$$

and its average value *entropy* as

$$H = \sum_i p(x_i) \log_2 \frac{1}{p(x_i)}. \tag{8.4-1}$$

It is proved in Gallagher [6] and Cover and Thomas [3] that any invertable coding of such a source into binary digits must have average codeword length greater than or equal to this entropy. The so-called *source coding theorem* states that, for the encoding of one message at a time,

$$H \leqslant \bar{l} < H + 1, \tag{8.4-2}$$

where $\bar{l}$ denotes the average codeword length $\bar{l} \triangleq \sum_i p(x_i) l(x_i)$. If we were to go to the expense of jointly coding M messages at a time, theory says that this bound becomes

$$H \leqslant \bar{l} < H + 1/M,$$

---

3. This is the so-called quantizer *index set*. We assume that the receiver knows the quantizer design and can map the set of quantizer indices back into the actual representation values. Such a mapping is often given the misnomer *inverse quantization*, and in practice would be communicated to the receiver as the default values for an international standard or alternatively as header information in a packet or file.

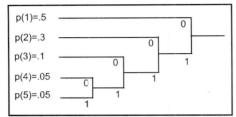

FIGURE 8.8 *Example of Huffman coding tree*

due to the assumed independence of the random source. Unfortunately, such coding of the so-called *extension source* is not often practical. In both of these equations, H is the entropy per single message.

8.4.1 HUFFMAN CODING

In the 1950s, David Huffman came up with an optimal variable-length encoding procedure, which came to be known as *Huffman coding* [3.6]. It is a method for coding a finite number of messages into variable-length binary codewords in such a way that the average codeword length is minimized. Huffman coding creates a binary tree by first joining together the two messages with the lowest probabilities, thereby creating a reduced source with one fewer messages. It then proceeds recursively to join together the two messages with lowest probability for this reduced source, proceeding down to a root node whose probability is 1. When the branches of this tree are populated with 1s and 0s, a variable-length code can be read off in the reverse order. This is perhaps best illustrated by example.

EXAMPLE 8.4-1 (*Huffman coding*)
Let the source have $M = 5$ messages, with probabilities $p_1 = 0.5$, $p_2 = 0.3$, $p_3 = 0.1$, $p_4 = 0.05$, $p_5 = 0.05$. Then we can construct the Huffman code as shown in Figure 8.8. Here, starting from the left, messages 4 and 5 have the lowest probabilities, and so, are combined first. Their total probability is then 0.1. It happens that this node is also a smallest probability in the reduced source, and it is joined with message 3 with $p(3) = 0.1$ at the second level. We continue on to complete the tree. The variable-length codewords are now read from right to left and become, reading from message 1 to message 5: $0, 10, 110, 1110$, and 1111, with lengths $l_i = 1, 2, 3, 4$, and 5, respectively. We note that this variable-length code is uniquely decodable, because no code word is the *prefix* of another codeword. It can be seen that the Huffman code tree guarantees that this will always be true. Computing the average

codeword length, we get

$$\bar{l} \triangleq \sum_i p(x_i)l(x_i)$$

$$= 0.5 \times 1 + 0.3 \times 2 + 0.1 \times 3 + 0.05 \times 4 + 0.05 \times 4$$

$$= 1.8 \text{ bits/message.}$$

In this case the entropy from (8.4-1) is $H = 1.786$. We note that the average codeword length for this example is quite close to the entropy of the source. This is not always the case, especially when the entropy per message is much less than one, as forewarned in (8.4-2).

In passing, we note that the Huffman variable-length tree is complete, i.e., all branches terminate in valid messages. Hence the resulting variable-length code will be sensitive to channel errors.

8.4.2 ARITHMETIC CODING

The Huffman coding procedure codes a fixed number of messages at a time, usually one, into a variable-length binary string. Another popular coding method, in contrast, codes a variable number of messages into a fixed-length binary string. This method, called *arithmetic coding* (AC), can usually achieve a very close match to the entropy of the source, even when the entropy is small. Arithmetic coding works with subintervals of [0, 1], with lengths equal to the probabilities of the messages to be coded.

With reference to Figure 8.9, as messages are processed, the subinterval is successively split according to the message probabilities. First, in this example, we "send" message 2 by centering attention on the middle interval of length p_2. Then we send message 1 by centering attention on the subinterval p_1. We then send the following messages 2 and 1 by the same approach. This continues until a fixed, very small subinterval size is reached, or is about to be exceeded. If we convey the final subinterval to the receiver, then the decoder can decode the string of messages that lead up to this final subinterval of [0, 1]. We note in passing that there is nothing to prevent the message probabilities from changing in some prescribed manner, as the successive messages are encoded into subintervals. The final subinterval is actually conveyed to the receiver by sending a fixed-length binary string of sufficient number of digits to point uniquely to the prechosen fixed but small subinterval. This is indicated by the arrow pointer in Figure 8.9. A typical pointer value is 14 bits or so. Note that decoding can proceed while the data is transmitted in a first-in/first-out fashion, because as the subinterval shrinks down in size, the leading digits of the fixed-length binary string are successively determined. A lot of practical details are left out of the simple argument here, including the

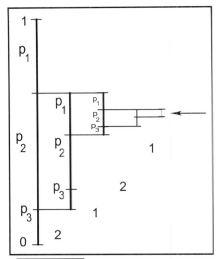

FIGURE 8.9 Illustration of an essential aspect of arithmetic coding

important question of numerical significance for a fixed-length computer word implementation; however, the basic idea is as shown.

The question remains: how efficient is this method? To provide a crude answer to this question, we note that the final subinterval size will be the product of the message probabilities that are encoded, $\prod_i p(x_i)$, and that the approximate number of binary digits to uniquely point to such an interval is

$$L = -\log_2(2^{-L}) \simeq -\log_2\left[\prod_i p(x_i)\right] = \sum_i -\log_2\left[p(x_i)\right],$$

where the term on the right is the sample total, the average value of which is the entropy of this independent string of messages from the source. By the Law of Large Numbers of probability theory, for an ergodic source, we can expect convergence of the average rate to the entropy (8.4-1). So we can expect to get very close to the entropy of a stationary ergodic source in this way. A more thorough derivation of arithmetic coding, including algorithms, can be found in Sayood [22] and Gersho and Gray [7].

8.4.3 ECSQ AND ECVQ

Sometimes it is useful to combine quantization and entropy coding into one operation. *Entropy-coded scalar quantization* (ECSQ) and *entropy-coded vector quantization* (ECVQ) accomplish just that. For a unstructured VQ(SQ) we can start out with the LBG algorithm (Section 8.3.4) and augment the error criteria with the Lagrange error

$$E = D + \lambda H(R)$$

$$= \sum_{l=1}^{L} \int_{C_l} d^2(\mathbf{x}, \mathbf{r}_l) f_x(\mathbf{x}) \, d\mathbf{x} + \lambda H(R),$$

where $H(R)$ is the resulting entropy of the quantized source X with representation vectors $\mathbf{r}_l$, and λ is a Lagrange multiplier. The parameter λ then becomes a variable controlling the rate $H(R)$ versus the distortion D. Here the entropy is computed as

$$H(R) = \sum_{l=1}^{L} p_l \log_2 \frac{1}{p_l},$$

with p_l being the pmf of the $\mathbf{r}_l$. To get the complete D-R function of the resulting ECVQ (ECSQ) system, we merely run this modified LBG algorithm for a range of λ values, and then plot the resulting points. These points can then be connected together with straight lines to get the final estimate of the joint quantizer–entropy model. Since conditional arithmetic coding has been found to be extremely efficient for image coding applications, one can effectively take the resulting D-R characteristic and regard it as virtually the joint quantizer–AC model also. This is called an operational distortion-rate function.

Taking a structured scalar quantizer, such as a uniform threshold quantizer (UTQ) with step-size Δ and deadzone 2Δ, and number of levels L, we can plot the D-R characteristics of an ECSQ directly as a function of parameter Δ, typically getting plots such as shown in Figure 8.10, where $L_1 < L_2 < L_3$. We see that large

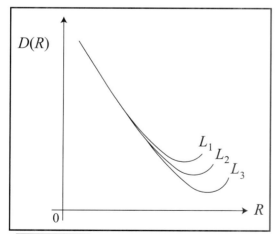

FIGURE 8.10   *Illustration of ECSQ joint quantizer and entropy encoder*

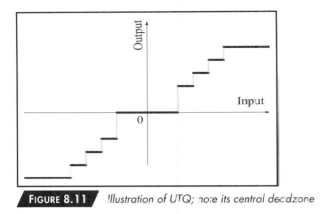

FIGURE 8.11 *Illustration of UTQ; note its central deadzone*

values of L do not penalize results below a certain maximum rate, above which point quantizer saturation begins to occur. This is mainly a consequence of the fact that messages with very small probability do not add much to the entropy of a source, via $\varepsilon \log_2 \varepsilon \searrow 0$ as $\varepsilon \searrow 0$. By making sure to operate at lower rates, then quantizer saturation can be pushed out of the range by using a large number of levels, without hurting the lower rate performance. (More on ECVQ and ECSQ is contained in Kim and Modestino [13].)

ERROR SENSITIVITY

Both Huffman and arithmetic coding are examples of *variable-length coding* (VLC). In any coding system that uses VLC, there is an increased error sensitivity, because a bit error will probably cause loss of synchronization, with resultant incorrect decoding, until a fortuitous chance event occurs later in the bitstream, i.e., until by chance a correct decoding occurs. This is not the case in fixed-length coding. As a result, the extra compression efficiency that can be obtained using VLC leads to increased sensitivity to errors. In practice, synch words and limitations on the length of VLC bitstreams can effectively deal with this increased sensitivity problem, but only with added complexity. A VLC coder that does not have such error resilience features is not really practical, as a single bit error anywhere in the image would most likely terminate correct decoding for the rest of the image. So, all but the most pristine channels or storage media will need some kind of error resilience features. We talk more about this in Chapter 12 for the image sequence or video case.

8.5 DCT CODER

DCT coders, the most representative of these being the international standard JPEG, use the tools presented above. The input image frame is first split into its Y, U, and V components. The Y (or luma) component is then subjected to

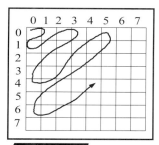

FIGURE 8.12 *Illustration of zig-zag or serpentine scan of AC coefficients of 8×8 DCT*

block-DCT with nonoverlapping 8×8 blocks. After this the DCT coefficients are split into one DC and 63 AC coefficients. They are then quantized by scalar uniform quantizers, with possibly a central deadzone, with step-size determined by a *quantization matrix* that gives a preassigned step-size to each of the 64 DCT coefficients. A quantizer with a central deadzone is called a uniform threshold quantizer (UTQ), as seen in Figure 8.11. It can be specified by its number of levels L, step-size Δ, and the width of its deadzone.[4] After quantization the DC coefficient commonly undergoes lossless DPCM coding across the blocks, while the AC coefficients are scanned in a certain zig-zag or serpentine scan (seen in Figure 8.12) and input to a *2-D run-value* or *2-D Huffman code*. This code assigns codewords to pairs of messages, each indicating a run of *xx* zeros terminated by a nonzero DCT value *yy*. This 2-D VLC is stored in a table in the coder and decoder.

There remains the problem of bit assignment to the DCT coefficients, which can be done using a quantizer model as follows. We first assume a quantizer model for DCT coefficient with index k_1, k_2, say,

$$D_{k_1,k_2} = g\sigma_{k_1,k_2}^2 2^{-2R_{k_1,k_2}}, \qquad (8.5\text{-}1)$$

where σ_{k_1,k_2}^2 is the coefficient variance (we assume zero mean) and R_{k_1,k_2} is the bitrate assigned to this coefficient. Due to orthogonality of the unitary DCT, the total MSE in the reconstruction will be

$$D = \sum_{k_1,k_2} D_{k_1,k_2},$$

and the average bitrate will be

$$R = \frac{1}{N^2} \sum_{k_1,k_2} R_{k_1,k_2},$$

4. The central deadzone is often useful for discriminating against noise in the coefficients to be quantized. Also, a central deadzone of 2Δ arises naturally in so-called bit plane coding in embedded scalable coding (see Section 8.6).

FIGURE 8.13   *Original 252 × 256 Cameraman image*

for an $N \times N$ DCT. To minimize the MSE, we introduce the Lagrangian parameter λ, and seek to minimize the function

$$D + \lambda R = \sum_{k_1, k_2} D_{k_1, k_2} + \lambda R_{k_1, k_2}.$$

Taking the partial derivatives with respect to each of R_{k_1, k_2} and setting them to zero, we can solve for the parameter λ and eventually obtain the following solution:

$$R_{k_1, k_2} = R + \frac{1}{2} \log_2 \left(\sigma^2_{k_1, k_2} / \sigma^2_{wgm} \right), \tag{8.5-2}$$

with

$$\sigma^2_{wgm} \triangleq \left(\prod \sigma^2_{k_1, k_2} \right)^{1/N^2}.$$

The resulting MSE then becomes simply

$$D = g \sigma^2_{wgm} 2^{-2R}. \tag{8.5-3}$$

It is surprising that the result depends only on the average bitrate R and the geometric mean of the subband variances σ^2_{wgm}.

EXAMPLE 8.5-1 (*block-DCT coding example*)
The 252×256 monochrome *Cameraman* image was coded by the DCT coder in *VcDemo* [14] at around 1 bit per pixel (bpp). An 8×8 block-DCT was used, with lossless DPCM coding of the quantized DC coefficient from block to block, and PCM coding of the AC coefficients. Variable-length coding was then applied to the AC coefficient quantized indices. The original image is

FIGURE 8.14 *DCT-coded image at 0.90 bpp*

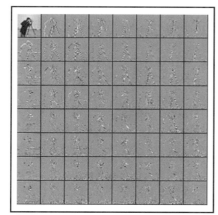

FIGURE 8.15 *DCT coefficients of Cameraman image*

shown in Figure 8.13 and the coded image at actual bit rate 0.90 bpp is shown in Figure 8.14.

The original block-DCT coefficients are shown in Figure 8.15 and the quantized coefficients are shown in Figure 8.16. We can see that about half of the DCT coefficients have been set to zero, yet the coded image still looks quite good. A close-up of Figure 8.14, though, reveals coding distortion, especially around edges in the image. The PSNR is 29.4 dB.

8.6 SWT CODER

Due to their advantages of no blocking artifacts and somewhat better compression efficiency, subband/wavelet coders have been extensively studied in recent

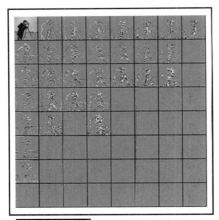

FIGURE 8.16 *Quantized DCT coefficients of Cameraman*

years. Here we consider the popular dyadic or recursive subband decomposition, also called the wavelet decomposition, which was mentioned previously. This transformation is followed by scalar quantization of each coefficient. Rate assignment then splits up the total bits for the image into bit assignments for each subband. Assuming an orthogonal analysis/synthesis system is used, then the total mean-square error distortion D due to the individual quantizations with distortion $D_m(R_m)$, with R_m bits assigned to the mth subband quantizer, can be written as the sum of the mean-square distortions over a total of M subbands as

$$D = \sum_{m=1}^{M} \frac{N_m}{N} D_m(R_m), \qquad (8.6\text{-}1)$$

where N_m is the number of samples in the mth subband, and the total number of pixels (samples) is N. Note that the weighted average in (8.6-1) over the M subbands is equal to an unweighted average over the N subband samples. Then modeling the individual quantizer as $D_m = g\sigma_m^2 2^{-2R_m}$, all with the same quantizer efficiency factor g, the optimal bit assignment for an average bitrate of R bits/pixel can be found with some assumptions, via the Lagrange multiplier method [25], to yield

$$R_m = R + \frac{1}{2} \log_2 \left(\frac{\sigma_m^2}{\sigma_{\text{wgm}}^2} \right), \quad m = 1, \ldots, M, \qquad (8.6\text{-}2)$$

where the weighted geometric mean σ_{wgm}^2 is defined as

$$\sigma_{\text{wgm}}^2 \triangleq \prod_{m=1}^{M} \left(\sigma_m^2 \right)^{N_m/N},$$

Table 8.1. Bit assignments

4.64	3.41	2.74	1.81
1.81	0.00	1.50	0.00
0.00	0.00	0.00	0.00
0.00	0.00	0.00	0.00

FIGURE 8.17 *Original* 512 × 125 *Lena (8 bits)*

where N is the total number of samples, i.e., $N = \sum_{m=1}^{M} N_m$. The resulting MSE in the reconstruction then becomes

$$D = g\sigma_{\text{wgm}}^2 2^{-2R},$$

which is analogous to (8.5-3) of DCT coding.

If we were to compare a DCT coder to an SWT coder, using this bit assignment method, all that would be necessary is to compare the corresponding weighted geometric means of the variances. Note that there are a number of assumptions here. First, we assume the quantizer model of (8.5-1), which is strictly only valid at high bitrates. Second, rate-distortion theory says this type of coding is only optimum for jointly Gaussian distributions. In practice, the quantizer model of (8.5-1) is useful only as a first approximation. More exact and complicated quantizer models have been developed for specific applications.

EXAMPLE 8.6-1 (*subband coding*)
The original 512 × 512 *Lena* image of Figure 8.17 was coded by the SBC coding module of the freely available software VcDemo [14] at 1 bpp. Two levels of subband decomposition were obtained using 16-tap separable fil-

FIGURE 8.18 *512 × 512 Lena coded by SBC at 1 bpp*

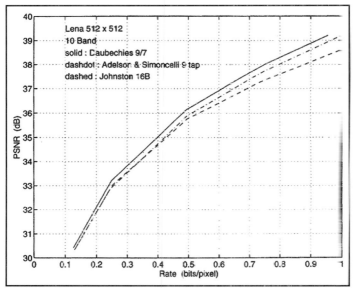

FIGURE 8.19 *Comparison of three subband/wavelet filters on Lena image*

ters, yielding 16 subbands. The DC subband was quantized by DPCM using a 1×1-order NSHP linear predictor, while the remaining AC subbands were coded by uniform quantization. Variable-length coding was then applied to the quantizer outputs. Bit assignment was done according to the above pro-

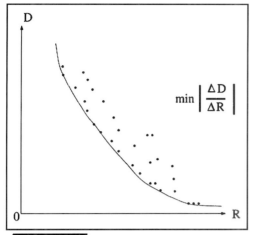

FIGURE 8.20   *Illustration of $D(R)$ in case of a discrete number of quantizer step-sizes*

cedure and resulted in the bit assignments shown in Table 8.1, resulting in an overall average bitrate of 0.99 bpp. The result is shown in Figure 8.18, where the actual bits used is 0.95 bpp and the PSNR is reported to be 37.4 dB equivalent to a standard deviation error of 3.4 on this 8-bit image with range [0, 255].

EXAMPLE 8.6-2 (*SWT coder performance*)
An example SWT image coder is reported in Choi [2], where a filter study is undertaken to evaluate the performance of various subband/wavelet filters for compression of the 512×512 *Lena* image. He used the subband finite-state scalar quantizer (SB-FSSQ)-embedded coder [17] and compared three filters using a 10-band dyadic subband decomposition. The filters considered were Daubechies 9/7, Johnston's 16B, and Adelson and Simoncelli's 9 tap. The results in Figure 8.19 show quite close performance with the top-performing PSNR curve corresponding to the Daubechies 9/7 filter. Interestingly enough, the 16B filter came out on top when we coded all 16 subbands rather than taking a dyadic or wavelet decomposition. One additional comment is that the Daubechies 9/7 has a smaller amount of ringing distortion when the LL subband is displayed alone, as happens in scalable coding. In fact, a key property of subband/wavelet coders is their inherent scalability in resolution to match display and communications-link bandwidth requirements.

A more precise approach to bit assignment may be done as follows. First we write the total rate as

$$R = \sum_{i=1}^{N} R_i,$$

where we assume N channels (or coefficients). We now wish to find the channel rate assignment $\mathbf{R} = (R_1, \ldots, R_N)^T$, such that the total distortion is minimized. Effectively, we seek the

$$\mathbf{R} \triangleq \arg\min_{\mathbf{R}} \sum_{i=1}^{N} D_i(R_i).$$

This problem of minimization with constraint can be formulated with a Lagrange multiplier as follows: $D(\mathbf{R}) \triangleq \sum_{i=1}^{N} D_i(R_i)$, and then

$$f(\mathbf{R}) = D(\mathbf{R}) + \lambda \left[R - \sum_{i=1}^{N} R_i \right].$$

Taking partial derivatives with respect to each R_i, we obtain the so-called constant slope conditions

$$\frac{\partial f}{\partial R_k} = \frac{\partial D_i}{\partial R_i} - \lambda = 0,$$

meaning that at optimality we should operate at points on the channel distortion-rate curves that all have the same slope, i.e.,

$$\frac{\partial D_i}{\partial R_i} = \lambda \qquad (8.6\text{-}3)$$

known as the *equal slope condition*.

As we vary the value λ we then can sweep out the entire operational distortion-rate curve $D(R)$.

EXAMPLE 8.6-3 *(discrete quantizer set)*
Often in practice we have only a discrete number of bit assignment points because a limited number of fixed quantizers can be used. Then there are no continuous distortion-rate curves $D_i(R_i)$ for the individual channels. In that case, if we try all the combinations of the various quantizers, we would get a large but finite set of data as sketched in Figure 8.20. We notice that the best points are to the left and downward, because for a given rate, these values have the lowest total distortion. This is called the *convex hull* of this set of data points. At high total rates, we just use the smallest step-size available among all the quantizers. Likewise, the optimal lowest rate would be obtained by using the largest step-size for each quantizer, sometimes even leaving that channel(s) out altogether, i.e., assigning it zero bits. A method to solve for the intervening points based on tree optimization is the Breiman, Friedman, Olshen, and Stone (BFOS) algorithm [20], and this is discussed in a problem at the end of this chapter.

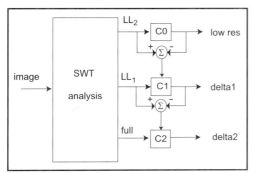

FIGURE 8.21   *Multiresolution image coder with three decoded image resolutions (full, 1/2, and 1/4)*

8.6.1 MULTIRESOLUTION SWT CODING

Figure 8.4 shows a dyadic SWT with three recursive applications of the basic separable 4-band subband/wavelet transform. We thus generate three lower resolution images. We call the first generated LL subband LL_1 and the second generated LL subband LL_2 and so on. For completeness we call the original image LL_0. Here we consider only two levels of decomposition giving a lowest resolution of LL_2, medium resolution of LL_1, and full resolution LL_0. A scalable coder, embedded in resolution, can then be constructed by first coding the lowest resolution data LL_2, and then an enhancement band for the subbands included in LL_1, and finally a further enhancement band consisting of the remaining subbands LH, HL, and HH, making up LL_0. If we simply transmitted these three scales, LL_i, $i = 0, 3$, we would have what is called *simulcast*, an inherently inefficient method of transmission. To make this scheme more efficient, a residual or error from the lower scale can be passed on to the next higher scale and coded. A diagram illustrating this residual coding approach is shown in Figure 8.21.

Following partition based on scale, the low-resolution coder C0 is a conventional SWT image coder. The medium-resolution coder C1 gets all the data up to the LL_2 or midspatial resolution. However, the coded error from C0 is also fed to the intermediate coder C1, which internally codes this error in place of the low-resolution subbands in LL_2, thereby providing a refinement of this data as well as coding with the new midfrequency data. This process is repeated again for coder C2. Decoding proceeds as follows. To decode low-resolution data, there is no change from conventional coding. To decode medium-resolution data, the delta1 enhancement layer is decoded and used to supplement the low-resolution layer. A similar method then uses bitstream delta2 to achieve the full spatial resolution.

One might ask why the coder error has to be propagated across the scales in a scalable coder; in other words, "Why not just code the high-frequency subbands ($-LH$, $-HL$, and $-HH$) at the new scale?" The reason is that the rate

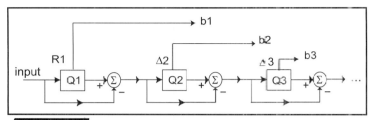

FIGURE 8.22 *Cascade of three quantizers provides a quality scalable coding of input signal*

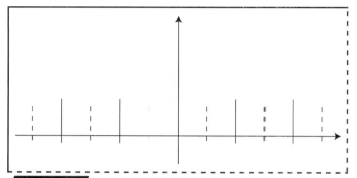

FIGURE 8.23 *Illustration of scalar quantizer embedding. Solid lines are coarse quantizer decision levels. Dashed and solid lines together constitute decision levels of the combined embedded quantizer*

assignment (8.6-2) will give different numbers of bits to the LL_2 subband depending on whether it is part of the low-, the medium-, or the high-resolution image. This assigned number of bits will generally increase with resolution, if the desired PSNR is around the same number or is increasing with resolution, which is often the case. So recoding the coding error of the bands already transmitted effectively increases the bit assignment for that band. Focusing on the LL_2 subband, for instance, it is coded once, then recoded for the medium-resolution image, and then recoded a second time for the high- or full-resolution image. For a given subband component of the LL_2 image, we can see a concatenation of quantizers, as shown in Figure 8.22. The input signal is quantized at rate R1 to produce bitstream b1. The rate $\Delta 2$ is then used to refine the representation, with corresponding enhancement bitstream b2. The process is repeated again with enhancement rate $\Delta 3$ and corresponding bitstream b3. A decoder receiving only b1 can decode a coarse version of the input. With the addition of b2, a decoder can refine this coarse representation, with information about the initial quantization error. Reception of the second enhancement layer can then refine the representation even more. This is then a rate-scalable way to transmit a pixel source, giving three possible bitrates.

Such quantizers are said to be embedded, because the successively higher accuracy quantizers defined in this way have the property that each stage refines the decision levels of the previous stage. This is illustrated in Figure 8.23, which shows the overall quantizer bins for two stages of embedding. Here the solid vertical lines are decision levels for the coarse quantizer Q1 and the dashed-line decision levels are provided by Q2, yielding an equivalent quantizer consisting of all the decision levels.

In an embedded quantizer, the first stage can be a mean-square error optimal quantizer, but the second and succeeding stages are restricted to using these same decision levels plus some more, which would not normally be optimal. A key paper by Equitz and Cover [5] considered the optimality of such cascaded or refined quantizers and concluded that the source must have a certain Markov property holding across the refinements for optimality. In particular, successive refinement bits from the source must be conditionally dependent only on the next higher significance bit. In general, a refinement bit would be dependent on all the prior more significant bits. Still, the greatest dependency would be expected to occur on the next higher significance bit. If all of the quantizers are uniform, and the step-sizes are formed as decreasing powers of two, i.e., 2^{-k}, then the outputs b_k become just the successively smaller bits from a binary representation of the input value. In this case all of the quantizers are uniform threshold and are fully embedded.

8.6.2 NONDYADIC SWT DECOMPOSITIONS

As mentioned earlier, the dyadic or wavelet decomposition does not always give the best compression performance; so-called *wavelet packets* optimize over the subband tree, splitting or not, to get the best coding quality for that frame [19]. When the splitting is allowed to change within an image (frame), they formulated a so-called *double tree algorithm* to find a spatially adaptive decomposition. Note, however, that this is not an easy problem, since the best subband/wavelet decomposition for a given image class will depend on the quantizing method used as well on the entropy coder.

8.6.3 FULLY EMBEDDED SWT CODERS

This class of coders follows the SWT by a variable-length coder that first codes the most significant subband samples/coefficients at their most significant bit (MSB) plane and then proceeds down in significance, bit plane by bit plane, both noting newly *significant* coefficients and also refining those coefficients already coded at higher MSB planes. The generally acknowledged first of these was the *embedded zero-tree wavelet* (EZW) by Shapiro [23]. The EZW was followed by the set partitioning in hierarchical trees (SPIHT) coder of Said and Pearlman [21]. The coding standard JPEG 2000, which can be considered as a kind of fruition of this type of coder, is covered thoroughly in the text by Taubman and Marcellin [26]. The

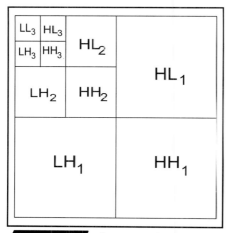

FIGURE 8.24   *Illustration of subband/wavelet structure for the dyadic (aka octave band and wavelet) decomposition, where the subbands XY_k are at the kth stage*

embedded zero block coder (EZBC) [10] is also presented. It has gained notice in scalable video coding research.

A common diagram of the subband structure of the SWT coder is shown in Figure 8.24, which represents three stages of subband/wavelet decomposition. The first stage results in subbands LL_1, LH_1, HL_1, and HH_1. The second stage then decomposes LL_1 to yield subbands LL_2, LH_2, HL_2, and HH_2. Finally, there is a third stage of decomposition that results in subbands LL_3, LH_3, HL_3, and HH_3. Subband LL_3 is not further decomposed here and plays the role of *baseband*. We can then look at the sets of three high-frequency subbands $\{LH_k, HL_k,$ and $HH_k\}$ for $k = 3, 2, 1$, as enhancement data that permits an increasing scale (resolution) over the coarse scale representation in the baseband. Please note that this diagram is in the spatial domain, in that each subband is displayed spatially within each subwindow XY_k. Equivalently, we are looking at wavelet coefficients.

As already mentioned, embedded coders proceed to code the data of Figure 8.24 by proceeding from *most significant bit* (MSB) to *least significant bit* (LSB), bit plane by bit plane. Of course, this implies a quantization, in fact it is a UTQ, specified by its step-size Δ and deadzone centered on the origin and of width 2Δ. Assuming, for the subband to be coded, that the LSB bit plane is 2^b, then $\Delta = 2^b$. This LSB bit plane is usually conservatively chosen at such a level that its quality will never to be exceeded.

8.6.4 EMBEDDED ZERO-TREE WAVELET (EZW) CODER

The EZW bit-plane coder proceeds from one quite significant observation: at low and even into medium bitrates, there are a lot of zeros in all but the most significant bit planes, and once a zero is encountered in a bit plane at a given position,

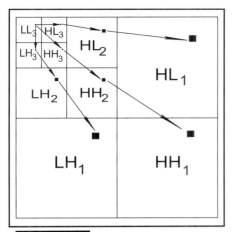

FIGURE 8.25   *Illustration of parent–child relationship for trees of coefficients in EZW. Here tree depth $N = 3$ levels*

then, for natural images, it is very likely accompanied by zeros at higher frequency subbands corresponding to the same location. Shapiro defined a special symbol, called a *zero-tree root* [23], to code all these zeros together. He regarded the quantized data in Figure 8.24 as composing separate data trees, each with roots in the base subband LL_N (here $N = 3$), where N is the number of levels of decomposition. Figure 8.25 shows the *parent–child relationships* for the three trees of coefficients (subband samples) growing out of one spatial location in the LL_3 subband.

In EZW coding, the first step is to find the most significant bit plane by computing the maximum over all the subband magnitude values,

$$2^T \leqslant \max|c(i,j)| < 2^{T+1}.$$

for subband data $c(i,j)$. Then we can normalize the data and introduce the binary notation

$$q[i,j] = s[i,j] + q_0[i,j] + q_1[i,j]2^{-1} + q_2[i,j]2^{-2} + \cdots,$$

where s is the sign bit, q_0 is the MSB, q_1 is the next MSB, etc. The scale factor would then be coded in a header, along with such factors as the image dimensions, pixel depth, and color space. Two new symbols are used in the coder:

- *Zero-tree root (ZTR)*: first zero that flows down to all zero remaining tree at current bit plane.
- *Isolated zero (IZ)*: a zero at this position in current bit plane, but not further down the data tree.

Other messages used in the EZW coder are − and − to indicate the sign of a subband value (wavelet coefficient), and then 0 and 1 to represent the binary values of the lower significance bits.

We proceed iteratively, with two passes through the bit-plane data corresponding to Figure 8.25. The first pass is called the *dominant pass* and creates a list of those coefficients not previously found to be dominant. The second *subordinate pass* refines those coefficients found to be significant in a previous dominant pass. The dominant pass scanning order is zig-zag, right-to-left, and then top-to-bottom within each scale (resolution), before proceeding to the next higher scale. On subsequent dominant passes, only those coefficients not yet found to be significant are scanned.

EZW ALGORITHM

A simplified version of the algorithm can then be written as follows:

1   Set threshold T_1 such that $\{2T_1 > \max |c(i, j)|$ and T_1 is smallest such$\}$, and normalize the data.

2   Set $k = 0$.

3   Conduct dominant pass by scanning through the data $q_k(i, j)$ and, for each newly significant pixel, output a message: $+$ $-$, IZ, and ZTR for *conditional adaptive arithmetic coder* (CAAC).

4   Conduct subordinate pass by scanning through the data $q_k(i, j)$ to refine pixels already known to be significant in the current bit plane. Such bit values may be output directly or also given to CAAC.

5   Set $k \longleftarrow k + 1$ and set $T_k = T_{k-1}/2$.

6   Stop (if *stopping criterion* met) or go to step 3.

The stopping criterion may be a fixed number of bit planes coded, but is usually a given numerical level. This is where the minimum step size of the quantizer comes in. If we want to code at highest quality with a step-size of Δ for each subband, then we will stop at $k = \arg\{T_k = \Delta\}$. If we decode the full bitstream, we get this quality. If we decode only the earlier part of the bitstream, we get a lesser quality, i.e., that corresponding to the quantizer step-size corresponding to whatever bit plane we stop at. Such a coder is thus fully embedded and permits fine-grain scalability in quality (PSNR) or, alternatively, in bitrate. To achieve a fixed bitrate, one simply stops decoding when that number of bits is reached.

We note that a scalable image coder really has three parts. First is the *pre-coder*, which codes the data into a coded image file archive with only a maximum bitrate or quality in mind. This is what the above algorithm accomplishes. The second part is the *extractor*, which pulls bits from the precoder's archive. These two steps together make up the scalable encoder. The third part is the decoder.

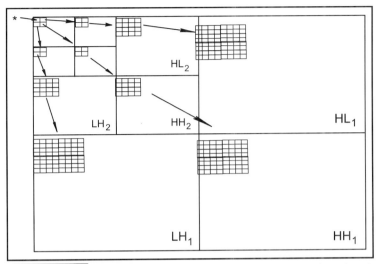

FIGURE 8.26 *Illustration of SPIHT parent–child dependencies for a three-level tree*

8.6.5 SET PARTITIONING IN HIERARCHICAL TREES (SPIHT) CODER

The SPIHT coder of Said and Pearlman [21] replaces the zero trees with three lists that indicate significance and insignificance of sets of pixels and their descendants. Its generally higher performance compared to EZW has made it the coder of choice for many applications and it is currently the one to beat in the research literature. The parent–child relation in SPIHT is somewhat different from EZW, as shown in Figure 8.26. In the coarsest subband LL_N we have sets of 2×2 pixels that each initiates three trees, in horizontal, diagonal, and vertical directions. Pixels in locations $(2i, 2j)$ do not have descendants.

SPIHT introduces three lists of subband pixels (wavelet coefficients):

1 *List of insignificant pixels* (LIP).
2 *List of significant pixels* (LSP).
3 *List of insignificant sets* (LIS).

LIS is further broken down into two types of sets of insignificant pixels: *type A* (all descendants are zero) and *type B* (all grandchildren and further descendants are zero). To describe the algorithm more easily, we define new sets with respect to a pixel at location $[i, j]$:

- $C[i, j]$ denotes its *children*.
- $D[i, j]$ denotes its *descendants*.
- $G[i, j]$ denotes its *grandchildren*.

Clearly we have $\mathcal{G} = \mathcal{D} - \mathcal{C}$, where the minus sign denotes set subtraction. Also, for a given set of pixels $\mathcal{B}$, we define the indicator function

$$S_k(\mathcal{B}) = \begin{cases} 1, & \mathcal{B} \text{ contains a pixel with } q_k[i,j] = 1 \\ 0, & \text{otherwise.} \end{cases}$$

SPIHT CODING ALGORITHM

We are now in a position to state the SPIHT coding algorithm.

1 Set $k = 0$, LSP $= \phi$, LIP $=$ {all coordinates $[i,j]$ of LL$_N$}, LIS $=$ {all coordinates $[i,j]$ of LL$_N$ that have children}.

2 *Significance pass*

- For each $[i,j] \in$ LIP, *output* $q_k[i,j]$. If $q_k = 1$, *output sign bit* $s[i,j]$ and move $[i,j]$ to end of LSP.
- For each $[i,j] \in$ LIS:
  - If set is of type A, *output* $S_k(\mathcal{D}[i,j])$. If $S_k(\mathcal{D}[i,j]) = 1$, then
    * For *each* $[l,m] \in \mathcal{C}[i,j]$, *output* $q_k[l,m]$. If $q_k[l,m] = 0$, add $[l,m]$ to the LIP. Else, *output* $s[l,m]$ and add $[l,m]$ to LSP.
    * If $\mathcal{G}[i,j] \neq \phi$, move $[i,j]$ to end of LIS as set of type B. Else delete $[i,j]$ from LIS.
  - If set is of type B, *output* $S_k(\mathcal{G}[i,j])$. If $S_k = 1$, then add each $[l,m] \in \mathcal{C}[i,j]$ to the end of the LIS as sets of type A and delete $[i,j]$ from LIS.

3 *Refinement pass*

- For each $[i,j] \in$ LSP, *output* $q_k[i,j]$ using *old* LSP.

4 Set $k \leftarrow k + 1$ and go to step 2.

Note that for sets of type A, we process all descendants, while for sets of type B, only grandchildren are processed. The SPIHT decoder is almost identical with the encoder. Initialization is the same; then the data are input from the binary codestream at each point in the coding algorithm where an output is indicated. In this way the decoder is able to process the data in the same sequence as the coder, and so no positional information has to be transmitted. While this is very efficient, one bit error somewhere in the SPIHT codestream can make the decoder lose track and start producing useless results. In particular, a bit error in the significance pass will cause such a problem. Of course, it can be said that any image coder that uses VLC potentially has this problem.

8.6.6 EMBEDDED ZERO BLOCK CODER (EZBC)

The main idea in EZBC is to replace the zero tree across scales (subbands) with separate quadtrees within each subband, to indicate significance of the data. This permits resolution scalability to accompany quality or SNR scalability in a more efficient manner than with zero trees, which necessarily carry information on all scales higher than the present one. We need some new notation to describe the quadtrees.

We denote quantized subband pixels as $c(i,j)$ with most significant bit $m(i,j)$, and pixel in subband k written as $c_k(i,j)$. Also note that as quadtree level l goes up, quadtree resolution goes down, with level $l = 0$ being full resolution.

DEFINITION 8.6-1 (*EZBC algorithm notation*)

$QT_k[l](i,j) \triangleq$ quadtree data at quadtree level l, in subband k, at position (i,j).

$K \triangleq$ number of subbands.

$D_k \triangleq$ quadtree depth for subband k.

$D_{\max} \triangleq$ maximum quadtree depth.

$\text{LIN}_k[l] \triangleq$ *list of insignificant nodes* for subband k and quadtree level l.

$\text{LSP}_k \triangleq$ *list of significant pixels* from subband k.

$S_n(i,j) \triangleq$ significance of node (i,j) at bit plane n.

$$S_n(i,j) \triangleq \begin{cases} 1, & \text{if } n \leqslant m(i,j) \\ 0, & \text{else.} \end{cases}$$

Coder initialization:

$$QT_k[0](i,j) = |c_k(i,j)|,$$
$$QT_k[l](i,j) = \max\{QT_k[l-1](2i,2j), QT_k[l-1](2i-1,2j),$$
$$QT_k[l-1](2i,2j-1), QT_k[l-1](2i-1,2j-1)\}.$$

$$\text{LIN}_k[l] = \begin{cases} \{(0,0)\}, & l = D_k \\ \phi, & \text{otherwise.} \end{cases}$$

EZBC CODING ALGORITHM

1   Initialization

- $\text{LIN}_k[l] = \begin{cases} \{(0,0)\}, & l = D_k \\ \phi, & \text{otherwise} \end{cases}$

- $LSP_k = \phi, \forall k$
- $n = \lfloor \log_2 \max_{k,i,j}\{|c_k(i,i)|\}\rfloor$

2 For $l = 0, D_{\max}$

- for $k = 0, K - 1$
  - CodeLIN(k, l)

3 For $k = 0, K - 1$

- — CodeLSP$_k$
- $n \leftarrow n - 1$ and go back to step 2. Stop when hit target bits

We now describe the functions CodeLIN(k, l) and CodeLSP$_k$:

CodeLIN(k, l)

- for each $(i, j) \in LIN_k[l]$
  - code $S_n(i, j)$
  - if $(S_n(i, j) = 0)$
    - (i, j) remains in $LIN_k[l]$
  - else
    - if $(l = 0)$, then **code** the sign bit of $c_k(i, j)$ and add (i, j) to LSP_k
    - else CodeDescendantNodes(k, l, i, j)

CodeDescendantNodes(k, l, i, j)

- for each node $(x, y) \in \{(2i, 2j), (2i, 2j + 1), (2i + 1, 2j), (2i + 1, 2j + 1)\}$ of quadtree level $l - 1$ in subband k
  - code $S_n(x, y)$
  - if $(S_n(x, y) = 0)$, add (x, y) to LSP_k
  - else CodeDescendantNodes($k - 1, l, i, j$)

CodeLSP$_k$

- for each pixel $(i, j) \in LSP_k$
  - code bit n of $|c_k(i, j)|$

The word code in this algorithm is like the word **output** in the SPIHT algorithm, in that they both specify algorithm output. Here, though, we take the algorithm output and send it to a CAAC rather than outputting it directly to the channel.

CONTEXT-BASED ARITHMETIC CODING IN EZBC

The context-based arithmetic coding used in EZBC is described next with reference to Figure 8.27. The context model is binary and is based on the quadtree

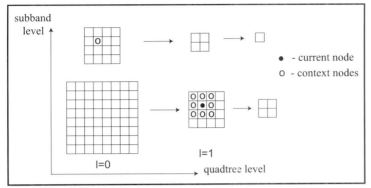

FIGURE 8.27 *Illustration of context modeling for AC in EZBC*

neighbors of the current datum. There are eight nearest neighbors from the current subband and quadtree levels, plus one context node from the next higher subband level and next lower quadtree level, which has the same position. The actual value of this 9-bit context is determined by the values encountered at the previous location, as scanned in a normal progressive raster. In this way, dependence between bit planes, and with nearby prior encoded neighbors, is not lost and so efficient coding results. Instead of using 2^9 states, states with similar conditional statistics are judiciously merged similarly to the method used in embedded block coding with optimal truncation (EBCOT) and JPEG 2000 [25,26]. Keeping the number of these states small permits more rapid and accurate adaptation to varying statistics and hence better compression.

8.7 JPEG 2000

The JPEG 2000 standard [26] is based on EBCOT [25] and is similar to the other embedded algorithms already discussed with the main exception that the sub-band values are separated into *code blocks* of typical size 64×64, which are coded separately. Individual R-D functions are established for each code block and the overall total bits is apportioned to the code blocks based on the equal slope condition (8.6-3), thereby funneling additional bits to wherever they are most needed to maximize the decrease in the total distortion, assumed to be approximately additive over the code blocks. So, some optimality is lost by breaking up the code blocks, but increased performance is obtained by the functionality of being able to send bits to code blocks where they are most needed.[5] The "optimal truncation" part of the EBCOT acronym refers to this bit assignment, since the code blocks' bit-streams are embedded, and thus can be truncated to get reduced

---

5. These relatively small code blocks also permit a region of interest (ROI) type of scalability, wherein higher accuracy representation may be given to certain high-interest areas of the image.

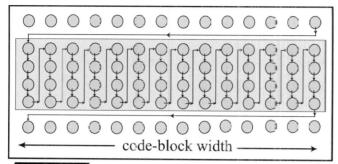

FIGURE 8.28 *An illustration of how a JPEG 2000 coder scans a 4-row stripe in a code block of wavelet coefficients (subband samples)*

quality points. It also follows that if each code block is bit-plane coded down to a sufficiently low bit plane, then the optimization can be done after the bit-plane coding in a "postcompression" pass. Then only occasionally, and at very high total bit values, would the binary version of the code block not be of sufficient accuracy.

EBCOT does not use any quadtree coding of significance at each bit-plane level. Rather, each bit plane is coded in three so-called *fractional* passes. With reference to Figure 8.28, wavelet coefficients (subband samples) are scanned in a horizontal stripe fashion, with a stripe width of 4 pixels vertically. A significance state is kept for each coefficient, and the first fractional pass *significant propagation* visits just those coefficients that neighbor a significant coefficient, since they have the highest likelihood of becoming significant, and when they do, will reduce the distortion the most for a given incremental number of bits. When a significant value is detected, its sign is coded with a context-based arithmetic coder. A second fractional pass, *magnitude refinement*, codes those coefficients already found to be significant in a previous bit plane. The last *cleanup pass* treats those coefficients that have not yet been coded, as they have the least expected contribution to expected distortion reduction. Here a special *run mode* is used to efficiently deal with the expected runs of zero bit-plane values.

In order to make this practical, there must be relatively efficient means to approximate the R-D functions of the code blocks, and also to do the optimized truncation. Means of doing both are included in the text of Taubman and Marcellin [26], which also treats certain simplificational differences between EBCOT and the JPEG 2000 standard.

8.8 COLOR IMAGE CODING

As we have seen in Chapter 6, color images are simply multicomponent vector images, commonly composed of red, green, and blue components. Normally, images are stored in the transformed YC_rC_b color space, where often the two chroma

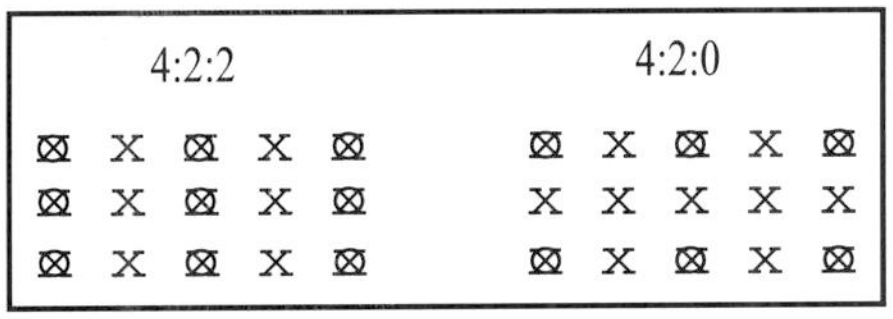

FIGURE 8.29 *Illustration of color subsampling strategies 4:2:2 and 4:2:0*

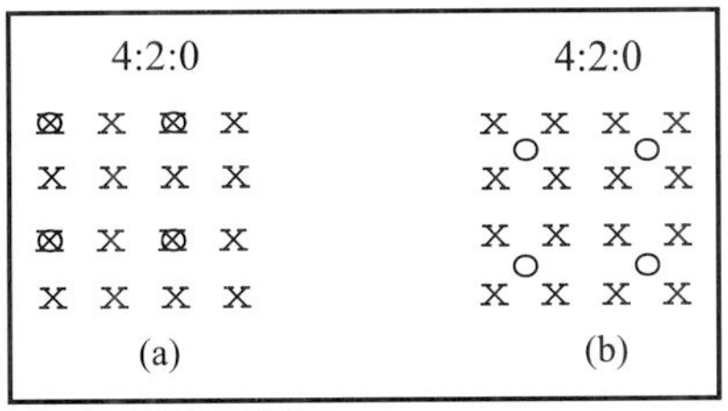

FIGURE 8.30 *Illustration of different possible sitings of chroma subsamples in 4:2:0*

channels C_r and C_b have been subsampled. Common subsampling strategies are labeled 4:2:2 and 4:2:0, as illustrated in Figure 8.29, with 4:2:2 referring to horizontal sampling rate of the Y, C_r, and C_b components, respectively, and meaning that there are only two C_r and C_b samples for each four luma Y samples, i.e., the chroma C_r and C_b signals are decimated horizontally by the factor 2.

The commonly used pattern 4:2:0 extends the chroma decimation into the vertical direction, again by the factor 2, i.e., chrominance components are decimated 2×2. The justification for this decimation is that the human visual system is less sensitive to spatial high frequencies in color than in luma. Sometimes, but not always, the chroma channels have been prefiltered to avoid aliasing error before being converted to 4:2:2 or 4:2:0. For definiteness, the full color space, with no chroma subsampling, is denoted 4:4:4.

In Figure 8.29, we show the chroma subsamples sited at the locations of a corresponding luma sample, but this is not commonly the case. Figure 8.30b shows an example of 4:2:0 with the chroma samples situated more symmetrically between luma samples, and is used in the DV digital video standard and in the H.263 video coding standard. Note in Figure 8.30 that conversion of video with color format (a) to that of (b) can be effected by a unity gain chroma filter approximating a half-sample delay in each dimension.

Normally, each of the three components Y, C_r, and C_b are separately coded by any of the methods of this chapter, just as though they were separate gray-level or monochrome images. The new issue thus is to apportion the total bitrate to the three color components. Due to color subsampling, which is almost universal, the number of luma samples is twice that of each chroma component in 4:2:2 and four times in 4:2:0. Now usually the chrominance components are easier to code than the luminance, because of their general lack of detail and often lowpass character. Efforts have been made to come up with a total distortion criteria for

Table 8.2. PSNR for some JPEG 2000 test set images at 0.55 bpp

PSNR (dB)	Lena	Barbara	GoldHill	Cafe	B ke	Hotel
SPIHT	37.2	31.4	33.1	26.5	33.0	33.6
EZBC	37.6	32.2	33.6	27.1	33.7	34.7
JP2K	37.2	32.1	33.2	26.7	33.4	34.0
EBCOT	37.3	32.3	33.2	26.9	33.5	34.1
SPECK	37.1	31.5	33.0	26.3	32.7	—

the three components [27]. A general good practice, though, is to assign bits to the components so as to make the quantizer step sizes about equal across the three components, with this then generating the most pleasing visual result, and also somewhat higher PSNR for the two chroma components as compared to that of the luma channel.

8.8.1 SCALABLE CODER RESULTS COMPARISON

Table 8.2 (from Hsiang [9]) lists some PSNR comparisons between some of the SWT coders, using a wide range of natural images that were part of the test set of the JPEG 2000 competition. If there is more than one version, we compare the only scalable version. While the PSNR performance levels of all of the coders are comparable, the advantage goes to the EZBC coder in this comparison. This is somewhat surprising in that the EBCOT and JPEG 2000 (JP2K) coders are rate-distortion optimized. However, the advantage of the R-D optimization is said to show up mainly in nonnatural or composite images composed of combined images, text, and graphics [26]. Of course, in that case, it may be more optimal still to decompose such a complex composite image into its constituents and code the graphics with a graphics coder and the text via recognition.

8.9 ROBUSTNESS CONSIDERATIONS

Almost all the specific image coding algorithms that we have discussed thus far are not robust to errors. This is because they use VLC. The Huffman code tree is complete, in the sense that all binary strings are possible. So, a decoder, upon encountering a bit error, will start decoding erroneous values from that point forward; this is loosely referred to as *decoder failure*. Practical coding algorithms additionally have markers such as *end of block* (EOB) and other synch words, which serve to terminate or at least control the error propagation. These synch words are inserted into the coder bitstream at appropriate points. Also, alternative non-Huffman VLCs may be employed, whose code tree, while reasonably efficient, is not complete. Then illegal code words can be detected and used to reset the decoder.

In practical coders, positional information and the block size in pixels are included in headers that start off a code block. Also, an overall header precedes the entire bitstream, giving image size, bit depth, color space, etc. Finally, the decoder needs to know how to act when it encounters a problem, e.g., if the code block is of fixed size, and the EOB comes too soon, indicating an error situation. A common response would be to discard the data in the current block, insert black into the now missing block, or perform some type of error-concealment strategy, and search for the next synch word, and start decoding again at that point. In essence, it is the *syntax* or structure of the coder, a structure that is created by the insertion of header and synch word information, that enables the decoder to detect errors. Then, if it knows how to react to these errors, it can be considered at least somewhat robust. Of course, if errors are expected to occur with some frequency, then error control coding and acknowledgment-retransmission (ACK/NACK) schemes can be used. More on this in the context of image sequence or video coding for networks will be presented in Chapter 12.

8.10 CONCLUSIONS

Image source compression is important for the transmission and storage of digital image and other 2-D data. Compression can be lossless, consisting of just an invertible transformation and variable-length coding, or lossy, employing a quantizer. Most practical application to natural images involves lossy coding due to the significantly higher compression ratios that can be achieved. Lossless coding is generally confined to medical images, works of art, and other high-value images. Its compression ratios are very modest by comparison, typically around two to one or so. There is also the category *visually lossless* that means the HVS cannot detect the loss (in many cases), i.e., the loss is below the JND, which is of interest for digital cinema compression.

The old standard image compression algorithm *JPEG* uses the DCT and is nonscalable. The new international standard *JPEG 2000* features an SWT and embedded scalable VLC. There is also the option for lossless coding in this recent coding algorithm. We briefly touched on color image coding.

Finally, we introduced the issue of coder robustness, achieved only by leaving in or reinserting some controlled redundancy that allows the decoder to detect and recover from even isolated single bit errors.

8.11 PROBLEMS

1   Interpret 8×8 block DCT image coding as a type of subband/wavelet coder with the DCT basis functions acting as the filters. What is the number of subbands? What is the 2-D decimation ratio? Do the filter supports of each subband overlap on the plane after subsampling? What is the passband of the filter used to produce the lowpass subband, i.e., the one including frequency $(0, 0)$? What is its minimum stopband attenuation?

2   In an optimal MSE scalar quantizer, the decision values must satisfy

$$d_i = \frac{1}{2}(r_i + r_{i-1}),$$

in terms of the neighboring representation values. This was derived as a stationary point, using partial derivatives, but here you are to conclude that it must be so, based on first principles. Hint: Consider the effect of moving the decision point d_i to the right or left. What does this do to the quantizer output?

3   An approximation has been derived (Panter and Dite, 1949) for so-called *fine scalar quantization* wherein the pdf of the signal x is approximately constant over a quantization bin, i.e.,

$$f_x(x) \approx f_x\!\left(\tfrac{1}{2}(d_{i-1} + d_i)\right), \quad \text{for } d_{i-1} < x \leqslant d_i, \text{ with } d_{i-1} \approx d_i,$$

resulting in the fine quantization approximation formula:

$$D = \frac{1}{12L^2}\left[\int_{-\infty}^{+\infty} f_x(x)^{1/3}\,dx\right]^3.$$

(a)   Evaluate this approximate distortion in the case where x is $N(0, \sigma^2)$.
(b)   Evaluate in the case where x is Laplacian distributed, i.e.,

$$f_x(x) = \frac{1}{\alpha}\exp{-\frac{2}{\alpha}|x|}, \quad \text{for } -\infty < x < +\infty, \text{ with } \alpha > 0,$$

and where $\alpha = \sqrt{2}\sigma$.

4   Work out the logarithmic bit assignment rule of (8.5-2) and (8.5-3). What are its shortcomings?

5   This problem concerns optimal bit assignment across quantizers after a unitary transform, i.e., the optimal bit allocation problem. We assume

$$R = \sum r_n$$

and

$$D = \sum d_n,$$

where n runs from 1 to N, the number of coefficients (channels). Here r_n and d_n are the corresponding rate and distortion pair for channel n. Assume the allowed bit allocations are M in number: $0 \leqslant b_1 < b_2 < \cdots < b_M$, and that the component distortion-rate functions $d_n = d_n(b_m)$ are given for all n and for all m, and are assumed to be convex.

(a)   Argue that the assignment $r_m = b_1$ for all m must be on the optimal assignment curve as the lowest bitrate point, at total rate $R = Nb_1$.

(b) Construct a straight line to all lower distortion solutions, and argue that the choice resulting in the lowest such line must also be on the optimal *R-D* curve.

(c) In part b, does it suffice to try switching in next higher bit assignment b_2 for each channel, one at a time?

6 Making use of the scalar quantizer model

$$D = g\sigma^2 2^{-2R},$$

find the approximate mean-square error for DPCM source coding of a random field satisfying the autoregressive (AR) model

$$x(n_1, n_2) = 0.8x(n_1 - 1, n_2) + 0.7x(n_1, n_2 - 1)$$
$$- 0.56x(n_1 - 1, n_2 - 1) + w(n_1, n_2),$$

where w is a white noise of variance $\sigma_w^2 = 3$. You can make use of the assumption that the coding is good enough so that $\hat{x}(n_1, n_2)$, the $(1, 0)$ step prediction based on $\tilde{x}(n_1 - 1, n_2)$, is approximately the same as that based on $x(n_1 - 1, n_2)$ itself.

7 Consider image coding with 2-D DPCM, but put the quantizer *outside* the prediction loop, i.e., complete the prediction-error transformation ahead of the quantizer operation. Discuss the effect of this coder modification. If you design the quantizer for this *open-loop* DPCM coder to minimize the quantizing noise power for the prediction error, what will be the effect on the reconstructed signal at the decoder? (You may assume the quantization error is independent from pixel to pixel.) What should the open-loop decoder be?

8 Consider using the logarithmic bit assignment rule, for Gaussian variables,

$$B_i = \frac{B}{N} + \frac{1}{2} \log_2 \frac{\sigma_i^2}{\sigma_{gm}^2}, \quad i = 1, \ldots, N,$$

with $\sigma_{gm}^2 \triangleq (\prod_i \sigma_i^2)^{1/N}$ and $N =$ number of channels (coefficients). Apply this rule to the 2×2 DCT output variance set given here:

$$\text{coef. map} = \begin{pmatrix} 00 & 10 \\ 01 & 11 \end{pmatrix} \quad \text{and} \quad \text{corresponding variances} = \begin{pmatrix} 22 & 4 \\ 8 & 2 \end{pmatrix}.$$

Assume the total number of bits to assign to these four pixels is $B = 16$. Please resolve any possible negative bit allocations by removing that pixel from the set and reassigning bits to those remaining. Noninteger bit assignments are OK since we plan to use variable-length coding. Please give the bits assigned to each coefficient and the total number of bits assigned to the four pixels.

9   Reconcile the constant slope condition (8.6-3) for bit assignment that we get from optimal channel (coefficient) rate-distortion theory in the continuous case, i.e., the constant slope condition $dD_i/dR_i = \lambda$, with the BFOS algorithm [20] for the discrete case, where we prune the quantizer tree of the branch with minimal distortion increase divided by the rate decrease: $\arg\min_i |\Delta D_i/\Delta R_i|$. Hint: Consider pruning a tree, as the number of rate values for each branch approaches infinity. Starting at a high rate, note that after a possible initial transient, we would expect to be repeatedly encountering all of the channels (coefficients, values) at near constant slope.

10 Show that the distortion and rate models in (8.6-1) and (8.6-2) imply that in scalable SWT coding of two resolution levels, the lower resolution level must always be refined at the higher resolution level if

$$\sigma^2_{\text{wgm}}(\text{base}) > \sigma^2_{\text{wgm}}(\text{enhancement}),$$

where $\sigma^2_{\text{wgm}}(\text{base})$ is the weighted geometric mean of the subband variances at the lower resolution and $\sigma^2_{\text{wgm}}(\text{enhancement})$ is the geometric mean of those subband variances in the enhancement subbands.

REFERENCES

[1] W. H. Chen and C. H. Smith, "Adaptive Coding of Monochrome and Color Images," *IEEE Trans. Commun.*, Nov. 1977.

[2] S.-J. Choi, *Three-Dimensional Subband/Wavelet Coding of Video with Motion Compensation*, Ph.D. thesis, ECSE Dept., Rensselaer Polytechnic Institute, Troy, NY, 1996.

[3] T. M. Cover and J. A. Thomas, *Elements of Information Theory*, Wiley, New York, NY, 1991.

[4] D. E. Dudgeon and R. M. Mersereau, *Multidimensional Digital Signal Processing*, Prentice-Hall, Englewood Cliffs, NJ, 1983

[5] W. Equitz and T. Cover, "Successive Refinement of Information," *IEEE Trans. Inform. Theory*, 37, 269–275, March 1991.

[6] R. G. Gallagher, *Information Theory and Reliable Communication*, John Wiley, New York, NY, 1968.

[7] R. Gersho and R. M. Gray, *Vector Quantization and Signal Compression*, Kluwer Academic Publishers, Boston, MA, 1992.

[8] H.-M. Hang and J. W. Woods, "Predictive Vector Quantization of Images," *IEEE Trans. Commun.*, COM-33, 1208–1219, Nov. 1985.

[9] S.-H. Hsiang, *Highly Scalable Subband/Wavelet Image and Video Coding*, Ph.D. thesis, ECSE Dept., Rensselaer Polytechnic Institute, Troy, NY, Jan. 2002.

[10] S.-T. Hsiang and J. W. Woods, "Embedded Image Coding using Zeroblocks of Subband/Wavelet Coefficients and Context Modeling," *Proc. ISCAS 2000*, Geneva, Switzerland, May 2000.

[11] A. K. Jain, *Fundamentals of Digital Image Processing*, Prentice-Hall, Englewood Cliffs, NJ, 1989.

[12] N. S. Jayant and P. Noll, *Digital Coding of Waveforms*, Prentice-Hall, Englewood Cliffs, NJ, 1984.

[13] Y. H. Kim and J. W. Modestino, "Adaptive Entropy Coded Subband Coding of Images," *IEEE Trans. Image Process.*, **1**, 31–48, Jan. 1992.

[14] R. L. Lagendijk, Information and Communication Theory Group, Delft University of Technology Delft, The Netherlands, Web site: http://ict.ewi.tudelft.nl [VcDemo available under "Software."]

[15] J. Lim, *Two-Dimensional Signal and Image Processing*, Prentice-Hall, Englewood Cliffs, NJ, 1990.

[16] H. S. Malvar and D. H. Staelin, "The LOT: Transform Coding without Blocking Effects," *IEEE Trans. Acoust., Speech, Signal Process.*, **37**, 553–559, April 1989.

[17] T. Naveen and J. W. Woods, "Subband Finite State Scalar Quantization," *IEEE Trans. Image Process.*, **5**, 150–155, Jan. 1996.

[18] W. A. Pearlman, In *Handbook of Visual Communications*, H.-M. Hang and J. W. Woods, eds., Academic Press, San Diego, CA, 1995. Chapter 2.

[19] K. Ramchandran and M. Vetterli, "Best Wavelet Packet Bases in a Rate-Distortion Sense," *IEEE Trans. Image Process.*, **2**, 160–175, April 1993.

[20] E. A. Riskin, "Optimal Bit Allocation via the Generalized BFOS Algorithm," *IEEE Trans. Inform. Theory*, **37**, 400–402, March 1991.

[21] A. Said and W. A. Pearlman, "A New, Fast, and Efficient Image Codec Based on Set Partitioning in Hierarchical Trees," *IEEE Trans. Circ. Syst. Video Technol.*, **6**, 243–250, June 1996.

[22] K. Sayood, *Introduction to Data Compression*, Morgan Kaufmann, San Francisco, CA, 1996.

[23] J. M. Shapiro, "Embedded Image Coding using Zerotrees of Wavelet Coefficients," *IEEE Trans. Signal Process.*, **41**, 3445–3462, Dec. 1993.

[24] Y. Q. Shi and H. Sun, *Image and Video Compression for Multimedia Engineering*, CRC Press, Boca Raton, FL, 2000.

[25] D. S. Taubman, "High Performance Scalable Image Coding with EBCOT," *IEEE Trans. Image Process.*, **9**, 1158–1170, July 2000.

[26] D. S. Taubman and M. W. Marcellin, *JPEG 2000 Image Compression Fundamentals, Standards, and Practice*, Kluwer Academic Publishers, Norwell, MA, 2002.

[27] P. H. Westerink, *Subband Coding of Images*, Ph.D. thesis, Delft Univ. of Technology, The Netherlands, Oct. 1989.

[28] Z. Zhang and J. W. Woods, "Large Block VQ for Image Sequences," *Proc. IEE IPA-99*, Manchester, UK, July 1999.

THREE-DIMENSIONAL AND SPATIOTEMPORAL PROCESSING

9

In Chapter 2 we looked at 2-D signals and systems. Here we look at the 3-D case, where often the application is to spatiotemporal processing, i.e., two spatial dimensions and one time dimension. Of course, 3-D more commonly refers to the three orthogonal spatial dimensions. Here we will be mostly concerned with convolution and filtering, and so need a regular grid for our signal processing. While regular grids are commonplace in spatiotemporal 3-D, they are often missing in spatial 3-D applications. In any event, the theory developed in this section can apply to any 3-D signals on a regular grid. In many cases, when a regular grid is missing or not dense enough (often the case in video processing), then interpolation may be used to infer values on a regular *super*grid, maybe a refinement of an existing grid. Often such an interpolative resampling will be done anyway for display purposes.

9.1 3-D SIGNALS AND SYSTEMS

We start off with a definition of three-dimensional linear system.

DEFINITION 9.1-1 (*3-D linear system*)
A system is linear when its operator **T** satisfies the equation

$$\mathbf{L}\{a_1 x_1(n_1, n_2, n_3) + a_2 x_2(n_1, n_2, n_3)\}$$
$$= a_1 \mathbf{L}\{x_1(n_1, n_2, n_3)\} + a_2 \mathbf{L}\{x_2(n_1, n_2, n_3)\}$$

for all (complex) scalars a_1 and a_2, and for all signals x_1 and x_2. When the system T is linear we usually denote this operator as **L**.

DEFINITION 9.1-2 (*linear shift-invariant system*)
Let the 3-D linear system **L** have output $y(n_1, n_2, n_3)$ when the input is $x(n_1, n_2, n_3)$, i.e., $y(n_1, n_2, n_3) = \mathbf{L}\{x(n_1, n_2, n_3)\}$. Then the system is 3-D LSI if it satisfies the equation

$$y(n_1 - k_1, n_2 - k_2) = \mathbf{L}\{x(n_1 - k_1, n_2 - k_2)\}$$

for all integer shift values k_1, k_2, and k_3.

Often such systems are called linear *constant parameter* or *constant coefficient*. An example would be a multidimensional filter. In the 1-D temporal case, we generally need to specify initial conditions and/or final conditions. In the 2-D spatial case, we often have some kind of boundary conditions. For 3-D systems, the solution space is a 3-D region, and we need to specify boundary conditions generally on all of the boundary surfaces. This is particularly so in the 3-D spatial case. In the 3-D spatiotemporal case, where the third parameter is time, then initial conditions often suffice in that dimension. However, we generally need both initial and boundary conditions to completely determine a filter output.

DEFINITION 9.1-3 (*3-D convolution representation*)
For an LSI system, with impulse response, the input and output are related by the 3-D convolution:

$$y(n_1, n_2, n_3) = \sum\sum\sum h(k_1, k_2, k_3)x(n_1 - k_1, n_2 - k_2, n_3 - k_3)$$

$$= (h * x)(n_1, n_2, n_3),$$

where the triple sums are over all values of k_1, k_2, and k_3, i.e., $-\infty < k_1, k_2, k_3 < +\infty$.

We can see that 3-D convolution is a commutative operation, so that $h * x = x * h$ as usual.

DEFINITION 9.1-4 (*3-D separability*)
A 3-D LSI system is a *separable system* if its impulse response factors or separates, e.g., $h(n_1, n_2, n_3) = h_1(n_1)h_2(n_2)h_3(n_3)$ or $h(n_1, n_2, n_3) = h_1(n_1, n_2)h_2(n_3)$. A signal is a *separable signal* if it separates into two or three factors. A *separable operator* factors into the concatenation of two or three operators, e.g.,

$$T_{n_3}\{T_{n_2}\{T_{n_1}[\cdot]\}\} \quad \text{or} \quad T_{n_3}\{T_{n_1, n_2}[\cdot]\}.$$

Examples of separable operators in 3-D are the familiar transforms, i.e., Fourier, DFT, DCT, and rectangular SWT, which, in this way, extend to the 3-D case naturally.

EXAMPLE 9.1-1 (*separable system and separable convolution*)
A given LSI system has input x and output y and separable impulse response $h(n_1, n_2, n_3) = h_1(n_1, n_2)h_2(n_3)$. Expressing the 3-D convolution, we have

$$y(n_1, n_2, n_3)$$

$$= \sum\sum\sum h_1(k_1, k_2)h_2(k_3)x(n_1 - k_1, n_2 - k_2, n_3 - k_3) \quad (9.1\text{-}1)$$

$$= \sum\sum h_1(k_1, k_2)\left[\sum_{k_3} h_2(k_3)x(n_1 - k_1, n_2 - k_2, n_3 - k_3)\right]$$

$$= h_1(n_1, n_2) * \left[h_2(n_3) * x(n_1, n_2, n_3)\right]$$

$$= h_2(n_3) * \left[h_1(n_1, n_2) * x(n_1, n_2, n_3)\right].^1 \quad (9.1\text{-}2)$$

---

1. Beware the notation here! The first convolution is just on the n_3 parameter. The second one is just on n_1, n_2.

The result is a 2-D convolution with the 2-D signal consisting of x for each fixed value of n_3, perhaps denoted as

$$x_{n_3}(n_1, n_2) \triangleq x(n_1, n_2, n_3),$$

followed by 1-D convolutions in n_3 for the 1-D output signal obtained by fixing n_1, n_2 and then ranging through all such n_1 and n_2, i.e., $x_{n_1, n_2}(n_3) \triangleq x(n_1, n_2, n_3)$. Of course, and by linearity, the 1-D and 2-D convolutions in (9.1-1) can be done in either order.

DEFINITION 9.1-5 (*3-D Fourier transform*)

$$X(\omega_1, \omega_2, \omega_3) \triangleq \sum_{n_1=-\infty}^{+\infty} \sum_{n_2=-\infty}^{+\infty} \sum_{n_3=-\infty}^{+\infty} x(n_1, n_2, n_3) \exp -j(\omega_1 n_1 + \omega_2 n_2 + \omega_3 n_3).$$

Note that X is continuous and triply periodic with period $2\pi \times 2\pi \times 2\pi$.

DEFINITION 9.1-6 (*inverse 3-D Fourier transform*)

$$x(n_1, n_2, n_3) = \frac{1}{(2\pi)^3} \int_{-\pi}^{+\pi} \int_{-\pi}^{+\pi} \int_{-\pi}^{+\pi} X(\omega_1, \omega_2, \omega_3)$$
$$\times \exp +j(\omega_1 n_1 + \omega_2 n_2 + \omega_3 n_3) \, d\omega_1 \, d\omega_2 \, d\omega_3,$$

which amounts to a 3-D Fourier series, with the roles of "transform" and "signal" reversed, as was the case in 2-D and 1-D also.

9.1.1 PROPERTIES OF 3-D FOURIER TRANSFORM

- *Ideal shift property*—shift by integers k_1, k_2, k_3:

$$\text{FT}\{x(n_1 - k_1, n_2 - k_2, n_3 - k_3)\}$$
$$= X(\omega_1, \omega_2, \omega_3) \exp -j(\omega_1 k_1 + \omega_2 k_2 + \omega_3 k_3).$$

- *Linearity*:

$$\text{FT}\{a_1 x_1(n_1, n_2, n_3) + a_2 x_2(n_1, n_2, n_3)\}$$
$$= a_1 X_1(\omega_1, \omega_2, \omega_3) + a_2 X_2(\omega_1, \omega_2, \omega_3).$$

- *3-D convolution*:

$$\text{FT}\{x_1(n_1, n_2, n_3) * x_2(n_1, n_2, n_3)\} = X_1(\omega_1, \omega_2, \omega_3) X_2(\omega_1, \omega_2, \omega_3).$$

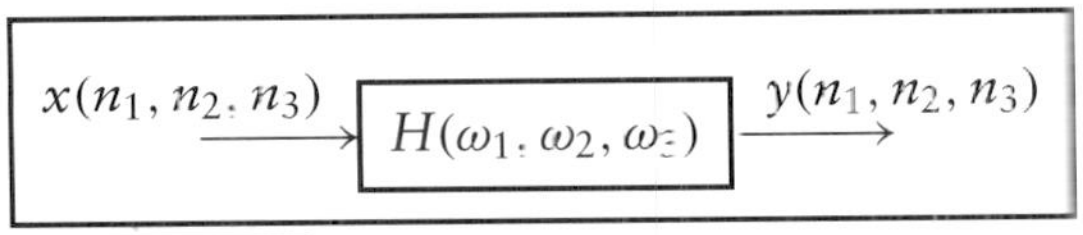

$$x(n_1, n_2, n_3) \xrightarrow{} \boxed{H(\omega_1, \omega_2, \omega_3)} \xrightarrow{} y(n_1, n_2, n_3)$$

FIGURE 9.1 *3-D system characterized by its frequency response*

- *Separable operator:*

$$\mathbf{FT}\{h_1(n_1, n_2) * [h_2(n_3) * x(n_1, n_2, n_3)]\} = H_1(\omega_1, \omega_2)H_2(\omega_3)X(\omega_1, \omega_2, \omega_3).$$

The last property can be interpreted as a 2-D filtering cascaded with a 1-D filtering along the remaining axis. While this is not general, it can be expedient and sometimes is all that is required to implement a given transformation.

9.1.2 3–D FILTERS

Consider a constant coefficient 3-D difference equation

$$y(n_1, n_2, n_3) = - \sum_{(k_1, k_2, k_3) \in \mathcal{R}_a - (0,0,0)} a_{k_1, k_2, k_3} y(n_1 - k_1, n_2 - k_2, n_3 - k_3)$$

$$+ \sum_{(k_1, k_2, k_3) \in \mathcal{R}_b} b_{k_1, k_2, k_3} x(n_1 - k_1, n_2 - k_2, n_3 - k_3),$$

where $\mathcal{R}_a$ and $\mathcal{R}_b$ denote the 3-D filter coefficient support regions. By using linear and shift-invariant properties of 3-D FTs, we can transform the above relation to multiplication in the frequency domain with a system function $H(\omega_1, \omega_2, \omega_3)$,

$$H(\omega_1, \omega_2, \omega_3) = \frac{B(\omega_1, \omega_2, \omega_3)}{A(\omega_1, \omega_2, \omega_3)},$$

given in terms of the 3-D Fourier transform $B(\omega_1, \omega_2, \omega_3)$ of the filter feed forward coefficients $b(k_1, k_2, k_3) \triangleq b_{k_1, k_2, k_3}$ and the Fourier transform $A(\omega_1, \omega_2, \omega_3)$ of the filter feedback coefficients $a(k_1, k_2, k_3) \triangleq a_{k_1, k_2, k_3}$, where $a(0, 0, 0) = +1$. A 3-D LSI system diagram is shown in Figure 9.1.

9.2 3-D SAMPLING AND RECONSTRUCTION

The 3-D sampling theorem extends easily from the corresponding 2-D theorem in Chapter 2. The method of proof also extends, so that the same method can be used. As before, we use the variables t_i for the continuous parameters of the function $x_c(t_1, t_2, t_3)$ to be sampled, with its corresponding continuous-parameter FT denoted $X_c(\Omega_1, \Omega_2, \Omega_3)$.

THEOREM 9.2-1 (*3-D sampling theorem*)

Let $x(n_1, n_2, n_3) \triangleq x_c(n_1 T_1, n_2 T_2, n_3 T_3)$, with sampling periods T_i, $i = 1, 2, 3$, then:

$$X(\omega_1, \omega_2, \omega_3) = \frac{1}{T_1 T_2 T_3} \sum_{k_1, k_2, k_3} X_c\left(\frac{\omega_1 - 2\pi k_1}{T_1}, \frac{\omega_2 - 2\pi k_2}{T_2}, \frac{\omega_3 - 2\pi k_3}{T_3}\right).$$

Aliasing occurs when the shifted FTs on the right overlap. This will occur when the T_i's are not small enough. If X_c is rectangularly bandlimited in three dimensions, then $\frac{\pi}{T_i} = \Omega_{c_i}$, for signal bandwidth limits Ω_{c_i}, will work fine.

Proof: We start by writing $x(n_1, n_2, n_3)$ in terms of the samples of the inverse Fourier transform of $X_c(\Omega_1, \Omega_2, \Omega_3)$:

$$x(n_1, n_2, n_3) = \frac{1}{(2\pi)^3} \int_{-\infty}^{+\infty} \int_{-\infty}^{+\infty} \int_{-\infty}^{+\infty} X_c(\Omega_1, \Omega_2, \Omega_3)$$
$$\times \exp +j(\Omega_1 n_1 T_1 + \Omega_2 n_2 T_2 + \Omega_3 n_3 T_3) \, d\Omega_1 \, d\Omega_2 \, d\Omega_3.$$

Next we let $\omega_i \triangleq \Omega_i T_i$ in this triple integral to obtain

$$x(n_1, n_2, n_3) = \frac{1}{(2\pi)^3} \int_{-\infty}^{+\infty} \int_{-\infty}^{+\infty} \int_{-\infty}^{+\infty} \frac{1}{T_1 T_2 T_3} X_c\left(\frac{\omega_1}{T_1}, \frac{\omega_2}{T_2}, \frac{\omega_3}{T_3}\right)$$
$$\times \exp +j(\omega_1 n_1 + \omega_2 n_2 + \omega_3 n_3) \, d\omega_1 \, d\omega_2 \, d\omega_3$$
$$= \frac{1}{(2\pi)^3} \sum_{\text{all } k_1, k_2, k_3} \int_{SQ(k_1, k_2, k_3)} \frac{1}{T_1 T_2 T_3} X_c\left(\frac{\omega_1}{T_1}, \frac{\omega_2}{T_2}, \frac{\omega_3}{T_3}\right)$$
$$\times \exp +j(\omega_1 n_1 + \omega_2 n_2 + \omega_3 n_3) \, d\omega_1 \, d\omega_2 \, d\omega_3, \qquad (9.2\text{-}1)$$

where $SQ(k_1, k_2, k_3)$ is a $2\pi \times 2\pi \times 2\pi$ cube centered at position $(2\pi k_1, 2\pi k_2, 2\pi k_3)$, i.e.,

$$SQ(k_1, k_2, k_3) \triangleq [-\pi + 2\pi k_1, +\pi + 2\pi k_1] \times [-\pi + 2\pi k_2, +\pi + 2\pi k_2]$$
$$\times [-\pi + 2\pi k_3, +\pi + 2\pi k_3].$$

Then, making the change of variables $\omega_i' = \omega_i - 2\pi k_i$ *inside* each of the preceding integrals, we get

$$
\begin{aligned}
x(n_1, n_2, n_3) &= \frac{1}{(2\pi)^2} \sum_{\text{all } k_1, k_2, k_3} \int_{-\pi}^{+\pi} \int_{-\pi}^{+\pi} \int_{-\pi}^{+\pi} \frac{1}{T_1 T_2 T_3} \\
&\quad \times X_c\left(\frac{\omega_1' + 2\pi k_1}{T_1}, \frac{\omega_2' + 2\pi k_2}{T_2}, \frac{\omega_3' + 2\pi k_3}{T_3} \right) \\
&\quad \times \exp +j(\omega_1' n_1 + \omega_2' n_2 + \omega_3' n_3) \, d\omega_1' \, d\omega_2' \, d\omega_3' \\
&= \frac{1}{(2\pi)^2} \int_{-\pi}^{+\pi} \int_{-\pi}^{+\pi} \int_{-\pi}^{+\pi} \left(\sum_{\text{all } k_1, k_2} \frac{1}{T_1 T_2 T_3} \right. \\
&\quad \times X_c\left(\frac{\omega_1' + 2\pi k_1}{T_1}, \frac{\omega_2' + 2\pi k_2}{T_2}, \frac{\omega_3' + 2\pi k_3}{T_3} \right) \Bigg) \\
&\quad \times \exp +j(\omega_1' n_1 + \omega_2' n_2 + \omega_3' n_3) \, d\omega_1' \, d\omega_2' \, d\omega_3' \\
&= \text{IFT}\left(\sum_{\text{all } k_1, k_2, k_3} \frac{1}{T_1 T_2 T_3} X_c\left(\frac{\omega_1' + 2\pi k_1}{T_1}, \frac{\omega_2' + 2\pi k_2}{T_2}, \frac{\omega_3' + 2\pi k_3}{T_3} \right) \right),
\end{aligned}
$$

as was to be shown. $\square$

This last equation says that x is the 3-D discrete-space IFT of X. If any T_i is too large, then this linear sampling theory says we should prefilter before sampling to avoid aliasing caused by the offendingly coarse T_is.

9.2.1 GENERAL 3-D SAMPLING

Consider more general, but still regular, sampling at locations

$$
\mathbf{t} = \begin{bmatrix} t_1 \\ t_2 \\ t_3 \end{bmatrix} = \mathbf{V} \begin{bmatrix} n_1 \\ n_2 \\ n_3 \end{bmatrix}
$$

$$
= n_1 \mathbf{v}_1 + n_2 \mathbf{v}_2 + n_3 \mathbf{v}_3,
$$

where $\mathbf{V} \triangleq [\mathbf{v}_1, \mathbf{v}_2, \mathbf{v}_3]$ is the 3-D *sampling matrix* formed from three noncolinear sampling basis vectors $\mathbf{v}_1, \mathbf{v}_2,$ and $\mathbf{v}_3$. The corresponding aliased representation in the discrete-space frequency domain is expressed in terms of the *periodicity matrix* $\mathbf{U}$, which must satisfy

$$
\mathbf{U}^T \mathbf{V} = 2\pi \mathbf{I}.
$$

Just as in the 2-D case, one can show that:

$$X(\omega) = \frac{1}{|\det \mathbf{V}|} \sum_{\mathbf{k}} X_c\left(\frac{1}{2\pi} \mathbf{U}(\omega - 2\pi \mathbf{k})\right), \tag{9.2-2}$$

where X_c is the 3-D continuous-space Fourier transform, and X is the 3-D Fourier transform of the samples. We can also express (9.2-2) in terms of vector analog frequency $\boldsymbol{\Omega}$ as

$$X(\mathbf{V}^T \boldsymbol{\Omega}) = \frac{1}{|\det \mathbf{V}|} \sum_{\mathbf{k}} X_c(\boldsymbol{\Omega} - \mathbf{U}\mathbf{k}),$$

analogously to the 2-D general sampling case in Chapter 2.

When X_c is bandlimited, and in the absence of aliasing, we have

$$X_c(\boldsymbol{\Omega}) = |\det \mathbf{V}| X(\mathbf{V}^T \boldsymbol{\Omega}),$$

which we can use for reconstruction from the 3-D samples. We start with the 3-D continuous-space IFT relation

$$x_c(\mathbf{t}) = \frac{1}{(2\pi)^3} \int\!\!\!\int\!\!\!\int_{-\infty}^{+\infty} X_c(\boldsymbol{\Omega}) \exp(j\boldsymbol{\Omega}^T \mathbf{t}) \, d\boldsymbol{\Omega},$$

and substitute to obtain

$$x_c(\mathbf{t}) = \frac{|\det \mathbf{V}|}{(2\pi)^3} \int\!\!\!\int\!\!\!\int X(\mathbf{V}^T \boldsymbol{\Omega}) \exp(j\boldsymbol{\Omega}^T \mathbf{t}) \, d\boldsymbol{\Omega}$$

$$= \frac{|\det \mathbf{V}|}{(2\pi)^3} \int\!\!\!\int\!\!\!\int \sum_{\mathbf{n}} x(\mathbf{n}) \exp(-j\boldsymbol{\Omega}^T \mathbf{V}\mathbf{n}) \exp(j\boldsymbol{\Omega}^T \mathbf{t}) \, d\boldsymbol{\Omega}$$

$$= \sum_{\mathbf{n}} x(\mathbf{n}) \frac{|\det \mathbf{V}|}{(2\pi)^3} \int\!\!\!\int\!\!\!\int \exp(j\boldsymbol{\Omega}^T (\mathbf{t} - \mathbf{V}\mathbf{n})) \, d\boldsymbol{\Omega}.$$

We thus see that the reconstruction interpolation filter is

$$h(\mathbf{t}) = \frac{|\det \mathbf{V}|}{(2\pi)^3} \int\!\!\!\int\!\!\!\int \exp(j\boldsymbol{\Omega}^T \mathbf{t}) \, d\boldsymbol{\Omega},$$

where the 3-D integral is taken over the bandlimited support of X_c, say $\|\boldsymbol{\Omega}\| \leqslant \Omega_c$. We then have the reconstruction formula

$$x_c(\mathbf{t}) = \sum_{\mathbf{n}} x(\mathbf{n}) h(\mathbf{t} - \mathbf{V}\mathbf{n}).$$

For more on general 3-D sampling and sample rate change, see the text by Vaidyanathan [3].

9.3 SPATIOTEMPORAL SIGNAL PROCESSING

In spatiotemporal processing, currently the data is typically undersampled in the time dimension AND there is no prefilter! The situation should improve as technology permits higher frame rates than 24–30 frames per second (fps) (48, 60, 72, etc.). Here we specialize 3-D processing to two spatial dimensions and one time dimension. We can then treat the filtering of image sequences or video. Notationally we replace n_3 by n, which will now denote discrete time. Also, the third frequency variable is written simply as ω radians or f when the unit is Hertz (Hz), giving the Fourier transform pair

$$x(n_1, n_2, n) \leftrightarrow X(\omega_1, \omega_2, \omega).$$

Time-domain causality now applies in the third dimension n. We can use a matrix notation:

$$\mathbf{x}(n) \triangleq \{x(\mathbf{n})\} \quad \text{with } \mathbf{n} \triangleq (n_1, n_2, n)^T,$$

to denote a frame of video data at time n. The video is then a sequence of these matrices, i.e.,

$$\mathbf{x}(0), \mathbf{x}(1), \mathbf{x}(2), \mathbf{x}(3), \mathbf{x}(4), \mathbf{x}(5), \ldots .$$

9.3.1 SPATIOTEMPORAL SAMPLING

EXAMPLE 9.3-1 (*progressive video*)
We apply rectangular sampling to the spatiotemporal continuous signal $x_c(t_1, t_2, t)$, sampled spatially with sample period $T_1 \times T_2$ and temporally with period $T = \Delta t$. We then have the sampling and periodicity matrices,

$$\mathbf{V} = \begin{bmatrix} T_1 & 0 & 0 \\ 0 & T_2 & 0 \\ 0 & 0 & \Delta t \end{bmatrix} \quad \text{and} \quad \mathbf{U} = \begin{bmatrix} 2\pi/T_1 & 0 & 0 \\ 0 & 2\pi/T_2 & 0 \\ 0 & 0 & \dfrac{2\pi}{\Delta t} \end{bmatrix}.$$

This spatiotemporal sampling is called *progressive* scanning or simply *noninterlaced* (NI) by video engineers.

Computer manufacturers like progressive (noninterlaced) because small characters do not flicker as much on a computer display. For natural video, progressive avoids the interline flicker of interlaced systems and is easier to interface with motion estimation and compensation schemes, which save channel bits. Nevertheless, interlaced CRT displays had been easier to build, because the pixel rate is lower for interlace given the same display resolution. Now solid-state displays are principally all progressive.

EXAMPLE **9.3-2** (*interlaced video*)

In Chapter 2, we introduced this concept for analog video as just 2-D sampling in the vertical and time directions $(v, t) = (t_2, t)$, with sampling and periodicity matrices,

$$V = \begin{bmatrix} 1 & 1 \\ 1 & -1 \end{bmatrix} \quad \text{and} \quad U = \begin{bmatrix} \pi & \pi \\ \pi & -\pi \end{bmatrix}.$$

This is only 2-D sampling, because in analog video there is no sampling horizontally. However, in digital video, there is horizontal sampling and so we need 3-D sampling. The interlaced digital video has a sampling matrix in space–time and periodicity matrix in spatiotemporal frequency given as

$$V = \begin{bmatrix} 1 & 0 & 0 \\ 0 & 1 & 1 \\ 0 & 1 & -1 \end{bmatrix} \quad \text{and} \quad U = \begin{bmatrix} 2\pi & 0 & 0 \\ 0 & \pi & \pi \\ 0 & \pi & -\pi \end{bmatrix},$$

in terms of sampling structure $(h, v, t) = (t_1, t_2, t)$. Here, we have normalized the sampling periods for simplicity, and the resulting simple sampling pattern is called *quincunx*. Of course, in a real system, the sample spacings would not be unity, and we would have sampling in terms of distances T_1 and T_2 on the frame, and T as the time between fields.[2]

Spatiotemporal aliasing is a real problem in video signal processing since optical lenses are poor bandlimiting filters and, further, the temporal dimension is usually grossly undersampled, i.e., 30 frames per second (fps) is often not high enough. A frame rate of 100 fps or more has been often claimed as necessary for common scenes. Only now with HD cameras (and beyond) is digital video beginning to become alias free in space. Alias free with respect to time sampling, however, remains elusive, without the use of special high-frame-rate capture devices. These are used in certain industrial inspection applications and sometimes in the movie industry to create slow-motion effects.

9.3.2 SPATIOTEMPORAL FILTERS

There is a variety of spatiotemporal filters, including finite impulse response (FIR), infinite impulse response (IIR), statistically designed filters, and adaptive filters. Additionally, there can be hybrids of FIR and IIR, e.g., the filter can be IIR in time and FIR in space. This kind of hybrid filter has become very popular due to the need to minimize memory in the temporal direction, thus minimizing use of

---

2. We remember from Chapter 2 that an interlaced frame is composed of two fields, the upper and lower.

frame memories. The frame predictor in a hybrid video coder (cf, Chapter 10) is an FIR spatiotemporal filter. The decoder must run the inverse of this filter, which is thus an IIR spatiotemporal filter. It must therefore be 3-D multidimensionally stable.

EXAMPLE 9.3-3 (*temporal averager*)
A very simple example of a useful spatiotemporal filter is the temporal averager, whose impulse response is given as

$$h(n_1, n_2, n) = \frac{1}{N}\delta(n_1, n_2)\big(u(n) - u(n - N)\big).$$

For input $x(n_1, n_2, n)$, the output $y(n_1, n_2, n)$ is given as

$$y(n_1, n_2, n) = \frac{1}{N}\sum_{k=0}^{N-1} x(n_1, n_2, n - k),$$

or just the average of the present and past $N - 1$ frames. This is a 3-D FIR filter that is causal in the time or n variable. Such a filter is often used to smooth noisy video data, most often due to image sensor noise, when the frames are very similar, i.e., have little interesting change, over the time scale of N frames.

If objects in the frame move much, then the temporal averager will blur their edges, so often such a device is combined with a suitable motion or change detector that will control the number of frames actually used. It can be based on some norm of the frame difference $\|x(n_1, n_2, n) - x(n_1, n_2, n - 1)\|$. Optimal filtering of noisy image sequences can be based on statistical models of the noise and data. One such model that is quite common is the spatiotemporal *autoregressive moving average* (ARMA) *random field sequence* (i.e., a sequence of random fields), which we discuss next.

SPATIOTEMPORAL ARMA MODEL

We start with the definition of an ARMA difference equation.

DEFINITION 9.3-1 (*autoregressive moving average*)

$$x(n_1, n_2, n) = \sum_{(k_1, k_2, k) \in \mathcal{R}_a - (0,0,0)} a_{k_1, k_2, k} x(n_1 - k_1, n_2 - k_2, n - k)$$

$$+ \sum_{(k_1, k_2, k) \in \mathcal{R}_b} b_{k_1, k_2, k} w(n_1 - k_1, n_2 - k_2, n - k). \quad (9.3-1)$$

Here the input spatiotemporal sequence w is taken as a white noise random field sequence. If $\mathcal{R}_b = \{(0, 0, 0)\}$, then the model is autoregressive (AR). We get a Markov random field sequence (cf., Section 9.4) if the input w is jointly independent as well as spectrally white. If additionally $\mathcal{R}_a$ only has coefficients for $n \geqslant 1$, then the model is said to be *frame based*. A frame-based model could use parallel computation capability to calculate a whole frame at a time.

9.3.3 INTRAFRAME FILTERING

The equation for *intraframe* filtering with filter impulse response $h(n_1, n_2)$ is

$$y(n_1, n_2, n) = \sum_{k_1, k_2} h(k_1, k_2) x(n_1 - k_1, n_2 - k_2, n).$$

Intraframe filtering is appropriate to restore mild spatial blur in the presence of good SNR, i.e., 30 dB or higher, and also when there is low interframe dependency (e.g., a very low frame rate, such as < 1 fps). Still other reasons are to retain frame identity (e.g., video editing, using so-called *I* frames in MPEG parlance[3]) and to reduce complexity and expense.

The *intraframe ARMA signal model* is given as

$$x(n_1, n_2, n) = \sum_{(k_1, k_2) \in \mathcal{R}_a - (0,0)} a_{k_1, k_2} x(n_1 - k_1, n_2 - k_2, n)$$

$$+ \sum_{(k_1, k_2) \in \mathcal{R}_b} b_{k_1, k_2} w(n_1 - k_1, n_2 - k_2, n). \qquad (9.3\text{-}2)$$

Normally the input term w is taken as spatiotemporal white noise. This ARMA amounts to spatial filtering of each input frame, with the 2-D recursive filter with coefficient arrays $\{a\}, \{b\}$. The filter coefficients are constant here across the frames, but could easily be time variant, making a time-varying intraframe ARMA model.

9.3.4 INTRAFRAME WIENER FILTER

Taking the 2-D Fourier transform of (9.3-2), we obtain

$$X(\omega_1, \omega_2; n) = \frac{B(\omega_1, \omega_2)}{1 - A(\omega_1, \omega_2)} W(\omega_1, \omega_2; n),$$

---

3. *I frame* is an MPEG compression term that denotes an image frame compressed without reference to other image frames, i.e., intraframe compression. MPEG compression is covered in Chapter 11.

for each frame (n). We could also write this in terms of 3-D Fourier transforms as

$$X(\omega_1, \omega_2, \omega) = \frac{B(\omega_1, \omega_2)}{1 - A(\omega_1, \omega_2)} W(\omega_1, \omega_2, \omega),$$

but will generally reserve 3-D transform notation for the interframe case to be treated in the following subsections.

For homogeneous random processes that possess power spectral densities (PSDs), we can write

$$S_x(\omega_1, \omega_2; n) = \left| \frac{B(\omega_1, \omega_2)}{1 - A(\omega_1, \omega_2)} \right|^2 S_w(\omega_1, \omega_2; n)$$

and

$$S_x(\omega_1, \omega_2, \omega) = \left| \frac{B(\omega_1, \omega_2)}{1 - A(\omega_1, \omega_2)} \right|^2 S_w(\omega_1, \omega_2, \omega),$$

where often $S_w = \sigma_w^2$, i.e., spatiotemporal white noise. Here A and B are the spatial or 2-D Fourier transforms of the spatial filter coefficients $c(k_1, k_2)$ and $b(k_1, k_2)$. We can then use these PSDs to design a 2-D or spatial Wiener filter for the observations

$$y(n_1, n_2, n) = h(n_1, n_2) * x(n_1, n_2, n) + v(n_1, n_2, n),$$

where the spatial blur h is known and the noise v is white with variance σ_v^2 and uncorrelated with the signal x. Then the *intraframe Wiener filter* is given as

$$G(\omega_1, \omega_2; n) = \frac{H^*(\omega_1, \omega_2) S_x(\omega_1, \omega_2; n)}{|H|^2(\omega_1, \omega_2) S_x(\omega_1, \omega_2; n) + \sigma_w^2} \tag{9.3-3}$$

at each frame n, where the PSDs S_x could be estimated via an AR or ARMA modeling procedure. Effectively, this is just image processing done for each frame (n). Note that the PDSs can vary with n, i.e., be time varying.

EXAMPLE 9.3-4 (*intraframe Wiener filter*)
Let $r = x + v$ with $x \perp v$, i.e., x and v orthogonal. Also assume that the random field sequences x and n are homogeneous in space with PSDs S_x and S_v, respectively. Then the spatially noncausal intraframe Wiener filter for frame n is given by

$$G(\omega_1, \omega_2; n) = \frac{S_x(\omega_1, \omega_2; n)}{S_x(\omega_1, \omega_2; n) + S_v(\omega_1, \omega_2; n)}$$

Additionally, assume a spatial ARMA model for the signal x as in (9.3-2) and also that the observation noise is white with PSD σ_v^2. Then the intraframe

Wiener filter becomes

$$G(\omega_1, \omega_2; n) = \frac{\left|\frac{B(\omega_1, \omega_2)}{1 - A(\omega_1, \omega_2)}\right|^2 \sigma_w^2}{\left|\frac{B(\omega_1, \omega_2)}{1 - A(\omega_1, \omega_2)}\right|^2 \sigma_w^2 + \sigma_v^2}$$

$$= \frac{|B(\omega_1, \omega_2)|^2 \sigma_w^2}{|B(\omega_1, \omega_2)|^2 \sigma_w^2 + |1 - A(\omega_1, \omega_2)|^2 \sigma_v^2}.$$

In the case of Wiener filtering, the optimality of this process would, of course, require that the frames be uncorrelated, both for the signal x and for the noise w. This is even though the blur h is confined to the individual frames, i.e., $h = h(n_1, n_2)$, which is also, of course, a necessary condition for the optimality of the linear filter (9.3-3).

9.3.5 INTERFRAME FILTERING

The equation for *interframe* filtering with filter impulse response $h(n_1, n_2, n)$ is

$$y(n_1, n_2, n) = \sum_{k_1, k_2, k} h(k_1, k_2, k) x(n_1 - k_1, n_2 - k_2, n - k).$$

Interframe filtering is the general case for spatiotemporal processing, and is the basis for the powerful motion-compensated models to be treated in Chapter 10.

The *interframe ARMA signal model* is given as

$$x(n_1, n_2, n) = \sum_{(k_1, k_2, k) \in \mathcal{R}_a - (0,0)} a_{k_1, k_2, k} x(n_1 - k_1, n_2 - k_2, n - k)$$

$$+ \sum_{(k_1, k_2, k) \in \mathcal{R}_b} b_{k_1, k_2, k} w(n_1 - k_1, n_2 - k_2, n - k). \quad (9.3\text{-}4)$$

Often these interframe models are categorized by how many prior frames they use. If we restrict to just one prior frame, we can write

$$x(n_1, n_2, n) = \sum_{(k_1, k_2) \in \mathcal{R}_a - (0,0)} a_{k_1, k_2, 0} x(n_1 - k_1, n_2 - k_2, n)$$

$$+ \sum_{(k_1, k_2) \in \mathcal{R}_a - (0,0)} a_{k_1, k_2, 1} x(n_1 - k_1, n_2 - k_2, n - 1)$$

$$+ \sum_{(k_1, k_2) \in \mathcal{R}_b} b_{k_1, k_2, 0} w(n_1 - k_1, n_2 - k_2, n)$$

$$+ \sum_{(k_1, k_2) \in \mathcal{R}_b} b_{k_1, k_2, 1} w(n_1 - k_1, n_2 - k_2, n - 1), \quad (9.3\text{-}5)$$

which explicitly shows the operation as FIR spatial filterings of the present and past input and output frames, and then their output summation to produce the next output point in the current frame.

In the frequency domain, we can write this filtering using 3-D Fourier transforms as

$$X(\omega_1, \omega_2, \omega) = \frac{B(\omega_1, \omega_2, \omega)}{1 - A(\omega_1, \omega_2, \omega)} W(\omega_1, \omega_2, \omega),$$

in the general case of (9.3-4), and

$$X(\omega_1, \omega_2, \omega) = \frac{B_0(\omega_1, \omega_2) + B_1(\omega_1, \omega_2) e^{-j\omega}}{1 - A_0(\omega_1, \omega_2) - A_1(\omega_1, \omega_2) e^{-j\omega}} W(\omega_1, \omega_2, \omega),$$

in the first-order temporal case of (9.3-5).

9.3.6 Interframe Wiener Filter

For the more general observation equation, with both blurring and observation noise,

$$y(n_1, n_2, n) = h(n_1, n_2, n) * x(n_1, n_2, n) + v(n_1, n_2, n).$$

we will need the *interframe Wiener filter*, expressed in 3-D Fourier transforms as

$$G(\omega_1, \omega_2, \omega) = \frac{H^*(\omega_1, \omega_2, \omega) S_x(\omega_1, \omega_2, \omega)}{|H|^2(\omega_1, \omega_2, \omega) S_x(\omega_1, \omega_2, \omega) + \sigma_v^2},$$

where we see the 3-D PSD $S_x(\omega_1, \omega_2, \omega)$, and the observation noise v is here assumed to be uncorrelated in time as well as in space, i.e., $R_v(m_1, m_2, m) = \sigma_v^2 \delta(m_1, m_2, m)$, with δ the 3-D Kronecker delta function.

EXAMPLE 9.3-5 (*first-order temporal interframe Wiener filter*)
For the case of the first-order temporal ARMA model of (9.3-5), we can compute the PSD as

$$S_x(\omega_1, \omega_2, \omega) = \left| \frac{B_0(\omega_1, \omega_2) + B_1(\omega_1, \omega_2) e^{-j\omega}}{1 - A_0(\omega_1, \omega_2) - A_1(\omega_1, \omega_2) e^{-j\omega}} \right|^2 \sigma_w^2,$$

so that the resulting interframe Wiener filter becomes

$G(\omega_1, \omega_2, \omega)$

$$= \frac{H^*(\omega_1, \omega_2, \omega) \left| \frac{B_0(\omega_1, \omega_2) + B_1(\omega_1, \omega_2) e^{-j\omega}}{1 - A_0(\omega_1, \omega_2) - A_1(\omega_1, \omega_2) e^{-j\omega}} \right|^2 \sigma_w^2}{|H|^2(\omega_1, \omega_2, \omega) \left| \frac{B_0(\omega_1, \omega_2) + B_1(\omega_1, \omega_2) e^{-j\omega}}{1 - A_0(\omega_1, \omega_2) - A_1(\omega_1, \omega_2) e^{-j\omega}} \right|^2 \sigma_w^2 + \sigma_v^2}$$

$$= \frac{H^*(\omega_1, \omega_2, \omega) |B_0(\omega_1, \omega_2) + B_1(\omega_1, \omega_2) e^{-j\omega}|^2 \sigma_w^2}{|H|^2(\omega_1, \omega_2, \omega) |B_0(\omega_1, \omega_2) + B_1(\omega_1, \omega_2) e^{-j\omega}|^2 \sigma_w^2 + |1 - A_0(\omega_1, \omega_2) - A_1(\omega_1, \omega_2) e^{-j\omega}|^2 \sigma_v^2}.$$

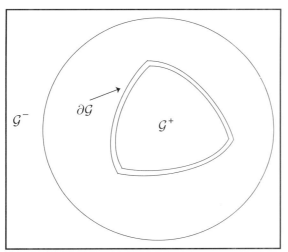

FIGURE 9.2 *Illustration of noncausal Markov field regions*

We notice that when $|H|^2(\omega_1, \omega_2, \omega)|B_0(\omega_1, \omega_2) + B_1(\omega_1, \omega_2) e^{-j\omega}|^2 \sigma_w^2 \gg |1 - A_0(\omega_1, \omega_2) - A_1(\omega_1, \omega_2) e^{-j\omega}|^2 \sigma_v^2$, then G is very close to an inverse filter. This condition is just the case where the blurred signal power $|H|^2 S_x \gg \sigma_v^2$, the power of the white observation noise.

More on intraframe and interframe Wiener filters, along with application to both estimation and restoration, is contained in the text by Tekalp [2].

In a 2-D ARMA model, if the input coefficient filtering through b_{k_1, k_2} is absent, we have just the 2-D NSHP AR model, which can be NSHP Markov in the case where the input random field is IID. In particular, if the input sequence w is white Gaussian, then the resulting random field is NSHP Markov. Now we turn to spatiotemporal or 3-D Markov.

9.4 SPATIOTEMPORAL MARKOV MODELS

We start out with the general definition of Markov random field sequence.

DEFINITION 9.4-1 (*a discrete Markov random field sequence*)
Let x be a random field sequence on $\mathcal{Z}^3$, the 3-D lattice. Let a band of minimum width p, $\partial \mathcal{G}$ ("the present"), separate $\mathcal{Z}^3$ into two regions: $\mathcal{G}^+$ ("the future") and $\mathcal{G}^-$ ("the past"). Then x is Markov-p if $\{x|_{\mathcal{G}^+}$ given $x|_{\partial \mathcal{G}}\}$ is independent of $\{x|_{\mathcal{G}^-}\}$ for all $\partial \mathcal{G}$.

These regions are illustrated in Figure 9.2, which shows a spherical region $\mathcal{G}^+$ inside a sphere of thickness $\partial \mathcal{G}$. Outside the sphere is the region $\mathcal{G}^-$.

For the homogeneous, zero-mean, Gaussian case, this noncausal Markov definition can be expressed by the following recursive equation model:

$$x(n_1, n_2, n) = \sum_{(k_1, k_2, k) \in D_p - (0,0,0)} c(k_1, k_2, k) x(n_1 - k_1, n_2 - k_2, n - k) + u(n_1, n_2, n),$$

$$(9.4\text{-}1)$$

where

$$E\{x(n_1, n_2, n) u(k_1, k_2, k)\} = \sigma_u^2 \delta_{n_1, k_1} \delta_{n_2, k_2} \delta_{n,k},$$

$$D_p \triangleq \{n_1, n_2, n \mid n_1^2 + n_2^2 + n^2 \leqslant p^2\}.$$

Here $u(n_1, n_2, n)$ is a Gaussian, zero-mean, homogeneous random field sequence with correlation function of bounded support,

$$R_u(n_1, n_2, n) = \begin{cases} \sigma_u^2, & (n_1, n_2, n) = 0 \\ -c_{n_1, n_2, n} \sigma_u^2, & (n_1, n_2, n) \in D_p - 0 \\ 0, & \text{elsewhere.} \end{cases}$$

The $c_{n_1, n_2, n}$ are the interpolation coefficients of the minimum mean-square error (MMSE) linear interpolation problem:

Given data x for $(n_1, n_2, n) \neq 0$, find the best linear estimate of $s(0)$. The solution is given as the conditional mean,

$$E\{x(0) \mid x \text{ on } (n_1, n_2, n) \neq 0\}$$

$$= \sum_{(k_1, k_2, k) \in D_p - (0,0,0)} c(k_1, k_2, k) x(-k_1, -k_2, -k).$$

In this formulation σ_u^2 is the mean-square interpolation error. Actually in Fig. 9.2, we could swap the past and the future, $\mathcal{G}^-$ and $\mathcal{G}^-$. However, this way when we shrink the sphere down to one point, we get the interpolative model (9.4-1).

9.4.1 CAUSAL AND SEMICAUSAL 3-D FIELD SEQUENCES

The future $\mathcal{G}^+$ can be defined in various ways that are constrained by the support of the coefficient array, supp(c):

- $\{n > 0\}$ in the *temporally causal case*.
- $\{(n_1, n_2) \neq (0, 0), n = 0\} \cup \{n > 0\}$ in the *temporally semicausal case*.
- $\{n_1 \geqslant 0, n_2 \geqslant 0, n = 0\} \cup \{n_1 < 0, n_2 > 0, n = 0\} \cup \{n > 0\}$, the *totally ordered temporally case*, also called a *nonsymmetric half-space* (NSHS).

The future–present–past diagram of the temporally causal model is shown in Figure 9.3. It can be obtained from the above noncausal model by stretching out the radius of the sphere in Figure 9.2 towards infinity. Then the thickness of the sphere becomes the plane shown in the temporally causal Fig. 9.3

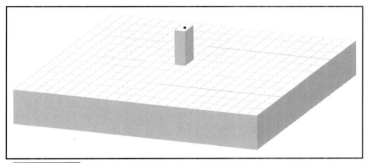

FIGURE 9.3   *Dot indicates the present, and the plane below is the immediate past*

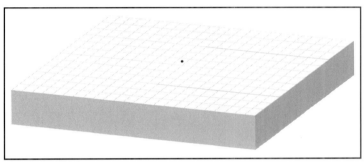

FIGURE 9.4   *Dot indicates the present, and it is surrounded by the immediate past*

for the Markov-1 case. A structured first-order temporally causal model can be constructed as follows:

$$x(n_1, n_2, n) = \sum_{k_1, k_2} c(k_1, k_2) x(n_1 - k_1, n_2 - k_2, n - 1) + u(n_1, n_2, n),$$

$$u(n_1, n_2, n) = \sum_{k_1, k_2} c^1(k_1, k_2, 1) u(n_1 - k_1, n_2 - k_2, n) + u^1(n_1, n_2, n).$$

Here $u(n_1, n_2, n)$ is uncorrelated, in the Gaussian case, in the time direction only, and $u^1(n_1, n_2, n)$ is strict near-neighbor correlated in space but completely uncorrelated in time. Effectively we have a sequence of 2-D Gauss Markov random fields u, driving a frame-wise prediction recursion for x.

The future–present–past diagram for a temporally semicausal model is shown in Figure 9.4. We can construct a Markov pth-order temporal example:

$$x(n_1, n_2, n) = \sum_{D_p - (0,0,0)} c(k_1, k_2, k) x(n_1 - k_1, n_2 - k_2, n - k) + u(n_1, n_2, n),$$

where

$$R_u(n_1, n_2, n) = \begin{cases} \sigma_u^2, & n = 0 \text{ and } (n_1, n_2) = 0 \\ -c_{n_1,n_2,n}\sigma_u^2, & n = 0 \text{ and } (n_1, n_2) \in D_p - 0 \\ 0, & n \neq 0 \text{ or } (n_1, n_2) \notin D_p. \end{cases}$$

9.4.2 REDUCED UPDATE SPATIOTEMPORAL KALMAN FILTER

For a 3-D recursive filter, we must separate the past from the future of the 3-D random field sequence. One way is to assume that the random field sequence is scanned in line-by-line and frame-by-frame,

$$x(n_1, n_2, n) = \sum_{(k_1, k_2, k) \in \mathcal{S}_c} c(k_1, k_2, k)x(n_1 - k_1, n_2 - k_2, n - k) + w(n_1, n_2, n),$$

where $w(n_1, n_2, n)$ is a white noise field sequence, with variance σ_w^2, and where

$$\mathcal{S}_c = \{n_1 \geqslant 0, \; n_2 \geqslant 0, \; n = 0\} \cup \{n_1 < 0, n_2 > 0, n = 0\} \cup \{n > 0\}.$$

Our observed image model is given by

$$r(n_1, n_2, n) = \sum_{(k_1, k_2, k) \in \mathcal{S}_h} h(k_1, k_2, k)x(n_1 - k_1, n_2 - k_2, n - k) + v(n_1, n_2, n).$$

The region $\mathcal{S}_h$ is the support of the spatiotemporal point spread function (psf) $h(k_1, k_2, k)$, which can model a blur or other distortion extending over multiple frames. The observation noise $v(n_1, n_2, n)$ is assumed to be an additive, zero-mean, homogeneous Gaussian field sequence with covariance $Q_v(n_1, n_2, n) = \sigma_v^2 \delta(n_1, n_2, n)$; i.e., a white Gaussian noise.

In recursive filtering, only a finite subset of the NSHS, called the *update region*, is updated at each step. This update region is slightly enlarged from the model support or local state for an (M_1, M_2, M)th-order $\oplus \oplus +$ model of the NSHS variety.

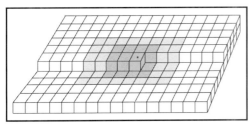

FIGURE 9.5 *Illustration of 3-D Kalman reduced support regions*

LOCAL STATE AND UPDATE REGION

The local state vector of a 3-D spatiotemporal RUKF that is of order $M = 1$ in the temporal direction, and spatial order $M_1 = M_2$, is given as

$$\mathbf{x}(n_1, n_2, n)$$

$$= \big[x(n_1, n_2, n), \ldots, x(n_1 - M_1 + 1, n_2, n);$$

$$x(n_1 + M_1 + 1, n_2 - 1, n), \ldots, s(n_1 - M_1 + 1, n_2 - 1, n);$$

$$\ldots; x(n_1 + M_1 + 1, n_2 - M_2, n), \ldots, x(n_1 - M_1 + 1, n_2 - M_2, n);$$

$$x(n_1 + M_1 + 1, n_2 + M_2, n - 1), \ldots, x(n_1 - M_1 + 1, n_2 + M_2, n - 1);$$

$$\ldots; x(n_1 + M_1 + 1, n_2 - M_2, n - 1), \ldots, x(n_1 - M_1 + 1, n_2 - M_2, n - 1)\big]^T.$$

Figure 9.5 shows a $1 \times 1 \times 1$-order *local state region* as the dark pixels, with a $2 \times 2 \times 1$-order *update region* $\mathcal{U}_{\oplus\oplus+}(n_1, n_2, n)$, to be updated recursively, shown as the medium-dark pixels in the figure. The total region shown is the *covariance update region* $\mathcal{T}_{\oplus\oplus+}(n_1, n_2, n)$. We see that these regions are NSHP in the current frame (the top layer) but of a symmetric support in the previous frame (lower layer). We only show two frames here, but, of course, there could be more, which would constitute a higher order recursive processing. The current pixel location in the current frame (n_1, n_2, n) is shown as the black dot in the figure.

The resulting approximate reduced update Kalman filter equations are given as

$$\hat{x}_b(n_1, n_2, n) = \sum_{(k_1, k_2, k) \in \mathcal{S}_c} c(k_1, k_2, k)\hat{x}_a(n_1 - k_1, n_2 - k_2, n - k),$$

$$\hat{x}_a(k_1, k_2, k) = \hat{x}_b(k_1, k_2, k) + g^{(n_1, n_2, n)}(n_1 - k_1, n_2 - k_2, n - k)$$

$$\times \left[r(n_1, n_2, n) - \sum_{(l_1, l_2, l) \in \mathcal{S}_h} h(l_1, l_2, l)\hat{x}_b(n_1 - l_1, n_2 - l_2, n - l) \right],$$

for all $(k_1, k_2, k) \in \mathcal{U}_{\oplus\oplus+}(n_1, n_2, n)$, the current update region.

The gain factors $g^{(n_1,n_2,n)}(n_1 - k_1, n_2 - k_2, n - k)$ are given in terms of error covariance functions $R_b^{(n_1,n_2,n)}$ by

$$g^{(n_1,n_2,n)}(k_1, k_2, k)$$

$$= \frac{\sum_{(l_1,l_2,l) \in S_h} h(l_1, l_2, l) R_b^{(n_1,n_2,n)}(n_1 - l_1, n_2 - l_2, n - l; n_1 - k_1, n_2 - k_2, n - k)}{\sum_{(l_1,l_2,l) \in S_h} \sum_{(o_1,o_2,o) \in S_h} h(l_1, l_2, l) R_b^{(n_1,n_2,n)}(n_1 - l_1, n_2 - l_2, n - l; n_1 - o_1, n_2 - o_2, n - o) + \sigma_v^2}$$

for all $(k_1, k_2, k) \in \mathcal{U}_{\oplus\oplus+}(n_1, n_2, n)$. The error covariance equations are given recursively as, *before update*,

$$R_b^{(n_1,n_2,n)}(n_1, n_2, n; k_1, k_2, k)$$

$$= \sum_{(l_1,l_2,l) \in S_c} c(l_1, l_2, l) R_a^{(n_1,n_2,n)}(n_1 - l_1, n_2 - l_2, n - l; k_1, k_2, k),$$

for all $(k_1, k_2, k) \in \mathcal{T}_{\oplus\oplus+}(n_1, n_2, n)$, and

$$R_b^{(n_1,n_2,n)}(n_1, n_2, n; n_1, n_2, n)$$

$$= \sum_{(l_1,l_2,l) \in S_c} c(l_1, l_2, l) R_a^{(n_1,n_2,n)}(n_1, n_2, n; n_1 - l_1, n_2 - l_2, n - l) + \sigma_w^2,$$

and, *after update*,

$$R_a^{(n_1,n_2,n)}(k_1, k_2, k; l_1, l_2, l)$$

$$= R_b^{(n_1,n_2,n)}(k_1, k_2, k; l_1, l_2, l) - g^{(n_1,n_2,n)}(n_1 - k_1, n_2 - k_2, n - k)$$

$$\times \sum_{(o_1,o_2,o) \in S_h} h(o_1, o_2, o) R_b^{(n_1,n_2,n)}(n_1 - o_1, n_2 - o_2, n - o; l_1, l_2, l),$$

for all $(k_1, k_2, k) \in \mathcal{U}_{\oplus\oplus+}(n_1, n_2, n)$ and for all $(l_1, l_2, l) \in \mathcal{T}_{\ominus\ominus+}(n_1, n_2, n)$.

As processing proceeds into the data, and away from any spatial boundaries, stability of the model generally has been found to lead to convergence of these equations to a steady state fairly rapidly, under the condition that the covariance update region $\mathcal{T}_{\oplus\oplus+}(n_1, n_2, n)$ is sufficiently large. In that case the superscripts on the error covariances and gain terms can be dropped and we have simply

$$g(k_1, k_2, k) =$$

$$= \frac{\sum_{(l_1,l_2,l) \in S_o} h(l_1, l_2, l) R_b(n_1 - l_1, n_2 - l_2, n - l; n_1 - k_1, n_2 - k_2, n - k)}{\sum_{(l_1,l_2,l) \in S_h} \sum_{(o_1,o_2,o) \in S_h} h(l_1, l_2, l) R_b(n_1 - l_1, n_2 - l_2, n - l; n_1 - o_1, n_2 - o_2, n - o) + \sigma_v^2}$$

for all $(k_1, k_2, k) \in \mathcal{U}_{\oplus\oplus+}(n_1, n_2, n)$, in which case the 3-D RUKF reduces to a simple linear spatiotemporal LSI filter.

As in the 2-D and 1-D cases, we see that the nonlinear error-covariance equations do not involve the data, just the signal model coefficients and variance parameters. A good way to design the steady-state 3D-RUKF is to run the error-covariance equations with the chosen stable signal model, not on the real image sequence, but just on a fictitious but much smaller sequence. Typically a sequence of 10 frames of 20×10 pixels is enough to obtain a quite accurate steady-state gain array g for commonly used image models. The 3-D RUKF has been extended to filter along motion trajectories [1,4] and this version will be covered in the next chapter after introducing motion estimation.

9.5 CONCLUSIONS

This chapter has provided extensions to three dimensions of some of the earlier two-dimensional signal processing methods. In the important spatiotemporal case, we have provided results of a general nature for both deterministic and random models. Chapter 10 will apply these methods to the processing of the 3-D spatiotemporal signal that is video. Chapter 11 will provide applications of 3-D processing in video compression and transmission.

9.6 PROBLEMS

1   Consider a $9 \times 9 \times 9$-order FIR filter, i.e., 10 taps in each dimension. How many multiplies are necessary per pixel to implement this filter: In general? If 3-D separable? If each separable factor is linear-phase?

2   In Example 9.1-1, we wrote the separable convolution representation,

$$y(n_1, n_2, n_3) = h_2(n_3) * [h_1(n_1, n_2) * x(n_1, n_2, n_3)],$$

of a system with two convolution operations. Carefully determine whether each one should be a 1-D or 2-D convolution by comparing to previous lines in Example 9.1-1.

3   Prove the 3-D Fourier transform property

$$\mathrm{FT}\{x(\mathbf{n}) \exp +j\boldsymbol{\omega}_0^T \mathbf{n}\} = X(\boldsymbol{\omega} - \boldsymbol{\omega}_0),$$

and specialize to the case

$$\mathrm{FT}\{(-1)^{n_1+n_2+n_3} x(\mathbf{n})\} = X(\omega_1 \pm \pi, \omega_2 \pm \pi, \omega_3 \pm \pi).$$

4   Extend the 2-D separable window filter design method of Chapter 5 to the 3-D case. Would there be any changes needed to the Kaiser window functions?

5   Extend Example 9.3-2 to the case of interlaced 3-D sampling with general sampling distances T_1 and T_2 in space, and T in time. Display both the sampling matrix $\mathbf{V}$ and the periodicity matrix $\mathbf{U}$.

6 If we sample the continuous parameter function $x_c(t_1, t_2, t_3)$ with sample vectors

$$\mathbf{v}_1 = \begin{pmatrix} 1 \\ 0 \\ 0 \end{pmatrix}, \qquad \mathbf{v}_2 = \begin{pmatrix} 1 \\ 1 \\ 0 \end{pmatrix}, \quad \text{and} \quad \mathbf{v}_3 = \begin{pmatrix} 0 \\ 0 \\ 1 \end{pmatrix},$$

express the resulting discrete parameter Fourier transform $X(\omega_1, \omega_2, \omega_3)$ in terms of the continuous parameter Fourier transform $X_c(\Omega_1, \Omega_2, \Omega_3)$. Specialize your result to the case where there is no aliasing.

7 Consider the first noncausal Wiener filter design method (Chapter 7, Section 7.2) and show that the method easily extends to the 3-D and spatiotemporal cases.

8 Estimate the memory size needed to run the 3D-RUKF equations on an image sequence of N frames, each frame being $N \times N$ pixels. Assume the model order is $1 \times 1 \times 1$, the update region is $2 \times 2 \times 1$, and the covariance update region is $4 \times 4 \times 1$.

REFERENCES

[1] J. Kim and J. W. Woods, "Spatiotemporal Adaptive 3-D Kalman Filter for Video," *IEEE Trans. Image Process.*, 6, 414–424, March 1997.

[2] A. M. Tekalp, *Digital Video Processing*. Prentice-Hall, Englewood Cliffs, NJ, 1995.

[3] P. P. Vaidyanathan, *Multirate Systems and Filter Banks*, Prentice-Hall, Englewood Cliffs, NJ, 1993.

[4] J. W. Woods and J. Kim, "Motion Compensated Spatiotemporal Kalman Filter." In *Motion Analysis and Image Sequence Processing*, I. Sezan and R. L. Lagendijk, eds. Kluwer, Boston, MA, 1993. Chapter 12.

DIGITAL VIDEO PROCESSING 10

In Chapter 9 we learned the generalization of multidimensional signal processing to the 3-D and spatiotemporal cases along with some relevant notation. In this chapter we apply and extend this material to processing the 3-D signal that is video, a sequence of 2-D images. We look at the general *interframe* or cross-frame processing of digital video. We study the various methods of motion estimation for use in motion-compensated temporal filtering and motion-compensated extensions of 3-D Wiener and Kalman filters. We then look at the video processing applications of deinterlacing and frame-rate conversion. Finally, we consider powerful Bayesian methods for motion estimation and segmentation.

10.1 INTERFRAME PROCESSING

EXAMPLE 10.1-1 *(spatiotemporal RUKF)*

The spatiotemporal RUKF introduced in Chapter 9 has application in noisy image sequence processing. In this example from [30], we show it processing a noisy version of the test clip *Salesman* at the spatial resolution of 360×280 pixels, and the frame rate of 15 fps. We created the noisy input by adding Gaussian white noise to set up an input SNR of 10 dB. After processing, the SNR improved by 6 to 7 dB, after an initial start-up transient of approximately 10 frames. We see a frame from the noise-free *Salesman* clip in Figure 10.1, followed by the noisy version in Figure 10.2. The smoothed result from the spatiotemporal (or 3D-) RUKF is shown in Figure 10.3. We see that the white noise has been reduced and the image frame has become somewhat blurred or oversmoothed.

FIGURE 10.1 *Frame from original CIF clip* Salesman

FIGURE 10.2   *Frame from* Salesman *with white noise added to achieve SNR = 10 dB*

FIGURE 10.3   *Frame from 3D-RUKF estimate*

The limit to what can be achieved by spatiotemporal filtering is determined by the overlap of the 3-D spectra of signal and noise. For fairly low-resolution signal and white noise, there will be a lot of overlap and hence a smoothing of the signal is unavoidable. For higher resolution video, such as obtained from HD cameras or the high-resolution scanning of 35 and 70 mm motion picture film, the overlap of the signal and noise spectra can be much less. In that case, more impressive results can be obtained.

The next example shows how deterministic multidimensional filters can be used in the processing of television signals to obtain what was called *improved definition*. A little known fact is that NTSC[1] color television lost some luma resolution that was present in the prior monochrome standard. Using multidimensional filters it was possible to largely get back the lost resolution [5].

EXAMPLE 10.1-2 (*multidimensional filter application to improved-definition television*)

Conventional analog NTSC broadcast television[2] has its luma power concentrated at low to mid frequencies and chroma data modulated up to spatial high-frequencies via the so-called chroma subcarrier. Compositing the chroma

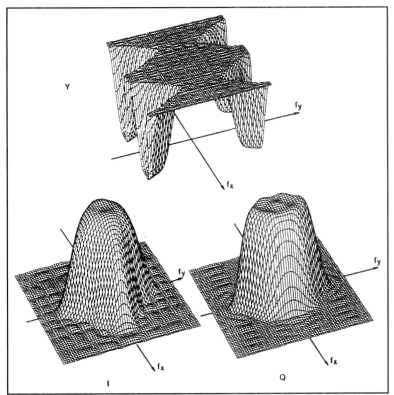

FIGURE 10.4   *2-D filter responses for generalized NTSC encoding of Y, I, and Q components. Reproduced with permission from Dubois and Schreiber [5]*

1. NTSC, National Television System Committee. Developed the Standard for television broadcast in 1953.

2. This is also known as a *composite signal*, since bandlimited chroma data is composited together with the bandlimited luma data.

data together with the luma data like this required, first, lowpass filtering of the luma data, but also rather severe lowpass filtering of the chroma data. This was the compromise solution that the NTSC decided upon to provide compatible color television in mid-20th century America (that is, compatible with the previously authorized monochrome, or luma-only, television). The downside of NTSC *compatible color television* was that the luma resolution went down slightly and the chroma (color) resolution was only about a sixth to an eighth that of the luma (monochrome) resolution, and that led to rather blurry color. Further complicating matters, typical receivers of the time were not able to completely separate the two signals, leading to *cross-color* and *cross-luma* artifacts. In the late 1980s, better solutions to separate the components at the transmitter were achieved through the use of multidimensional filtering. Perspective plots of three preprocessing filters are shown in Figure 10.4. The diamond band shape of these 11×31-point FIR filters was obtained using a nonseparable design method by Dubois and Schreiber [5]. These filters would be employed prior to forming an NTSC composite signal with very little to no overlap of the luma and quadrature-modulated chroma components. The filters in Figure 10.4 marked I and Q are used to bandlimit the chroma components prior to their modulation upwards in spectral frequency to fit in the spectral holes created by the filter marked Y that has created the Y component. Together with appropriate 2-D postprocessing filters, cross-color and cross-luma artifacts are effectively eliminated [5].

Receiver-processing multidimensional filters go by various names in the consumer electronics business, e.g., *comb* filter, line comb, and frame comb. Of course, the above transmitter filters should also be used in preprocessing the *YIQ* data prior to compositing, in order to reduce or eliminate any spectral overlap. Clearly the best way to avoid these cross-contamination problems is to keep the luma and chroma signals separate, as in *RGB* and *YIQ component format*. A compromise is *S-video*, where two channels are used, one for the luma and one for the two chroma signals. While S-video allows for a higher chroma bandwidth than does composite video, the best analog color bandwidth is still achieved by analog component format.

The next example shows modification of a designed spatial lowpass filter to meet needed step response characteristics.

EXAMPLE 10.1-3 [*video processing example* (adapted from Schröder and Blume [22])]
Television engineers are very concerned with the spatial impulse and step response of the filters used in video processing. In particular they want the first overshoot of the step response to be close in to the main rising section, and any further undershoots and overshoots to be small enough to be not visually noticed. One approach would be to use a general optimization program, ex-

tending the design methods of Chapter 5. First choose error criteria based on pth errors (p even) in both the frequency and spatial domains, with a weighting parameter λ to control the relative importance of frequency and spatial errors:

$$\left\| H_{\mathrm{I}}(\omega_1, \omega_2) - H(\omega_1, \omega_2) \right\|_p + \lambda \left\| s_{\mathrm{I}}(n_1, n_2) - s(n_1, n_2) \right\|_p,$$

where s_{I} and s are the ideal and acheived step responses, respectively. However, a simpler procedure, due to Schröder [21], is to cascade an enhancement network with a lowpass filter that has been conventionally designed. The goal of the enhancement circuit is to optimize the overall step response. The 3-tap enhancement filter with impulse response

$$h_{\mathrm{en}}(n_1) = -\frac{\alpha}{2}\delta(n_1 - 2) + (1 + \alpha)\delta(n_1) - \frac{\alpha}{2}\delta(n_1 + 2),$$

and frequency response

$$H_{\mathrm{en}}(\omega_1) = [1 + \alpha - \alpha \cos 2\omega] \exp{-j2\omega},$$

can optimize the horizontal step response of a 2-D lowpass filter. The parameter α is used to achieve the desired *visual definition* characteristics. Figure 10.5 shows a properly optimized step response on a vertical bar test pattern.

Figure 10.6 shows an under- or nonoptimized result and displays an overly smoothed appearance. Figure 10.7 shows the result of an overly optimized filter, i.e., too sharp a transition band, showing ringing around the

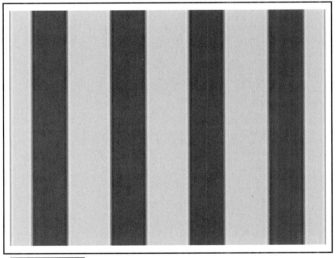

FIGURE 10.5   *Properly optimized or compensated lowpass filtering of test signal. Reproduced with permission from Schröder [21]*

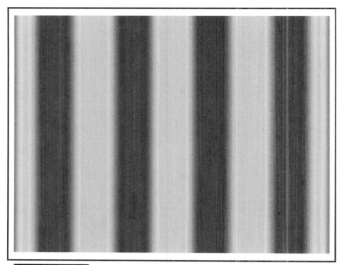

FIGURE 10.6 *Nonoptimized, showing oversmoothing. Reproduced with permission from Schröder [21]*

FIGURE 10.7 *Overoptimized, showing ringing. Reproduced with permission from Schröder [21]*

vertical edges. The same enhancement of visual definition can be done in the vertical direction by adding the needed vertical component to the enhancement filter, making the overall enhancement filter separable with 3×3 support.

10.2 MOTION ESTIMATION AND MOTION COMPENSATION

Motion compensation (MC) is very useful in video filtering to remove noise and enhance signal. It is also employed in all distribution-quality video coding formats. It is useful since it allows the filter or coder to process through the video on a path of near-maximum correlation based on following the motion trajectories across the frames making up the image sequence or video. Motion is usually characterized in terms of a motion or displacement vector, $\mathbf{d} = (d_1, d_2)^T$, that must be estimated from the data. Several methods of motion estimation are commonly used:

- Block matching
- Hierarchical block matching
- Pel-recursive methods
- Optical flow methods

Optical flow is the apparent motion vector field (d_1, d_2) we get from setting, i.e., forcing, equality in the so-called *constraint equation*

$$x(n_1, n_2, n) = x(n_1 - d_1, n_2 - d_2, n - 1). \qquad (10.2\text{-}1)$$

All four motion estimation methods start from this basic equation, which is just an idealization. Departures from the ideal are caused by covering and uncovering of objects in the viewed scene, lighting variation both in time and across the objects in the scene, movement toward or away from the camera, as well as rotation about an axis, i.e., spatially 3-D motion. Often this basic equation is only solved approximately in the least-squares sense. Also, the displacement is not expected to be an integer, as assumed in (10.2-1), often necessitating some type of interpolation to be used.

A basic problem with motion estimation is deciding the *aperture* or region over which we will estimate the motion, since we effectively assume that the motion is constant over this aperture. If the aperture is too large, then we will miss detailed motion and only get an average measure of the movement of objects in our scene. If the aperture is too small, the motion estimate may be poor to very wrong. In fact, the so-called *aperture problem* arises as illustrated in the square region shown in Figure 10.8. If the motion of the smooth dark region is parallel to its edge, then this motion cannot be detected. Since this situation would typically only hold for small regions in natural images, this aperture effect causes our choice of aperture size to avoid being too small. We are thus led to find a good aperture size for each problem.

An illustration of the problem of *covering* and *uncovering* is given in Figure 10.9, which shows a depiction of two successive frames n and $n - 1$, over which we wish to determine motion. We assume a simple object translating in the foreground over a fixed background, not an unreasonable local approximation of

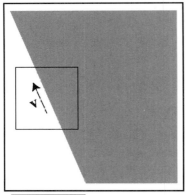

FIGURE 10.8   *Illustration of the aperture problem, with the square indicating the aperture size*

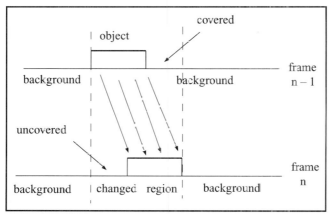

FIGURE 10.9   *Illustration of covering and uncovering of background by an object moving in the foreground*

video frames. We see that part of the background region in frame *n* is uncovered, while part of the background region from frame *n* − 1 is covered. Motion estimation that simply tries to match regions in the two frames will not be able to find good matches in either the covered or uncovered regions. However, within the other background regions, matches should be near perfect, and matching should be good within a textured object, at least if it moves in a trackable way and the pixel samples are dense enough. However, if we consider small regions around the object edges, we can expect problems finding good matches there.

10.2.1 BLOCK MATCHING METHOD

We intend to estimate a displacement vector at the location (n_1, n_2, n). In the *block matching* (BM) method [11], we match a block centered on this point to blocks in a specified *search area* in the previous frame, as illustrated in Figure 10.10.

Often the search area size is given as $(\pm M_1, \pm M_2)$, and if centered on the current pixel location (n_1, n_2) in the previous frame, then a total of $(2M_1 + 1) \times (2M_2 + 1)$ matches must be done in a so-called *full search*, and this must be done for each pixel where the motion is desired. Often the block matching is not conducted at every pixel and an interpolation method is used to estimate the motion in between these points. Common error criteria are *mean-square error* (MSE), *mean-absolute difference* (MAD), or even number of pixels in the block actually disagreeing for discrete-amplitude or digital data.

We express the MSE in block matching as

$$\mathcal{E}(\mathbf{d}) = \sum_{\mathbf{k}} \left(x(\mathbf{n} + \mathbf{k}, n) - x(\mathbf{n} + \mathbf{k} - \mathbf{d}, n - 1) \right)^2, \qquad (10.2\text{-}2)$$

for a block centered at position $\mathbf{n} = (n_1, n_2)^T$ as a function of the vector displacement $\mathbf{d} = (d_1, d_2)^T$. We seek the displacement vector

$$\mathbf{d}_o \triangleq \arg \min_{\mathbf{d}} \mathcal{E}(\mathbf{d}).$$

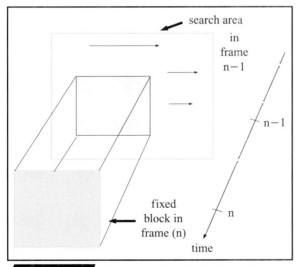

FIGURE 10.10   *Illustration of simple block matching*

Since the MSE in (10.2-2) is susceptible to outliers, often the MAD is used in applications. In addition to its being less sensitive to statistical outliers, another advantage of MAD is that it is simpler to compute.

In addition to the computationally demanding full-search method, there are simpler approximations involving a much reduced number of evaluations of the error. The methods either involve sampling the possible locations in the search region or sampling the elements in the block when calculating (10.2-2). Two examples of the former strategy are the 2D-log search and the three-step search (Figure 10.11); the performance of both is shown in Figure 10.12 [9]. The three-step search proceeds as follows: First the search window is broken up into four quadrants, and motion vectors are tested on a 3×3 grid with corners centered in the four quadrants of a square, as shown in Figure 10.11, which illustrates a case where the lower right corner of the square is the best match at the first step, the upper right corner of the once-reduced square is best at the second step, and the top right corner of the twice-reduced square is best at the third and final step.

We note that all three techniques perform much better than simple frame differencing, which equivalently treats all displacement vectors as zero. While both fast methods cut computation by a factor of 10 or more versus full-search block matching, they can lose up to 1–2 dB in *prediction PSNR*, measured in decibels as

$$PSNR = 10 \log_{10} \left[\frac{255^2}{\mathcal{E}(\mathbf{d})} \right],$$

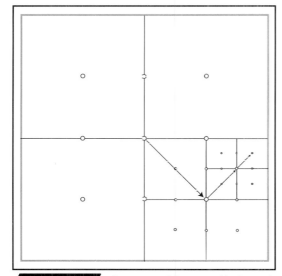

FIGURE 10.11 *Illustration of three-step block matching*

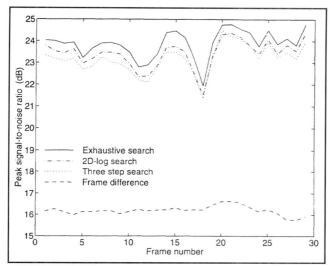

FIGURE 10.12   *Illustration of PSNR performance of exhaustive, 2D-log, three-step search, and simple frame difference. Reproduced with permission from Hang and Chou [9]*

for the 8-bit images being considered here, since their peak value would be 255. For 10-bit images, the formula would substitute 1023 for the peak value.

The other class of methods to speed up block matching involves sampling in the calculation of the error (10.2-2). So the amount of computation in evaluating the distortion for each searched location is reduced. Liu and Zaccarin [15] presented a method involving four phases of subsampling and alternated the subsampling patterns among the blocks, while using the MAD error criterion. This method achieved approximately a fourfold reduction in the amount of computation, but only a slight increase in the average prediction MSE.

There are some unavoidable problems with the block matching approach. A small blocksize can track small moving objects, but the resulting displacement estimate is then sensitive to image noise.[3] For example, a small block might just span a flat region in the image, where the displacement cannot be defined. A large blocksize is less sensitive to noise, but cannot track the motion of small objects. Similarly, a large search area can track fast motion, but is computationally intensive. A small search area may not be large enough to catch or track the real motion.

The best matching block is often good enough for a block-based video compression, where bits can be spent coding the prediction residual to cover up this error. However, in video filtering, when the estimated motion is not the true phys-

---

3. By the way, it turns out that some amount of noise is always present in real images. This comes from the common practice of setting the bit depth on sensors and scanners to reach the first bit or two bits of the physical noise level, caused by photons, grains, etc.

ical motion, visible artifacts will often be created. Hence, we need an improvement on the basic block matching method for the filtering application. The same is true for MC interpolation, frame-rate conversion, and pre- and postprocessing in video compression. Also, new highly scalable video coders use MC in a temporal filtering structure to generate lower frame-rate versions of the original video. Highly accurate motion vectors are important in this case too. A variation of block matching, called hierarchical block matching, can achieve a much better estimate of the true motion.

10.2.2 HIERARCHICAL BLOCK MATCHING

The basic idea of hierarchical algorithms is to, first, estimate a coarse motion vector at low spatial resolution. Then this estimated motion is refined by increasingly introducing higher spatial frequencies. Both subband/wavelet pyramids and Gaussian pyramids have been used for this purpose. An often cited early reference on *hierarchical block matching* (HBM) is Thoma and Bierling [26].

We start out by creating a spatial pyramid, with resolution levels set at a power of two. We start out at the lowest resolution level (highest pyramid level) and perform simple block matching. It has been found helpful to have the blocksize there agree with the average size of the largest moving areas in the video. Typically three or four stages of resolution are employed. We then start down the pyramid, increasing the resolution by the factor 2×2 at each level. We double the displacement vector from the previous level to get the initial search location at the present level. We finish up at the pyramid base level, which is full resolution. Both the blocksizes and the search regions can be chosen distinctly for each resolution, and are generally in the range from 4×4 to 32×32. The maximum search area is usually small at each resolution, i.e., ± 2, since only refinements are needed. The search area may also be small at the initial pyramid level because of the low resolution. Thus we can expect considerable savings in complexity for HBM. Some other improvements to block matching include subpixel accuracy, *variable-size block matching* (VSBM), overlapping blocks, block prediction of motion vectors, and *hierarchical VSBM* (HVSBM).

A diagram of HVSBM is shown in Figure 10.13. We start out with a spatial pyramid that can be obtained as a succession of LL subbands by subband/wavelet filtering and 2×2 decimation. Starting at the coarsest level at the top of the pyramid, a block matching motion estimation is performed. Then this displacement estimate d_0 is propagated down the pyramid one level, and $2d_0$ is used to initialize a search over a small region to *refine* the motion value to d_1. At this time, if the MC error measure is too large, the block is *split* into four, and the process of refining is repeated to generate d_1, and this process of refining and splitting is carried down the pyramid to the bottom, resulting in a variable-size block-based motion field. In a computationally more demanding variation of HVSBM, we start at the

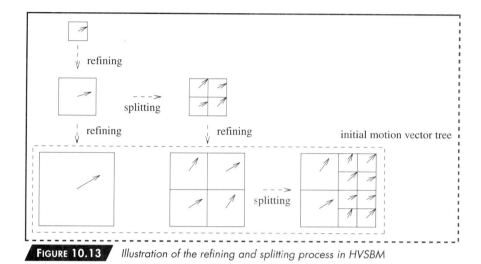

FIGURE 10.13 *Illustration of the refining and splitting process in HVSBM*

coarsest resolution with the smallest blocksize, and refine this motion field to the bottom of the pyramid, i.e., the highest resolution level. Then this resulting motion field is *pruned* back, by merging nodes, to a variable-size block-based motion field. This can be done using the BFOS algorithm, and this *bottom-up* approach generally results in a more accurate motion field than does the top-down method mentioned in the previous paragraph, but it is more computationally intensive.

In the video coding application, the error criteria can be composed of motion field error, either *mean-square error* (MSE) or *mean absolute difference* (MAD) plus $\lambda\Delta R_{mv}$, to measure the increase in motion-vector rate R_{mv} due to the split (top-down) or the decrease in motion-vector rate due to the merge (bottom-up). More on coding motion vectors for video compression is contained in Chapter 11.

10.2.3 OVERLAPPED BLOCK MOTION COMPENSATION

The motivation to overlap the blocks used in a conventional block matching estimate is to increase the smoothness of the resulting velocity field. This can be considered as a method to reduce the spatial aliasing in sampling the underlying velocity field. While one could simply overlap the blocks used in a simple block matching estimate, this could mean much more computation. For example, if the blocks were overlapped by 50% horizontally and vertically, it would be four times more computation if the block matching estimation were done independently, as well as four times more velocity information to transmit in the video compression application of Chapter 11. So, effectively we are more interested in smoothing than in alias reduction, and the *overlapped block motion compensation* (OBMC) technique [17,18] simply weights each velocity vector with four neighbor velocity estimates from the four nearest neighbor nonoverlapping blocks. Thus we effectively overlap the velocity vectors without overlapping the blocks. This is usually done with a few prescribed weighting windows.

A theoretical motivation for the overlapping can be obtained from (4) of Orchard and Sullivan [13],

$$\hat{x}(\mathbf{n}, n) = E\{x(\mathbf{n},n)|\mathbf{X}(n-1), \mathcal{V}_n\}$$

$$= \int f_n(\mathbf{v}|\mathcal{V}_n)x(\mathbf{n} - \mathbf{v}\Delta t, n-1)\, d\mathbf{v},$$

only slightly changed for our notation. Here we are performing a motion-compensated estimate of the intensity of frame $\mathbf{X}(n)$, as a conditional mean over shifted versions of frame $\mathbf{X}(n-1)$, with interframe interval Δt, making use of the conditional pdf $f_n(\mathbf{v}|\mathcal{V}_n)$, which depends on $\mathcal{V}_n$, the motion vector data sent in this and neighboring blocks. Assuming linear weights, i.e., no dependence on the data values in $\mathcal{V}_n$, they obtain

$$\hat{x}_n(\mathbf{n}) = \sum_{N_b(\mathbf{n})} w_b(\mathbf{n})x(\mathbf{x} - v_b\Delta t, n-1)\,, \qquad (10.2\text{-}3)$$

where the sum is over velocity vectors in the neighboring blocks $N_b(\mathbf{n})$. A formula for obtaining an optimized block weighting function $w_b(\mathbf{n})$ (simple weighting windows) is also given.

The initial estimate obtained in this way can be improved upon by iteratively updating the velocity estimates from the various blocks, one at a time, by utilizing the resulting overlapped estimate (10.2-3) in the error calculation (10.2-2). OBMC is used in the H.263 video compression standard for visual conversation and has also been adapted for use in some SWT video coders. In the compression application, it can smooth the velocity field without the need to transmit additional motion vector bits, since the block overlapping can be done separately at the receiver given the transmitted motion vectors still only one for each non-overlapped block. The overlapping of the estimates makes the velocity field smoother and removes the artificial blocked structure. This is especially important for SWT coders, where a blocky motion vector field could lead, through motion compensation, to a blocky prediction residual that would have false and excessively high spatial frequency information.

10.2.4 PEL-RECURSIVE MOTION ESTIMATION

This iterative method recursively calculates a displacement vector for each pixel (aka pel) in the current frame. We start with an estimate $\mathbf{d} = (d_1, d_2)^T$ for the current displacement. Then we use the iterative method:

$$\hat{d}_1^{(k+1)} = \hat{d}_1^{(k)} - \varepsilon \left.\frac{\partial \mathcal{E}}{\partial d_1}\right|_{\mathbf{d}=\hat{\mathbf{d}}^{(k)}},$$

$$\hat{d}_2^{(k+1)} = \hat{d}_2^{(k)} - \varepsilon \left.\frac{\partial \mathcal{E}}{\partial d_2}\right|_{\mathbf{d}=\hat{\mathbf{d}}^{(k)}},$$

with initial value supplied by the final value at the previously scanned pixel,

$$\hat{\mathbf{d}}^{(0)}(n_1, n_2) = \hat{\mathbf{d}}^{(\text{final})}(n_1 - 1, n_2),$$

where $\mathcal{E}$ is a local measure of average motion estimation error.

A key reference is Netravalli and Robbins [16] (see also Hang and Chou [9]). The method works well with just a few iterations when the motion is small, but often fails to converge when the displacements are large. In Bierling and Thoma [1], this differential displacement approach was extended to hierarchically structured motion estimation, with application to image sequence frame interpolation.

10.2.5 OPTICAL FLOW METHODS

This method is a least-squares approximation to the *spatiotemporal constraint equation*,

$$v_x \frac{\partial f}{\partial x} + v_y \frac{\partial f}{\partial y} + \frac{\partial f}{\partial t} = 0, \tag{10.2-4}$$

which is derived by partial differentiation of the optical flow equation (10.2-1) rewritten as a function of real variables (x, y) with velocity parameters v_x and v_y,

$$f(x, y, t) = f(x - v_x \, dx, y - v_y \, dy, t - \Delta t).$$

Equation (10.2-4) is then approximated in a least-squares sense over small local regions to get a velocity estimate. Specifically, we form the error criteria

$$\iiint_{\mathcal{R}(x,y,t)} \left(v_x \frac{\partial f}{\partial x} + v_y \frac{\partial f}{\partial y} + \frac{\partial f}{\partial t} \right)^2 dx \, dy \, dt,$$

where $\mathcal{R}(x, y, t)$ is a suitably small spatiotemporal region (aperture) centered at the position (x, y, t). Thus optical flow is a differential method that works by approximating the derivatives rather than the function error itself, as in block matching. In practice, a smoothing term must be added to this error term to *regularize* the estimate, which otherwise would be much too rough, i.e., there would be too much high-frequency energy in the velocity estimate $\hat{\mathbf{v}}(x, y, t)$. In the Horn and Schunck method [10] a smoothness term is introduced via a Lagrange multiplier as

$$\lambda \iiint_{\mathcal{R}(x,y,t)} \left(\nabla^2 v_z + \nabla^2 v_y\right)^2 dx\,dy\,dt,$$

which, for large values of the positive parameter λ, makes the velocity estimate change slope only gradually as a function of the spatial variables x and y. The overall error criterion then becomes

$$\mathcal{E}^2(\mathbf{v}) = \iiint_{\mathcal{R}(x,y,t)} \left(v_z \frac{\partial f}{\partial x} + v_y \frac{\partial f}{\partial y} + \frac{\partial f}{\partial t}\right)^2 dx\,dy\,dt$$

$$+ \lambda \iiint_{\mathcal{R}(x,y,t)} \left(\nabla^2 v_x + \nabla^2 v_y\right)^2 dx\,dy\,dt, \tag{10.2-5}$$

and we seek the minimizing velocity vector

$$\hat{\mathbf{v}} \triangleq \arg\min \mathcal{E}^2(\mathbf{v}),$$

at each (x, y, t) position. While this estimate has been used a lot in computer vision, it is not often used in video compression because of its dense velocity estimate. However, the optical flow estimate has been used in video (image sequence) filtering, where the need to transmit the resulting motion vectors does not occur.

In practice, we must approximate these integrals from the sampled data available, i.e., from frames of pixels. Thus the integrals in (10.2-5) have to be approximated on a dense grid of pixels, with the integrals replaced by summations over local neighborhoods $\mathcal{R}(n_1, n_2, n)$, and the partial derivatives also approximated. Often the approximation of derivatives is just done by first-order difference, i.e.,

$$\frac{\partial f}{\partial x} \approx \frac{f(x, y, t) - f(x - \Delta x, y, t)}{\Delta x},$$

but better results can be obtained using longer digital filters to provide improved estimates of these derivatives of the, assumed bandlimited, analog image frames [7].

The actual estimate is then obtained via least squares applied to the approximate version of (10.2-5). We can differentiate this equation with respect to both v_x and v_y and set the derivatives to zero, resulting in the by-now familiar set of linear equations (Normal equations) to solve. Assuming almost any spatial activity or texture in the aperture, the determinant of the coefficients will be nonsingular and a unique velocity estimate will emerge.

Having estimated motion, the next section moves on to motion compensated filtering.

10.3 MOTION-COMPENSATED FILTERING

If the motion is slow or there is no motion at all, then simple temporal filtering can be very effective for estimation, restoration, frame-rate change, interpolation, and pre/postprocessing. In the presence of strong motion, however, artifacts begin to appear in the simple temporal filter outputs, and the needed coherence in the input signal begins to break down. Such coherence is needed to distinguish it from the noise, distortion, interference, and artifacts that may also be present in the input frame data. A solution is to modify the filter trajectory so that it follows along the trajectory of motion. In this way signal coherence is maintained, even with moderate to fast motion.

The basic idea is to modify an LSI filter as follows:

$$y(n_1, n_2, n) = \sum_{k_1, k_2} h(k_1, k_2, 0) x(n_1 - k_1, n_2 - k_2, n)$$

$$+ \sum_{k_1, k_2} h(k_1, k_2, 1) x(n_1 - d_1 - k_1, n_2 - d_2 - k_2, n - 1) \quad (10.3\text{-}1)$$

$$+ \text{etc.}$$

Here $d_1 = d_1(n_1, n_2, n)$ is the horizontal component of the displacement vector between frames n and $n - 1$, and d_2 is the vertical component of displacement. In order to get the corresponding terms for frame $n - 2$, we must add the displacement vectors from the frame pair $n - 1$ and $n - 2$, to get the correct displacement. We should add them vectorially, i.e.,

$$d_1'(n_1, n_2) = d_1(n_1, n_2, n) + d_1\big(n_1 - d_1(n_1, n_2, n), n_2 - d_2(n_1, n_2, n), n - 1\big),$$

$$d_2'(n_1, n_2) = d_2(n_1, n_2, n) + d_2\big(n_1 - d_1(n_1, n_2, n), n_2 - d_2(n_1, n_2, n), n - 1\big).$$

Here we assume that the displacement vectors are known or estimated previously, and that they are integer valued. In the case where they are not integer valued, which is most of the time, then the corresponding signal value, e.g., $x(n_1 - d_1 - k_1, n_2 - d_2 - k_2, n - 1)$, must be itself estimated via interpolation. Most often spatial interpolation is used based on various lowpass filters. Effectively, in (10.3-1) we are filtering along the motion paths rather than simply filtering straightforward (or backward) in time (n) at each spatial location. One way to conceptualize this is via the diagram in Figure 10.14.

10.3.1 MC–WIENER FILTER

In Figure 10.15, the operator MC denotes *motion-compensated warping* performed on the noisy observations $y(n_1, n_2, n)$. The Wiener filtering is then done

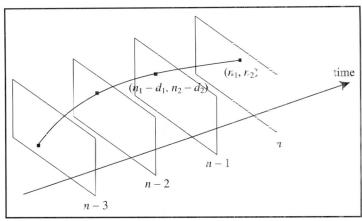

FIGURE 10.14 *Illustration of motion-compensated filtering along a motion path*

FIGURE 10.15 *Illustration of MC warping followed by Wiener filter, followed by IMC warping*

on the warped data in the MC domain, with signal and noise PSDs calculated from some similar MC data called the prototype. At the output, the *inverse motion compensation* (IMC) operator dewarps the frames back to original shape, to produce the output estimate $\hat{x}(n_1, n_2, n)$. Three-dimensional MC–Wiener filtering was introduced in Ozakan *et al.* [19]. The concept of IMC depends on the motion field being one-to-one. In a real image sequence, there are a relatively small number of pixels where this is not true, due to coverings and uncoverings of objects in the scene, the so-called *occlusion problem*. In these areas some approximation must be used in Figure 10.15 in order to avoid introducing artifacts into the final video estimate. It is common to resort to intraframe filtering in these occluded areas.

Because of the strong correlation in the temporal direction in most video, there is often a lot to gain by processing the video jointly in both the temporal and spatial directions. Since a Wiener filter is usually implemented as an FIR filter, exploiting the temporal direction in this way means that several to many frames must be kept in active memory. An alternative to this method is the spatiotemporal Kalman filter, which can use just one frame of memory to perform its recursive estimate, a motion-compensated version of which is presented next. Of course, both methods require the use of a suitable image-sequence model and noise model. The signal model spectra can be obtained via spectral estimation on similar noise-free data for the Wiener filter, while the Kalman filter needs parameter estimation of an autoregressive model. Both models must be trained or estimated on data that has been warped by the motion compensator, since this warped domain is where their estimate is performed.

10.3.2 MC–KALMAN FILTER

The basic idea here is that we can apply the 3-D RUKF of Chapter 9 along the motion trajectory using a good motion estimator that approximates true motion. As before, we can use multiple models for both motion and image estimation. To reduce object blurring and sometimes even double images, we effectively shift the temporal axis of the filter to be aligned with motion trajectories. When a moderate-sized moving object is so aligned, we can then apply the filter along the object's trajectory of motion, by filtering the MC video. Since the MC video has a strong temporal correlation, its image sequence model will have a small prediction error variance. This suggests that high spatial frequencies can be retained even at low input SNRs via motion-compensated Kalman filtering. The overall block diagram of a motion-compensated 3-D Kalman filter [30], or MC–RUKF, is shown in Figure 10.16. This motion-compensated spatiotemporal filter consists of three major parts: the motion estimator, the motion compensator, and the 3-D RUKF. While filtering a video, two different previous frames could be used for motion estimation together with the present noisy frame $y(\mathbf{n})$; one could use either the previous smoothed frame $E\{\mathbf{x}(n-1)|\mathbf{y}(n), \mathbf{y}(n-1), \ldots\}$, or the previous noisy frame $\mathbf{y}(n-1)$. We have found it best to use the smoothed previous frame, the best estimate currently available. For this motion estimation, a hierarchical block matching method was used [30].

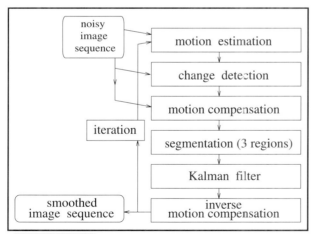

FIGURE 10.16   *System diagram for motion-compensated spatiotemporal Kalman filter*

The motion estimate is used to align a set of local frames along the motion trajectory. To effect this local alignment, the smoothed previous frame estimate is MC displaced with respect to the current frame. In an iterative extension to this basic method, shown in Figure 10.16, two smoothed frames are used to improve on this initial motion estimate. These smoothed frames retain spatial high frequencies and have reduced noise, so that these frames can now be used for motion estimation with a smaller size block. A motion vector field is then estimated from these two frames, i.e., $E\{x(n-1)|y(n), y(n-1), \ldots\}$ and $E\{x(n)|y(n+1), y(n+2), \ldots\}$, followed by a second application of the steady-state 3-D RUKF. A few iterations suffice.

MULTIMODEL MC–RUKF

The local correlation between two MC frames depends on the accuracy of the motion estimation. Since the SNR of the noisy observed video will be low, the motion estimation has a further limitation on its accuracy, due to statistical variations. To deal with the resulting limited motion vector accuracy, we can use a variable number of models to match the various motion regions in the image; for example, we can use three motion models: *still*, *predictable*, and *unpredictable*. The motivation here is that in the still region, we can perform unlimited temporal smoothing at each pixel location. In the predictable region, there is motion but it is motion that can be tracked well by our motion estimator. Here we can smooth along the found motion trajectory with confidence. Finally, in the unpredictable region, we find that our motion estimate is unreliable and so we fall back on the spatial or 2-D RUKF there. This *multiple model* version (MM MC–RUKF) results in a very high temporal coherence in the still region, high temporal coherence in the predictable region, and no motion blurring in the unpredictable region. This segmentation is based on local variance of the *displaced frame difference* (DFD).

As mentioned earlier, we employ a block matching method for motion estimation. Even when there is no local correspondence between two motion-compensated frames, the block matching method still chooses a pixel in the search area that minimizes the displaced frame difference measure. However, the resulting estimate will probably not have much to do with the real motion, and this then can lead to low temporal correlation. This is the so-called *noisy motion vectors* problem. We can compensate for this, in the case of still regions, by detecting them with an extra step, based on the straight frame difference. The frame difference is filtered and a simple 7×7 box filter is used to reduce the effect of the observation noise. Also, a 3×3 box filter is used on the MC output to detect the predictable region. The outputs are fed into local variance detectors. When a pixel in a still region is miss-classified to be as in the predictable region, a visible error in the filtered image sequence is noticeable, while in the opposite case, the error is not noticeable. Hence, we detect the still region again in the filtering step. Three spatiotemporal AR models are obtained from the residual video of the original sequence for our simulation. (For more details, please see [30].)

EXAMPLE 10.3-1 (*experimental result*)

We used the *Salesman*, monochrome and of CIF size, 360×280 pixels, at the frame rate 15 fps. We then added white Gaussian noise to achieve a 10 dB SNR. The processing parameters of the MC–RUKF were image model order $1 \times 1 \times 1$, update region $2 \times 2 \times 1$, and final MC blocksizes of both 9×9 and 5×5. The 3-D autoregressive model obtained from the original (modified) video was used. This model could also have been obtained from the

FIGURE 10.17   *Frame from MC–RUKF*

FIGURE 10.18   *Frame from MM MC–RUKF*

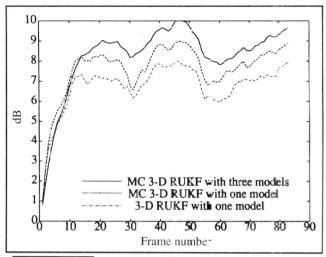

FIGURE 10.19   *Plot of PSNRs versus frame number for MM MC–RUKF, MC–RUKF, and 3-D RUKF on noisy Salesman clip*

noisy video or from a prototype noise-free, with some loss of performance. (Based on existing work in the identification of 2-D image models, it is our feeling that the additional loss would not be great.) We used a steady-state gain array, calculated off-line on a small fictitious image sequence. The SNR improvement achieved was up to 8 dB with the 3-D RUKF alone, with an additional MC–RUKF further improvement of about 1 dB. Using the multi-model feature, a further MM MC–RUKF improvement of about 1 dB was achieved. The restored video in Figures 10.17 and 10.18 shows motion artifacts visible in some motion areas, but is generally quite visually pleasing.

The resulting PSNR curves are given in Figure 10.19. We notice that the MM MC–RUKF provides the best objective performance by this MSE-based measure. We can see the initial start-up transient of about 10 frames. We notice also how the up to 10 dB improvement varies with frame number; this is caused by the motion of objects and moving shadows in the scene.

10.3.3 FRAME-RATE CONVERSION

TV standard or format conversion is needed today between the analog world standards NTSC (30 fps) and PAL and SECAM (25 fps)[4] and also leads to a convenient separation of acquisition format, transmission standard, and viewing or

4. PAL, Phase Alternating Line standard; introduced in most of Europe in the 1960s. SECAM, Sequential Couleur Avec Memoire (Sequential Color with Memory) standard; introduced in France in the 1960s.

display format. Frame-rate up-conversion is also often used to double (quadruple) the frame rate for display, e.g., from 25 fps to 50 or 100 fps. There are various methods for increasing the frame rate, the simplest being frame repeat, i.e., sample and hold in time. Somewhat more complicated is making use of a straight temporal average, without motion compensation. The filtering is a 1-D interpolation, i.e., linear filtering, in the temporal direction, done for each pixel separately. A potentially more accurate method for frame-rate increase is MC-based frame interpolation, first suggested in Thoma and Bierling [26].

EXAMPLE 10.3-2 (*up-conversion*)

We applied the method of [26] to the test clip *Miss America*, which is color and Source Interchange Format (SIF) sized 352×240, with 150 frames at 30 fps. To perform our simulation, we first decimated it down to 5 fps, and used only this low frame rate as input. Then we used the following strategies to interpolate the missing frames, i.e., raise the frame rate back up to 30 fps: frame replication, linear averaging, and motion-compensated interpolation. As a motion estimation method, we employed HBM, with the result smoothed by a simple lowpass filter.

Figure 10.20 shows a frame from temporal up-conversion using a simple linear averaging filter. Note that during this time period there was motion, which has caused a double image effect on the up-converted frame. Figure 10.21 shows the result of using the motion-compensated up-conversion at a frame number near to that of the linear result in Figure 10.20. We do not see any double image and the up-conversion result is generally artifact free.

FIGURE 10.20 *Frame from a linearly interpolated temporal up-conversion of Miss America from 5 to 30 fps*

FIGURE 10.21   *Frame from motion-compensated temporal up-conversion of Miss America from 5 to 30 fps (see the color insert)*

In this case our translational motion model worked very well, in part because of the rather simple motion displayed in the *Miss America* clip. It does not always work this well, and MC up-conversion remains a challenging research area.

10.3.4 DEINTERLACING

As mentioned in Example 2.2-5 of Chapter 2, deinterlacing is used to convert from a conventional interlaced *standard definition* (SD) format to one that is progressive, or noninterlaced. In so doing, the deinterlacer must estimate the missing data, i.e., the odd lines in the so-called *even frames* and the even lines in the *odd frames*. A conventional deinterlacer makes use of a diamond v–t multidimensional filter, in upsampling the data to progressive format. If the interlaced video had been prefiltered prior to its original sampling on this lattice, then, using an ideal filter, the progressive reconstruction can be exact, but still with the original spatiotemporal frequency response. If proper prefiltering had not been done at the original interlaced sampling, then aliasing error may be present in both the interlaced and progressive data. In this case, a nonideal frequency response for the conversion filter can help to suppress the alias energy that usually occurs at high spatiotemporal locations.

EXAMPLE 10.3-3 (*conventional deinterlacer*)
This example uses a 9×9 diamond-shaped support in the $v \times t$ plane. The filter coefficients, shown in Table 10.1, were obtained via window-based design. The frequency response of this filter is shown in Figure 10.22, where we

Table 10.1. v–t diamond filter coefficients

0	0	0	0	0.001247	0	0	0	0
0	0	0	0.004988	−0.005335	0.004988	0	0	0
0	0	0.007481	−0.016016	−0.013060	−0.016016	0.007481	0	0
0	0.004988	−0.016016	−0.036095	0.162371	−0.036095	−0.016016	0.004988	0
0.001247	−0.005339	−0.013060	0.162371	0.621808	0.162371	−0.013060	−0.005339	0.001247
0	0.004988	−0.016016	−0.036095	0.162371	−0.036095	−0.016016	0.004988	0
0	0	0.007481	−0.016016	−0.013060	−0.016016	0.007481	0	0
0	0	0	0.004988	−0.005339	0.004988	0	0	0
0	0	0	0	0.001247	0	0	0	0

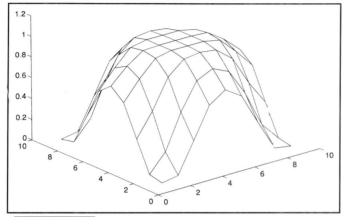

FIGURE 10.22 *Sketch of diamond filter response in v–t frequency domain (origin at center)*

can see a broader response along both temporal and vertical frequency axes than along diagonals, hence approximating a diamond pattern in the $v \times t$ ($n_2 \times n$) frequency domain.

Figure 10.23 shows one field from the interlaced *Salesman*, obtained by filtering and downsampling from the corresponding progressive clip. It serves as the starting point for our deinterlacing experiments. Figure 10.24 shows a frame from the resulting *progressive* (noninterlaced) output. We note that the result is generally pleasing, if somewhat soft (slightly blurred). From the frequency response in Figure 10.22, we can see that the image frame sharpness should be generally preserved for low temporal frequencies, i.e., slowly moving or stationary objects. Fast-moving objects, corresponding to diagonal support on the $v \times t$ filter frequency response function, will be blurred.

While blurring of fast-moving objects is generally consistent with the limitations of the human visual system response function, coherent motion can be tracked by the viewer. As such, it appears as low temporal frequency on the tracking viewer's retina, and hence the blurring of medium- to fast-moving objects can be detected for such *trackable motion*.

FIGURE 10.23 — *One field from the interlace version of Salesman*

FIGURE 10.24 — *Progressive frame from the diamond filter ($v \times t$) output for an interlaced input*

EXAMPLE 10.3-4 (*adaptive median deinterlacer*)

An alternative to the use of the classic multidimensional filter is the spatial adaptive deinterlacer, which first tries to decide whether the local region of the current frame is moving or not. This can be done by a local measure of the frame differences, either MAD or MSE. If the local area is determined as stationary, then the data from the previous field is used to fill the missing data in the current field, thus completing the current frame. Otherwise, some type of spatial interpolation is used to fill in the missing data. The most common method uses a 4-pixel median filter, with 1 pixel above and below the current pixel, and 1 pixel right behind it on the previous field. On the progressive lattice, this median operation can be written as follows: for odd frames,

$$\hat{x}(n_1, 2n_2, n) = \text{median}\{x(n_1, 2n_2 + 1, n), x(n_1, 2n_2 - 1, n),$$
$$x(n_1, 2n_2, n - 1), x(n_1, 2n_2, n + 1)\};$$

for even frames,

$$\hat{x}(n_1, 2n_2 + 1, n) = \text{median}\{x(n_1, 2(n_2 + 1), n), x(n_1, 2n_2, n),$$
$$x(n_1, 2n_2 + 1, n - 1), x(n_1, 2n_2 + 1, n + 1)\}.$$

In Figure 10.25, open circles indicate pixels (lines) present in a field; the × symbols represent missing pixels (lines). We see four input pixels (B–E) and one output pixel (A), which represent a missing pixel in frame n. Note that the input pixel E will require one field delay for hardware realization, and this can be avoided by simply leaving the pixel E out of the median filter, as a causal alternative. The statistical median of the four input values (B–E) will tend to favor D or E if there is no motion, but to favor B or C in the case of motion. Thus this simple median deinterlacer moves back and forth among temporal and spatial interpolation to fill in the missing pixel lines in the even and odd fields. An adaptive median deinterlaced frame is shown in Figure 10.26. While the result is sharper than that of the multidimensional filter, this sharpness is obtained at the cost of small artifacts occurring on fast-moving objects.

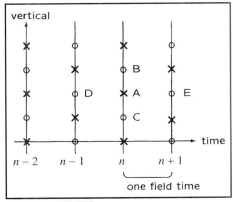

FIGURE 10.25 *Illustration of pixels input (B–E) to adaptive median filter deinterlacer*

FIGURE 10.26 *De-interlaced frame of Salesman by the adaptive median filter*

A more powerful alternative is motion-compensated de nterlacing. It uses motion estimation to find the best pixels in the previous field or fields for the prediction of the current pixel in the missing lines of the current field.

EXAMPLE 10.3-5 (*motion-compensated deinterlacer*)
In this example we try to detect and track motion [29] and then use it to perform the deinterlacing. Using an HBM motion estimator based on a QMF SWT, we first determine if the velocity is zero or not based on looking at a local average of the mean-square frame difference and comparing it to a threshold. A simple 3-tap vertical filter was used at this first stage. The motion is then said to be trackable if the local motion-compensated MSE is below a second threshold. The algorithm then proceeds as follows:

- When no motion is detected, we smooth in the temporal direction only, i.e., use the pixel at the same position in the previous field.
- When motion is detected, and with reference to Figure 10.27, we project the motion path onto the previous two fields, with a *cone of uncertainty* opening to 0.1–0.2 pixels at the just prior field. If the cone includes a pixel in the first prior field, then that pixel is copied to the missing pixel location A in the current field. Otherwise, we look to the second prior field. If no such previous pixels exist in the cone regions, we perform linear spatiotemporal interpolation.

The result in Figure 10.28 is potentially the best of these examples shown, albeit the most computationally demanding due to the motion estimation. It is sharp and clear, but unfortunately suffers from some motion artifacts,

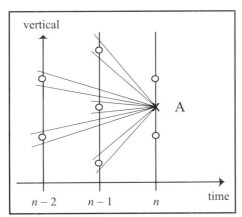

FIGURE 10.27   *Illustration of cone approach to motion-compensated deinterlacing*

FIGURE 10.28   *MC deinterlaced* Salesman *frame*

which could be ameliorated by more sophisticated motion estimation and compensation methods. The frame rate of *Salesman* was 15 fps.

One last comment on deinterlacing—of course, missing data not sampled cannot be recovered without some assumptions on the original continuous parameter data. In the specific case of deinterlacing, and in the worst case, one can imagine a small feature 1 pixel high, moving at certain critical velocities, such that it either always is or never is present on the interlaced grid. In real life, the velocity would not exactly match a critical velocity and so the feature would appear and disappear at a non-zero temporal frequency. If this feature is the edge of a rectangle or part of a straight line, the flickering can be noticed by the eye, but may be very hard to correct for, so an issue in MC deinterlacers is the consistency of their estimate. A recursive block estimate showing a high degree of visual consistency was given by de Haan *et al.* [4].

10.4 BAYESIAN METHOD FOR ESTIMATING MOTION

In Chapter 7 we introduced Bayesian methods for image estimation and restoration, which use a Gibbs Markov signal model together with some, most often iterative, solution method, e.g., simulated annealing (SA). Other methods used include deterministic iterative methods such as *iterated conditional mode* (ICM) and *mean field annealing* (MFA). ICM is a method that sweeps through the sites (pixels) and maximizes the individual conditional probability of each pixel in sequence. It is fast, but tends to get stuck at local optima of the joint conditional probability, rather than find the global maxima. MFA is an annealing technique that assumes that the effect of a neighboring clique's potential function can be well modeled by their mean value. While this changes the detailed global energy function, but the iteration proceeds much more quickly and reportedly provides very nice results, generally being somewhere between ICM and SA. A review of these approaches is contained in the *SP* magazine article by Stiller and Konrad [24].

To extend the Gibbs Markov spatial model, we need to model the displacement vector or motion field **d**, which can be done as

$$f_{\mathbf{d}}(\mathbf{D}) = K \exp - U_{\mathbf{d}}(\mathbf{D}),$$

where the matrix **D** contains the values of the displacement vector field **d** on the imaged region or frame. The energy function $U_{\mathbf{d}}(\mathbf{D})$ is a sum of potential functions over all of the pixels (sites) **n** and their corresponding cliques,

$$U_{\mathbf{d}}(\mathbf{D}) \triangleq \sum_{\mathbf{n}} \sum_{c_{\mathbf{n}} \in \mathcal{C}} V_{c_{\mathbf{n}}}(\mathbf{D}),$$

where $C_{\mathbf{n}}$ denotes the clique system for the displacement field **d**, over frame n.

A common setup calls for the estimation of the displacement between two frames $\mathbf{X}_n$ and $\mathbf{X}_{n-1}$, and using the MAP approach, we seek the estimate

$$\widehat{\mathbf{D}} = \arg \max_{\mathbf{D}} f(\mathbf{D}|\mathbf{X}_n, \mathbf{X}_{n-1}). \tag{10.4-1}$$

This simple model can be combined with a line field on an interpixel grid as in Chapter 7, to allow for smooth estimates of displacement that still respect the sharp boundaries of apparent motion that occur at object boundaries. The assumption is that moving objects generally have smoothly varying velocities, while different objects, and the background, are free to move at much different velocities. An early contribution to such estimates is the work of Konrad and Dubois [13].

Now, as was the case in Chapter 7, there are so many variables in (10.4-1) that simultaneous joint maximization is out of the question. On the other hand, iterative schemes that maximize the objective function one site at a time are practical. The nice thing about Gibbs models is that it is easy to find the local or marginal model. One simply gathers all of the potential functions that involve that site. All of the other terms go into a normalizing constant. So in the ICM approach, we seek the peak of the resulting conditional pdf. In the SA technique, we take a sample from this pdf, proceed through all of the sites, and then reduce the temperature a small amount per iteration. In MFA, we use the conditional mean of the conditional pdf, and iterate through all of the sites.

EXAMPLE 10.4-1 (*motion estimation for MC prediction*)
This example comes from the work of Stiller and Konrad [24] and shows both the predicted frame and the prediction error frame for three types of motion estimate: block, pixel, and region based, the latter segmenting moving areas in the scene in some way, e.g., using line fields. The test data comes from the monochrome videophone test clip *Carphone* in quarter CIF (QCIF) resolution. Figure 10.29 shows the result of block-based motion, using fairly large 16×16 blocks and 1/2-pixel accuracy, with the block structure being clearly evident, especially around the mouth. Its prediction MSE is reported as 31.8 dB. The predicted frame in Figure 10.30, resulting from a dense motion estimate (optical flow), is much better visually, with the much better prediction MSE of 35.9 dB. The region-based estimate, shown in Figure 10.31, is almost as good as the dense one, but has many fewer motion vectors. Its prediction MSE was reported at 35.4 dB. The corresponding prediction error frames are shown in Figures 10.32–10.34. Regarding the prediction error

FIGURE 10.29   *Predicted frame via block-based motion. Reproduced with permission from Stiller and Konrad* [24]

FIGURE 10.30   *Predicted frame via dense motion estimate (optical flow). Reproduced with permission from Stiller and Konrad* [24]

FIGURE 10.31   *Predicted frame via region-based estimate. Reproduced with permission from Stiller and Konrad* [24]

frames for the dense and region-based motion, we can see some blurring of object edges caused by the dense motion estimate, yet the object boundaries are respected better in the region-based estimate. Details on this example are contained in the paper by Dang *et al.* [3].

10.4.1 JOINT MOTION ESTIMATION AND SEGMENTATION

An alternative to motion estimation with a line field to prevent oversmoothing at object edges is to jointly estimate an object segmentation along with the motion [23]. The object boundary then serves to provide a linear feature over which motion should not be smoothed.

EXAMPLE 10.4-2 *(simple joint segmentation and displacement potential)*
In [24] Stiller and Konrad provide the following example of a joint segmenta-

FIGURE 10.32 *Prediction error frame for block-based motion. Reproduced with permission from Stiller and Konrad* [24]

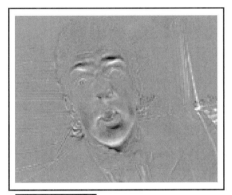

FIGURE 10.33 *Prediction error frame for dense motion (optical flow). Reproduced with permission from Stiller and Konrad* [24]

FIGURE 10.34 *Prediction error frame for region-based motion. Reproduced with permission from Stiller and Konrad* [24]

tion S and displacement D potential function. Here the segment label $s(\mathbf{n})$ can take on a fixed number of label values $1, \ldots, L$ corresponding to an assumed number of objects in the video frames. The potential for pairwise cliques was suggested as

$$V_{\mathbf{n},\mathbf{n}'}\big(\mathbf{d}(\mathbf{n}), \mathbf{d}(\mathbf{n}'), s(\mathbf{n}), s(\mathbf{n}')\big) = \lambda_d \|\mathbf{d}(\mathbf{n}) - \mathbf{d}(\mathbf{n}')\|^2 \delta\big(s(\mathbf{n}) - s(\mathbf{n}')\big)$$
$$+ \lambda_l\big(1 - \delta\big(s(\mathbf{n}) - s(\mathbf{n}')\big)\big), \qquad (10.4\text{-}2)$$

with all variables in frame n.

Here λ_d and λ_l are weights and δ is the discrete-time impulse function. We see that if the labels are the same, the first term in (10.4-2) penalizes the potential if the motion vectors are different at the two neighbor pixels (sites) $\mathbf{n}$ and $\mathbf{n}'$. The second term in the potential also penalizes the case where the labels are different at these two neighbor sites. The overall potential function then provides a compromise between these two effects.

One fairly complete formulation of the joint motion–segmentation problem is in the thesis of Han [8], where he applied the resulting estimates to video frame-rate increase and low-bitrate video coding. His energy function followed the approach of Stiller [23] and was given as

$$\big(\widehat{\mathbf{D}}_n, \widehat{\mathbf{S}}_n\big) = \arg \max_{\mathbf{D}_n, \mathbf{S}_n} f(\mathbf{D}_n, \mathbf{S}_n | \mathbf{X}_n, \mathbf{X}_{n-1})$$
$$= \arg \max_{\mathbf{D}_n, \mathbf{S}_n} f(\mathbf{X}_{n-1} | \mathbf{D}_n, \mathbf{S}_n, \mathbf{X}_n) f(\mathbf{D}_n | \mathbf{S}_n, \mathbf{X}_n) f(\mathbf{S}_n | \mathbf{X}_n), \quad (10.4\text{-}3)$$

where the three factors are the pdfs/pmf of *the video likelihood model, motion field* $\mathbf{D}_n$ *prior model,* and *segmentation field* $\mathbf{S}_n$ *prior model,* respectively.

The likelihood model was Gibbsian with energy function

$$U(\mathbf{X}_{n-1} | \mathbf{D}_n, \mathbf{X}_n) = \frac{1}{2\sigma^2} \sum_{\mathbf{n}} \big(x(\mathbf{n}, n) - x(\mathbf{n} - \mathbf{d}(\mathbf{n}, n), n - 1)\big)^2,$$

which is simplified to not depend on the segmentation.

The motion field prior model is also Gibbsian and is given in terms of energy function

$$U_\mathbf{d}(\mathbf{D}_n | \mathbf{S}_n) = \lambda_1 \sum_{\mathbf{n}} \sum_{\mathbf{m} \in N_\mathbf{n}} \|\mathbf{d}(\mathbf{n}, n) - \mathbf{d}(\mathbf{m}, n)\|^2 \delta\big(s(\mathbf{n}, n) - s(\mathbf{m}, n)\big)$$
$$+ \lambda_2 \sum_{\mathbf{n}} \|\mathbf{d}(\mathbf{n}, n) - \mathbf{d}(\mathbf{n} - \mathbf{d}(\mathbf{n}, n), n - 1)\|^2$$
$$- \lambda_3 \sum_{\mathbf{n}} \delta\big(s(\mathbf{n}, n) - s(\mathbf{n} - \mathbf{d}(\mathbf{n}, n), n - 1)\big).$$

Here he made the assumption that given the segmentation of the current frame, the motion field does not depend on the frame data. The first term in U_d sums over the local neighborhoods of each pixel and raises the motion field prior energy function if the labels agree and the l^2 norms of the motions are different. The second term increases this energy if the motion vector has changed along the motion path to the previous frame. The third term *lowers* the energy function if the labels agree along the motion paths. Of course, careful selection of the lambda parameters is necessary to achieve the best trade-off among these various factors, which try to enforce spatial smoothness of motion along with temporal smoothness of motion and segmentation along the motion path.

The segmentation field prior model is given in terms of the energy function

$$U_s(\mathbf{S}_n | \mathbf{X}_n) = \sum_n \sum_{\mathbf{m} \in N_n} V_m\big(s(\mathbf{n}, n), s(\mathbf{m}, n) | \mathbf{X}_n\big),$$

where

$$V_m\big(s(\mathbf{n}), s(\mathbf{m}) | \mathbf{X}_n\big) = \begin{cases} -\gamma, & s(\mathbf{n}) = s(\mathbf{m}) \text{ and } l(\mathbf{n}) = l(\mathbf{m}) \\ 0, & s(\mathbf{n}) = s(\mathbf{m}) \text{ and } l(\mathbf{n}) \neq l(\mathbf{m}) \\ +\gamma, & s(\mathbf{n}) \neq s(\mathbf{m}) \text{ and } l(\mathbf{n}) = l(\mathbf{m}) \\ 0, & s(\mathbf{n}) \neq s(\mathbf{m}) \text{ and } l(\mathbf{n}) \neq l(\mathbf{m}), \end{cases}$$

where $l(\mathbf{n})$ is a (deterministic) label field determined from $\mathbf{X}_n$ alone via a standard region growing algorithm [8]. The overall intent is to encourage smooth segmentation labels that agree with the 2-D label field determined just by color (gray level), penalize changes in segmentation field that do not agree with changes in color, and treat neutrally the remaining discrepancies. This is because the motion segmentation can change when an object starts or stops moving, and usually it should correspond to a color segmentation.

It is common to try to iteratively maximize (10.4-3) via an alternating method, where one sweep of all of the sites tries to improve the criterion by updating $\widehat{\mathbf{D}}_n$ given the current $\widehat{\mathbf{S}}_n$, while the next or alternate sweep updates $\widehat{\mathbf{S}}_n$ given the current $\widehat{\mathbf{D}}_n$. The procedure starts with a segmentation determined by region growing on the first color frame. Figure 10.35 shows results of the alternating motion–segmentation optimization from Han [8]. We notice that the determined motion in Figure 10.35d is very smooth within the objects found in Figure 10.35b, unlike the generally more noisy motion vectors in Figure 10.35c determined by block matching.

In the next chapter, we will show how joint motion estimation and segmentation can be used in an object-based low-bitrate video coder. This application can use the preceding formulation that assumes clean data is available. Another class of applications tries to use Bayesian methods to restore degraded image sequence data, such as the restoration of old movie film and videotape. A comprehensive

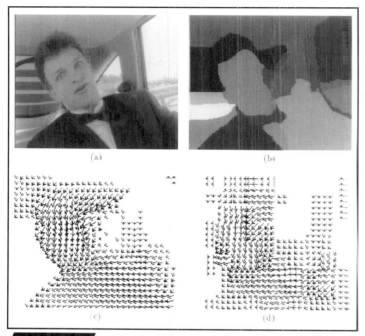

FIGURE 10.35   (a) *Original*, (b) *joint segmentation*, (c) *block-based motion*, (d) *joint segmentation motion vectors*

review of this area appears in the research paper by Kokaram [12], where the formulation also includes estimation of a binary masking function that indicates the areas in each frame with missing data, due to dirt, scratches, chemical deterioration, tape dropout, etc. (see also Section 3.11 in Chapter 3 on video enhancement and restoration by Lagendijk *et al.* and Section 4.10 in Chapter 4 on video segmentation by Tekalp in the handbook by Bovik [2]).

10.5 CONCLUSIONS

Multidimensional filtering plays an important role in video format conversion. Whenever an image or video is resized to fit a given display, or is frame-rate converted, a properly designed multidimensional filter will improve results over commonly employed decimation and 1-D filtering alone. Motion-compensated filtering turns out to be very effective for restoration of noisy or distorted video data. Motion-compensated multidimensional filtering also offers increased sharpness in video standards conversion. However, unless the MC estimates are very good, artifacts can result, and this is called *motion model failure*. Finally, we looked at Bayesian methods for joint motion estimation and segmentation. In the next chapter, MC plays a big role in video compression.

10.6 PROBLEMS

1 How many pixels per second must be processed in D1[5] video? How much storage space (in gigabytes) for 10 minutes of D1?

2 Use MATLAB to design a spatial lowpass filter and then optimize it visually with the approach of Example 10.1-3. Can you see the difference? What value of α did you determine? How does it relate to the transition bandwidth of the original lowpass filter?

3 Compare the computation (multiplies and adds) of full-search versus three-step search block matching, both using the MSE method. Assume an $M \times M$ block and $N \times N$ search window, and that block matching estimates are calculated on an $M \times M$ decimated grid. Take the image size as CIF, i.e., 352×288.

4 In overlapped block motion compensation (OBMC), the weighting matrices of Orchard and Sullivan [18] are used to smooth the motion vectors over the fixed block boundaries. Comment on how this would be expected to affect the computational complexity of OBMC versus regular block matching.

5 In this problem you will show the connection between the spatiotemporal constraint equation, (10.2-4), and direct minimization of the error

$$E(d_x, d_y) \triangleq \sum_{R(x,y)} \left[f(x, y, t) - f(x - d_x, y - d_y, t - \Delta t) \right]^2.$$

First calculate $\partial E/\partial d_x$ and $\partial E/\partial d_y$ and set them both to zero. Then argue that if at least one of the spatial gradients $\partial f/\partial x$ and $\partial f/\partial x$ is not zero, then, for sufficiently small region $R(x, y)$ and time interval $(t, t + \Delta t)$, the spatiotemporal constraint equation must hold, with $d_x = v_x \Delta t$ and $d_y = v_y \Delta t$. Hint: expand the function $f(x, y, t)$ in a Taylor series and keep only the first-order terms.

6 Show in block diagram form a system realizing the diamond filter deinterlacing method of Example 10.3-3. Your diagram should explicitly include the upsamplers and the filter. For ATSC 1080i format, approximately how many multiplies per second will be required?

7 Extending Example 10.3-5 on motion-compensated deinterlacing, we can look back over several field pairs (frames) and find multiple references for the missing pixel. Then the estimate for the missing pixel can be a weighted average of the references from the different frames. Suggest a good way of weighting these multiple references. (You may care to look at Wiegand *et al.* [28] for some ideas.)

8 Can mean field annealing (MFA) be used to estimate the segment labels in a joint motion–segmentation approach to Bayesian motion estimation? How would an iterated conditional mode (ICM) estimate be obtained?

---

5. A description of the D1 format is provided in the appendix at the end of this chapter.

9 Separate the three energy functions corresponding to the Gibbs pdfs/pmf factors in (10.4-3), into those involving $\mathbf{s}(\mathbf{n}, n)$ and those involving $\mathbf{d}(\mathbf{n}, n)$.

REFERENCES

[1] M. Bierling and R. Thoma, "Motion Compensating Field Interpolation using a Hierarchically Structured Displacement Estimator," *Signal Process.*, **11**, 387–404, Dec. 1986.

[2] A. C. Bovik, ed., *Handbook of Image and Video Processing*, 2nd edn., Elsevier Academic Press, Burlington, MA, 2005.

[3] V.-B. Dang, A. R. Mansuric, and J. Konrad, "Motion Estimation for Regional-Based Video Codings," *Proc. IEEE ICIP*, Washington, DC, Oct. 1995.

[4] G. de Haan, P. W. A. C. Biezen, H. Huigen, and O. A. Ojo, "True-Motion Estimation with 3-D Recursive Search Block Matching," *IEEE Trans. CSVT*, **3**, 368–379, Oct. 1993.

[5] E. Dubois and W. Schreiber, "Improvements to NTSC by Multidimensional Filtering," *SMPTE J.*, **97**, 446–463, June 1988.

[6] D. E. Dudgeon and R. M. Mersereau, *Multidimensional Digital Signal Processing*, Prentice-Hall, Englewood Cliffs, NJ, 1983.

[7] H. Farid and E. P. Simoncelli, "Differentiation of Discrete Multidimensional Signals," *IEEE Trans. Image Process.*, **13**, 496–508, April 2004.

[8] S.-C. Han, *Object-Based Representation and Compression of Image Sequences*, Ph.D. thesis, ECSE Dept., Rensselaer Polytechnic Institute, Troy, NY, 1997.

[9] H.-M. Hang and Y.-M. Chou, "Motion Estimation for Image Sequence Compression," In *Handbook of Visual Communication*, H.-M. Hang and J. W. Woods, eds., Academic Press, New York, NY, 1995. Chapter 5.

[10] B. K. P. Horn and B. G. Schunck, "Determining Optical Flow," *Artificial Intell.*, **17**, 309–354, 1984.

[11] J. R. Jain and A. K. Jain, "Displacement Estimation and its Application in Interframe Image Coding," *IEEE Trans. Commun.*, **COM-29**, 1799–1808, Dec. 1981.

[12] A. C. Kokaram, "On Missing Data Treatment for Degraded Video and Film Archives: A Survey and a New Bayesian Approach," *IEEE Trans. Image Process.*, **13**, 397–415, March 2004.

[13] J. Konrad and E. Dubois, "Estimation of Image Motion Fields: Bayesian Formulation and Stochastic Solution," *Proc. IEEE ICASSP*, pp. 1072–1075, April 1988.

[14] J. Lim, *Two-Dimensional Signal and Image Processing*, Prentice-Hall, Englewood Cliffs, NJ, 1990.

[15] B. Liu and A. Zaccarin, "New Fast Algorithms for the Estimating of Block Motion Vectors," *IEEE Trans. CSVT*, **3**, 148–157, April 1993.

[16] A. N. Netravalli and J. D. Robbins, "Motion Compensated Coding: Some New Results," *Bell Syst. Tech. J.*, **59**, 1735–1745, Nov. 1980.

[17] S. Nogaki and M. Ohta, "An Overlapped Block Motion Compensation for High Quality Motion Picture Coding," *Proc. IEEE Int. Sympos. Circuits Syst.*, pp. 184–187, May 1992.

[18] M. T. Orchard and G. J. Sullivan, "Overlapped Block Motion Compensation: An Estimation-Theoretic Approach," *IEEE Trans. Image Process.*, **3**, 693–699, Sept. 1994.

[19] M. K. Ozkan, I. Sezan, A. T. Erdam, and A. M. Tekalp, "Motion Compensated Multiframe Wiener Restoration of Blurred and Noisy Image Sequences," *Proc. IEEE ICASSP*, San Francisco, CA, Mar. 1992.

[20] Y. Q. Shi and H. Sun, *Image and Video Compression for Multimedia Engineering*, CRC Press, Boca Raton, FL, 2000.

[21] H. Schröder, "On Vertical Filtering for Flicker Free Television Reproduction," *IEEE Trans. Circuits, Syst. Signal Process.*, **3**, 161–176, 1984.

[22] H. Schröder and H. Blume, *One- and Multidimensional Signal Processing*, John Wiley and Sons, New York, NY, 2000.

[23] C. Stiller, "Object Based Estimation of Dense Motion Fields," *IEEE Trans. Image Process.*, **6**, 234–250, Feb. 1997.

[24] C. Stiller and J. Konrad, "Estimating Motion in Image Sequences," *IEEE Signal Process. Magazine*, **16**, 70–91, July 1999.

[25] A. M. Tekalp, *Digital Video Processing*, Prentice-Hall PRTR, Englewood Cliffs, NJ, 1995.

[26] R. Thoma and M. Bierling, "Motion-Compensated Interpolation Considering Covered and Uncovered Background," *Signal Process.: Image Commun.*, **1**, 191–212, Oct. 1989.

[27] Y. Wang, J. Ostermann, and Y.-Q. Zhang, *Video Processing and Communications*, Prentice-Hall, Englewood Cliffs, NJ, 2001.

[28] T. Wiegand, X. Zhang, and B. Girod, "Long-Term Memory Motion-Compensated Prediction," *IEEE Trans. CSVT*, **9**, 70–84, Feb. 1999.

[29] J. W. Woods and S.-C. Han, "Hierarchical Motion Compensated De-Interlacing," *Proc. SPIE VCIP VI*, Boston, Nov. 1991.

[30] J. W. Woods and J. Kim, "Motion Compensated Spatiotemporal Kalman Filter," In *Motion Analysis and Image Sequence Processing*, I. Sezan and R. L. Lagendijk, eds., Kluwer, Boston, MA, 1993. Chapter 12.

10.7 APPENDIX: DIGITAL VIDEO FORMATS

This appendix presents some common video formats. We start with the half-resolution formats SIF and CIF and then discuss the standard definition (SD) format, with the official name ITU 601. In passing, we consider a smaller quarter-size format called QCIF used for low-bitrate video conferencing. We also list the more recent high definition (HD) formats specified by the Advanced Television Systems Committee (ATSC).

SIF

The Source Interchange Format (SIF) is the format of MPEG 1. It has 352×240 pixels at 30 frames per second (fps) in a noninterlaced or progressive lattice. This is the U.S. version, more formally called SIF-525. The European version, SIF-625, is 352×288. The MPEG committee has given recommendations for the conversion of ITU 601 to SIF and for the conversion of SIF to ITU 601 using certain suggested filters. These filters are commonly called *MPEG half-band filters*.

Decimation filter: −29   0   88   138   88   0   −29   //256
Interpolation filter: −12   0   140   256   140   0   −12   //256

CIF

The Common Intermediate Format (CIF) has 352×288-pixel progressive image frames, with a sometimes variable frame rate. CIF is used in the H.263 video conferencing standard of the International Telecommunications Union (ITU). CIF gets its dimensions from the larger of SIF-525 and SIF-625. Also commonly used is quarter CIF (QCIF), the format of the H.263 video telephone 176×144 standard that runs at typically 5–15 fps. Finally, going in the direction of higher resolution, there is 4CIF, with resolution 704×576. The set 4CIF, CIF, and QCIF makes a nice set of resolutions for scalability in spatial resolution.

ITU 601 DIGITAL TV (AKA SMPTE D1 AND D5)

This is the component digital standard of the D1 (3/4 inch) and D5 (1/2 inch) digital TV tape recorders. The signal is sampled in 4:2:2 YC_RC_B format, and interlaced 2:1. The field rate is 59.94 Hz (in the United States), with the frame rate 29.97 Hz (in the United States). The aspect ratio is 4:3, with the bit depth of 8 or 10 bits per pixel.

The intent was to digitize standard-resolution television as seen in the United States and Europe. This standard is meant for a viewing distance of five to six times the picture height. There are various members of the ITU 601 family, the most prominent being the so-called *4:2:2 member*. The sampling frequency was standardized at 13.5 MHz for both 525 and 625 line systems, i.e., "4" means 13.5 MHz. This choice results in a static orthogonal sampling pattern for both 525/60 and 625/50 systems, since 13.5 and 6.75 MHz are integer multiples of 2.25 MHz, the least common multiple of the scan line frequencies of these systems. There are 720 active luma samples/line and 486 active lines, i.e., 720×486

pixels in the U.S. version at 60 fields/sec, and in Europe, 720×576 pixels at 50 fields/sec. The quantization is uniform PCM, again to either 8 or 10 bits. The standard provides filter specs both for the sampling of luma and for the subsampling of the chroma channels. The overall data rate was specified as 27 MHz.

From the digital point of view, there are 486 lines, not 525 lines, in the U.S. system. The 525 lines come about when analog standard-definition TV is digitized, since analog TV left time for the CRT to "retrace" or return to the beginning of a line (horizontal retrace interval) and to return to the top left corner for display of the next field or frame (vertical retrace).

The Society of Motion Picture and Television Engineers (SMPTE) recommended the following color difference components, also called a color space:

$$\begin{bmatrix} Y \\ C_R \\ C_B \end{bmatrix} = \begin{bmatrix} 0.299 & 0.587 & 0.114 \\ 0.500 & -0.418 & -0.081 \\ -0.169 & -0.331 & 0.500 \end{bmatrix} \begin{bmatrix} R \\ G \\ B \end{bmatrix}.$$

In the common 4:2:2 mode, the C_R and C_B samples are co-sited with odd samples of Y on each line, i.e., first, third, fifth, etc.

WHAT IS 4:2:0?

The 4:2:0 specification means that the C_R and C_B chroma signals are subsampled by two vertically as well. This gives a combined chroma sample rate of 50% of the luma sample rate. Whenever there is subsampling of the color information, as in 4:2:2 and 4:2:0 rasters, there is need to specify any offsets between the chroma and luma rasters. Some information on this is given in Chapter 8, Section 8.8 (Color Image Coding). In converting from one representation to another, it may thus be necessary to apply a 2-D delay operator to the data.

ATSC FORMATS

The digital formats in the United States were set up by the Advanced Television System Committee (ATSC) in the early 1990s. They mainly consist of two HD formats and two standard definition (SD) formats. The SD formats are 720×486 interlaced[6] at 60 fields/sec and 720×486 progressive at 60 fps. The main HD formats are 1280×720 at 60 fps, often referred to as 720p, and 1920×1080 interlaced at 60 fields/sec, often referred to as 1080i. In recent years, a 1080p spec has also emerged, with an uncompressed data rate of 3 Gbps, twice that of 1080i. Two bit depths are supported for the components, 8 and 10 bits per sample. The ATSC maintains a web site at www.atsc.org.

High definition television as in the ATSC standards has elicited a new ITU recommendation 709 that specifies an HD CIF format of 1920×1080 square pixels

---

6. See Example 2.2-5 and Section 10.3.4 for the "interlacing" concept.

at 24p, 25p, 30p, and 60p as well as 50i and 60i. The colorimetry is changed too from the ITU 601. The ITU 709 color space is specified as

$$Y = 0.2126R + 0.7152G + 0.0722B,$$

with color difference signals

$$C_R = (R - Y)/1.5748 \quad \text{and} \quad C_B = (B - Y)/1.8556.$$

Color is specified as 4:2:2 and components can have both 8 and 10 bit depth.

Common HD tape formats presently include: D5-HD, DVCPRO HD, HD-CAM, and HDCAM SR. These formats all include intraframe compression of the ITU 709 signal. Both D5-HD and HDCAM SR record the full 1920 × 1080 frame in 4:2:2 format at 10 bit pixel depth. DVCPRO HD and HDCAM record 1440 × 1080 pixels at 8 bit depth, and convert to/from 1920 pixels/line on input/output. The bitrate to tape of HDCAM SR is 440 Mbps, HDCAM is 180 Mbps, and DVCPRO HD is 100 Mbps. The prosumer format HDV also records 1440 × 1008 at 8 bits, but uses interframe coding and 4:2:0 color space to compress the data rate down to 25 Mbps on tape, using the MPEG 2 main profile at high 1440 level. ITU recommendations can be obtained at web site www.itu.int.

DIGITAL VIDEO COMPRESSION 11

Video coders compress a sequence of images, so this chapter is closely related to Chapter 8 on image compression. Many of the techniques introduced there will be used and expanded upon in this chapter. At the highest level of abstraction, video coders comprise two classes: *interframe* and *intraframe*, based on whether they use statistical dependency between frames or are restricted to using dependencies only within a frame. The most common intraframe video coders use the DCT and are very close to JPEG; the video version is called M-JPEG, wherein the "M" can be thought of as standing for "motion," as in "motion picture," but does not involve any motion compensation. Another common intraframe video coder is that used in consumer DV camcorders. By contrast, interframe coders exploit dependencies across frame boundaries to gain increased efficiency. The most efficient of these coders make use of the apparent motion between video frames to achieve their generally significantly larger compression ratio. An interframe coder will code the first frame using intraframe coding, but then use predictive coding on the following frames. The new HDV format uses interframe coding for HD video.

MPEG coders restrict their interframe coding to a *group of pictures* (GOP) of relatively small size, say 12–60 frames, to prevent error propagation. These MPEG coders use a transform compression method for both the frames and the predictive residual data, thus exemplifying *hybrid coding*, since the coder is a hybrid of the block-based transform spatial coder of Chapter 8 and a predictive or DPCM temporal coder. Sometimes transforms are used in both spatial and time domains, the coder is then called a 3-D *transform coder*.

All video coders share the need for a source buffer to smooth the output of the variable length coder for each frame. The overall video coder can be *constant bitrate* (CBR) or *variable bitrate* (VBR), depending on whether the buffer output bitrate is constant. Often, intraframe video coders are CBR, in that they assign or use a fixed number of bits for each frame. However, for interframe video coders, the bitrate is more highly variable. For example, an action sequence may need a much higher bitrate to achieve a good quality than would a so-called "talking head," but this is only apparent from the interframe viewpoint.

Table 11.1 shows various types of digital video, along with frame size in pixels and frame rate in frames per second (fps). Uncompressed bitrate assumes an 8-bit pixel depth, except for 12-bit digital cinema (DC), and includes a factor of three for RGB color. The last column gives an estimate of compressed bitrates using

Table 11.1. Types of video with uncompressed and compressed bitrates

Video	Pixel size	Frame rate	Rate (uncomp.)	Rate (comp.)
Teleconference (QCIF)	176×144	5–10 fps	1.6–3 Mbps	32–64 Kbps
Multimedia (CIF)	352×288	30	36 Mbps	200–300 Kbps
Standard definition (SD)	720×486	30	168 Mbps	4–6 Mbps
High definition (HD)	1920×1080	30	1.2 Gbps	20 Mbps
Digital cinema (DC)	4096×2160	24	7.6 Gbps	100 Mbps

technology such as H.263 and MPEG 2. We see fairly impressive compression ratios comparing the last two columns of the table, upwards of 50 in several cases. While the given "pixel size" may not relate directly to a chosen display size, it does give an indication of recommended display size for general visual content, with SD and above generally displayed at full screen height, multimedia at half screen height, and teleconference displayed at one quarter screen height, on a normal display terminal. In terms of distance from a viewing screen, it is conservatively recommended to view SD at 6 times the picture height, HD at 3 times the picture height, and DC at 1.5 times the picture height (see Chapter 6).

11.1 Intraframe Coding

In this section we look at three popular intraframe coding methods for video. They are block-DCT, motion JPEG (M-JPEG), and subband/wavelet transform (SWT). The new aspect over the image coding problem is the need for rate control, which arises because we may need variable bit assignment across the frames to get more uniform quality. In fact, if we would like to have constant quality across the frames that make up our video, then the bit assignment must adapt to frame complexity, resulting in a *variable bitrate* (VBR) output of the coder.

Figure 11.1 shows the system diagram of a transform-based intraframe video coder. We see the familiar transform, quantize, and VLC structure. What is new is the buffer on the output of the VLC and the feedback of a rate control from the buffer. Here we have shown the feedback as controlling the quantizer. If we need *constant bitrate* (CBR) video, then the buffer output bitrate must be constant, so its input bitrate must be controlled so as to avoid overflow or underflow, the latter corresponding to the case where this output buffer is empty and so unable to supply the required CBR. The common way of controlling the bitrate is to monitor buffer fullness, and then feed back this information to the quantizer. Usually the step-size of the quantizer is adjusted to keep the buffer around the midpoint, or half full.

As mentioned earlier, a uniform quantizer has a constant step-size, so that if the number of output levels is an odd number and the input domain is symmetric around zero, then zero will be an output value. If we enlarge the step-size around zero, say by a factor of two, then the quantizer is said to be a *uniform threshold quantizer* (UTQ) (see Figure 11.2). The bin that gets mapped into zero

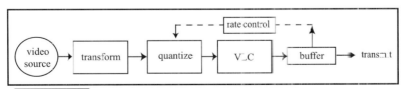

FIGURE 11.1 *Illustration of intraframe video coding with rate control*

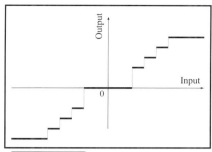

FIGURE 11.2   *Illustration of uniform threshold quantizer, named for its deadzone at the origin*

is called the *deadzone*, and can be very helpful to reduce noise and "insignificant" details. Another nice property of such UTQs is that they can be easily nested for scalable coding, as mentioned in Chapter 8, and which we discuss later in this chapter.

If we let the reconstruction levels be bin-wise conditional means and then entropy code this output, then UTQs are known to be close to optimal for common image data pdfs, such as Gaussian and Laplacian. In order to control the generated bitrate via a buffer placed at output of the coder, a quality factor has been introduced. Both CBR and VBR strategies are of interest. We next describe the M-JPEG procedure.

11.1.1 M-JPEG PSEUDO ALGORITHM

1   Input a new frame.
2   Scan to next 8×8 image block.
3   Take 8×8 DCT of this block.
4   Quantize AC coefficients in each DCT block, making use of a *quantization matrix* $Q = \{Q(k_1, k_2)\}$,

$$\widehat{\mathrm{DCT}}(k_1, k_2) = \mathrm{Int}\left[\frac{\mathrm{DCT}(k_1, k_2)}{sQ(k_1, k_2)}\right], \qquad (11.1\text{-}1)$$

where s is a scale factor used to provide a rate control. As s increases, more coefficients will be quantized to zero and hence the bitrate will decrease. As s decreases toward zero, the bitrate will tend to go up. This scaling is inverted by multiplication by $sQ(k_1, k_2)$ at the receiver. The JPEG quality factor Q is inversely related to the scale factor s.

5   Quantize the DC coefficient DCT(0,0), and form the difference of successive DC terms ΔDC (effectively a noiseless spatial DPCM coding of the DC terms).
6   Scan the 63 AC coefficients in conventional zig-zag scan.

7 Variable-length code this zig-zag sequence as follows:
    (a) Obtain run length (RL) to next nonzero symbol.
    (b) Code the pair (RL, nonzero symbol value) using a *2-D Huffman* table.
8 Transmit (store, packetize) bitstream with end-of-block (EOB) markers and headers containing quantization matrix Q, Huffman table, image size in pixels, color space used, bit depth, frames per second, etc.
9 If more data in frame, return to step 2.
10 If more frames in sequence, return to step 1.

Variations on this M-JPEG algorithm are used in desktop SD video editing systems, consumer digital video camcorders (25 Mbps), and professional SD video camcorders such as the Sony Digital Betacam (90 Mbps) and Panasonic DVCPro 50 (50 Mbps), and others. A key aspect is rate control, which calculates the scale factor s in this algorithm. The following example shows a simple type of rate control with the goal of achieving CBR on a frame-to-frame basis, i.e., intraframe rate control. It should be mentioned that, while JPEG is an image coding standard, M-JPEG is not a video coding standard, so various "flavors" of M-JPEG exist that are not fully compatible.

EXAMPLE 11.1-1 (*M-JPEG*)
Here we present two nearly constant bitrate examples obtained with a buffer and a simple feedback control to adjust the JPEG quality factor Q. Our control strategy is to first fit a straight line of slope γ to plot the Q factor versus $\log R$ of a sample frame in the clip to estimate an appropriate step-size, and then employ a few iterations of steepest descent. Further, we use the Q factor of last frame as first guess for the present frame. The update equation becomes

$$Q_{\text{new}} = Q_{\text{prev}} + \gamma(\log R_{\text{target}} - \log R_{\text{prev}}).$$

We ended up with 1.2–1.4 iterations per frame on average to produce the results. We have done two SIF examples: *Susie* at 1.3 Mbps and *Table Tennis* at 2.3 Mbps. Figure 11.3 shows the results for the *Susie* clip, from top to bottom: PSNR, rate in KB/frame, and Q factor. We see a big variation in rate without rate control, but a fairly constant bitrate using our simple rate control algorithm. Results from the second example, *Tennis*, are shown in Figure 11.4; note that the rate is controlled to around 76 KB/frame quite well.

Notice that in both of these cases the constraint of constant rate has given rise to perhaps excessive increases in PSNR in the easy-to-code frames that contain blurring due to fast camera and/or object motion.

Most SD video has been, and remains at this time interlaced. However, in the presence of fast motion, grouping of two fields into a single frame prior to 8×8 DCT transformation is not very efficient. As a result, when M-JPEG is used on

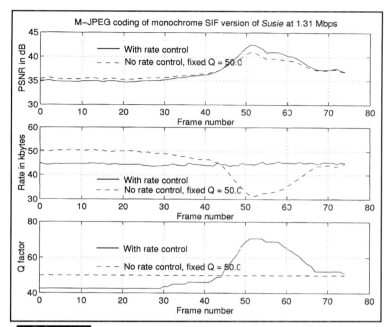

FIGURE 11.3 *Rate control of M-JPEG on* Susie *clip*

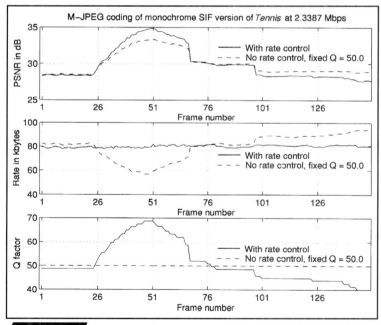

FIGURE 11.4 *Illustration of rate control performance for M-JPEG coding of* Tennis *clip*

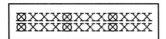

FIGURE 11.5 4:1:1 color subsampling pattern of DV in the so-called 525/30 (NTSC) system

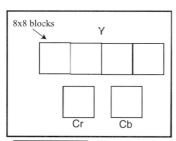

FIGURE 11.6 DV macroblock for the NTSC (525/30) system

interlaced SD video, usually each field is coded separately, and there is no provision for combining fields into frames, which would be more efficient in the case of low motion. The common DV coder, still based on DCT, has such an adaptive feature, giving it an efficiency edge over M-JPEG.

11.1.2 DV CODEC

Currently the source coders in consumer SD video camcorders are called *DV coders* [9], and there is some variation in the details of how they work. These intraframe coders first subsample the chroma components in the NTSC system in 4:1:1 fashion,[1] as shown in Figure 11.5, where the x symbols indicate luma samples and the circles represent chroma (both *Cr* and *Cb* sample locations in a field. Note that the sites of the two chroma samples occur at a luma sample site. This is not always the case for 4:1:1, but is true for NTSC DV coders.

There is provision for combining the two fields of the interlaced NTSC system into a frame or not, depending on a mode decision (not standardized, but left up to the DV coder designer). Then 8 × 8 DCTs of either fields or frames are performed and grouped into *macroblocks* (MBs), as shown in Figure 11.6.

Thus a DV macroblock consists of six blocks: four luma and two chroma blocks. These macroblocks will be quantized and variable-length coded, but first, they are mapped one-to-one onto *synch blocks* (SBs) and then stored *K* at a time into *segments*. Each segment is assigned a fixed code length, in order to permit robust reading of digital tape at high speed, for so-called "trick play," such as search in fastforward. Thus VLC is only used in DV coders within segments. The quantization is performed similar to (11.1-1) with the scaling variable *s* controlled

---

1. A PAL version is available that uses a 4:2:0 color space at 25 interlaced frames/sec.

to achieve a fixed target bitrate per segment. In what is called a *feedforward* or *look-ahead strategy*, 8 to 16 different values of *s* are tried within the segment to meet the goal for fixed bit assignment. The best one is then selected for the actual coding of the segment. The DV standard adopted $K = 30$ macroblocks per segment, and they are distributed "randomly" around the image frame to improve the uniformity of picture quality, which at 25 Mbps is generally between 35 and 40 dB for typical interlaced SD video content.

Both the M-JPEG and DV coder have a constant bitrate per frame and so are CBR. This means that the video quality will be variable, depending on the difficulty of the various image frames. They effectively are buffered at the frame level to achieve this CBR property. Use of a longer buffer would allow VBR operation, wherein we could approach constant quality per frame.

11.1.3 INTRAFRAME SWT CODING

Here we look at intraframe video coders that use SWT in place of DCT. Again, the need for some kind of control of the video bitrate is the new main issue over SWT image coding. First we consider the rate assignment over the spatial subbands within each frame.

We must split up the total bits for each frame into the bit assignment for each subband. Assuming an orthogonal analysis/synthesis SWT, then the total mean-square error distortion D due to the individual quantizations with assumed distortion $D_m(R_m)$, for R_m bits assigned to the mth subband quantizer, can be written as the weighted sum of the mean-square distortions over the M subbands as

$$D = \sum_{m=1}^{M} \frac{N_m}{N} D_m(R_m), \tag{11.1-2}$$

where N_m is the number of samples in the mth subband, as pointed out for the image compression case in Chapter 8. Then modeling the individual quantizers as $D_m = g\sigma_m^2 2^{-2R_m}$, all with the same quantizer efficiency factor g, the optimal bit assignment, for an average bitrate of R bits/pixel, was determined as

$$R_m = R + \frac{1}{2} \log_2 \left(\frac{\sigma_m^2}{\sigma_{\text{wgm}}^2} \right), \quad m = 1, \ldots, M, \tag{11.1-3}$$

and the weighted geometric mean σ_{wgm}^2 is defined as

$$\sigma_{\text{wgm}}^2 \triangleq \prod_{m=1}^{M} \left(\sigma_m^2 \right)^{N_m/N},$$

where N is the total number of samples, i.e., $N = \sum_{m=1}^{M} N_m$. The resulting MSE per frame in the reconstruction then becomes

$$D = g\sigma_{\mathrm{wgn}}^2 2^{-2R},$$

which is analogous to the case of SWT image coding (see Chapter 8).

If we choose to have CBR, then this is the solution. However, we may want to either achieve constant distortion or minimum average distortion over time. This then requires that we vary the bit assignment over the frames. Given the image frames' subband variances, we can compute the weighted geometric mean of each frame $\sigma_{\mathrm{wgm}}^2(n)$, and then the distortion at frame $D(n)$ will be

$$D(n) = g\sigma_{\mathrm{wgm}}^2(n)2^{-2R(n)},$$

so that the total distortion over all of the frames becomes

$$D_{\mathrm{T}} = \sum_{n=1}^{N} D(n)$$

$$= \sum_{n=1}^{N} g\sigma_{\mathrm{wgn}}^2(n)2^{-2R(n)}, \tag{11.1-4}$$

where $R(n)$ is the bit assignment to the nth frame, and the average bitrate is then

$$\frac{1}{N}R_{\mathrm{T}} = \frac{1}{N}\sum_{r=1}^{N} R(n). \tag{11.1-5}$$

This pair of operational rate and distortion functions, (11 1-5) and (11.1-4), can then be optimized for best average performance over time. Normally this minimum will not occur at the point where all of the frames have the same distortion $D(n) = D$. In that case, one has to make the choice of either optimizing the constant distortion performance, or optimizing for best average performance. They will generally be different.

One other issue in SWT coding is the choice of the subband/wavelet filter to use, and this is explored in the following example.

EXAMPLE 11.1-2 (*SWT filter study*)
A goal is to see the effect of different SWT filters on an efficient intraframe coder. Individual frames are first subband analyzed and then individually quantized by UTQ. The output of the quantizer is then separately arithmetic coded. The closed-form log bit-assignment rule of this section is used. We also used the arithmetic coding (AC) algorithm of Jones [21]. We employed generalized Gaussian models, whose parameters were estimated from

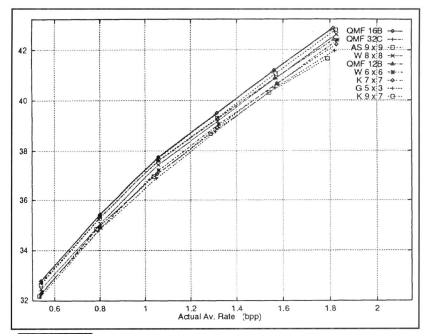

FIGURE 11.7   *Comparison of various SWT filters for intraframe video compression (reproduced from [30] with permission)*

the first frame. The AC output was grouped into synch blocks: LL–LL subband blocked to 2 lines, LL–LH, –HL, –HH subbands blocked to 4 lines, and the higher subbands not blocked. A 5-bit end-of-block marker was used to separate these synch blocks. The results from Naveen and Woods [30] for the HD test clip *MIT Sequence* are shown in Figure 11.7. We see that there is not a drastic difference in the performance of these filters, but here the longer QMFs do better. The popular CDF 9/7 wavelet filter was not included in this study, but was later found to have similar performance, slightly better than the QMFs for dyadic or wavelet decomposition, but slightly worse for full subband decomposition [44].

11.1.4 M-JPEG 2000

M-JPEG 2000 is officially JPEG 2000, Part 3, and amounts to a standardized file format or container for a sequence of JPEG 2000 image files. It can be CBR or VBR within the standard. Various methods of bitrate control can be applied to the resulting sequence of coded frames, making them either best constant quality, best average quality, HVS weighted quality, etc. The M-JPEG 2000 standard is just the standard for the decoder, thus permitting any variable numbers of bits per frame. As always, rate control is only an encoder problem.

As mentioned in Chapter 8, JPEG 2000 uses operational rate-distortion optimization over tiles on the image frame. These tiles are then coded using the *constant slope condition*, resulting in minimum average mean-square coding error. This constant slope matching of JPEG 2000 can be extended between frames to achieve a minimum mean-square coding error for the whole image sequence. Such an approach may mean an unacceptable delay, so a matching of slope points on a local frame-by-frame basis may be chosen as an alternative VBR solution that has been found much closer to constant quality than is CBR.

EXAMPLE 11.1-3 (*digital cinema*)
In 2004, Digital Cinema Initiatives (DCI) ran a test of compression algorithms for the digital cinema distribution at both 4K and 2K resolution, meaning horizontal number of pixels, and aspect ratios around 2.35. DCI had been created 2 years earlier by the major U.S. motion picture companies. A test suite was generated for the tests by the American Society of Cinematographers (ASC) and consisted of 4K RGB image frames at 24 fps. The bit depth of each component was 12 bits. After extensive testing, involving both expert and non-expert viewers, DCI made the recommendation of M-JPEG 2000 as the best choice for digital cinema. In part this was because of the 4K/2K scalability property, and in part because of its excellent compression performance. The tests used the CDF 9/7 filter or lossy version of the M-JPEG 2000 coder. As of this writing, the JPEG committee was standardizing a new profile of M-JPEG 2000 specially for digital cinema use. Bit rates are in the vicinity of 100 Mbps at the 4K resolution.

We now turn to the generally more efficient interframe coders. While these coders offer more compression efficiency than do the intraframe coders just discussed, they are not so artifact free, and are mainly used for *distribution quality* purposes. Intraframe coders, with the exception of the digital cinema distribution standard, have been mainly used for *contribution quality*, i.e., professional applications, because of their high quality at moderate bit rates and their ease of editing.

11.2 INTERFRAME CODING

Now we look at coders that make use of the dependence between frames or interframe dependence. We consider spatiotemporal (3-D) DPCM, basic MC-DCT concepts, MPEGx multimedia coding, H.26x visual conferencing, and MC-SWT coding. We start out with the 1-D DPCM coder and generalize it to a sequence of temporal frames.

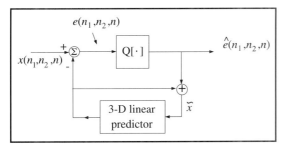

FIGURE 11.8   *Spatiotemporal generalization of DPCM with spatiotemporal predictor*

11.2.1 GENERALIZING 1-D DPCM TO INTERFRAME CODING

We replace the 1-D scalar value $x(n)$ with the video frame $x(n_1, n_2, n)$ and do the prediction from a *nonsymmetric half-space* (NSHS) region in general. However, in practice the prediction is usually based only on the prior frame or frames. Then we quantize the prediction error difference or residual. We thus have the system shown in Figure 11.8. We can perform *conditional replenishment* [4], which only transmits pixel significant differences, i.e., beyond a certain threshold value. These significant differences tend to be clustered in the frame so that we can efficiently transmit the cluster position with a context-adaptive VLC. The average bitrate at the buffer output can be controlled by a variable threshold to pass only significant differences. (For an excellent summary of this and other early contributions, see the 1980 review article by Netravali and Limb [32].)

Looking at Figure 11.8, we can generalize 3-D DPCM by putting any 2-D spatial or intraframe coder in place of the scalar quantizer $Q[\cdot]$. If we use block DCT for the spatial coder that replaced the spatial quantizer, we have a *hybrid coder*, being temporally DPCM and spatially transform-based. We typically use a frame-based predictor, whose most general case would be a 3-D nonlinear predictor operating on a number of past frames. Often though, just a frame delay is used, effectively assuming the current frame will be the same as the motion-warped version of the previous frame.

In some current coding standards, a spatial filter (called a *loop filter*) is used to shape the quantizer noise spectrum and to add temporal stability to this otherwise only marginally stable system. Another way to stabilize the DPCM loop is to put in an *Intra* frame every so often. Calling this frame an *I* frame, and the intercoded frames *P*, we can denote the coded sequence as *IPPP* · · · *PIPPP* · · · *PIPPP* · · · ·. Another idea, in the same context, is to randomly insert *I* blocks instead of complete *I* frames. While the former refresh method is used in the MPEG 2 entertainment standard, the latter random *I* block refresh method is used in the H.263 visual conferencing standard.

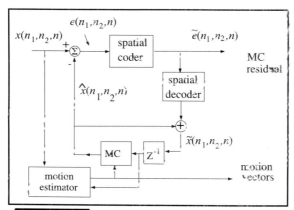

$e(n_1, n_2, n)$

$x(n_1, n_2, n)$

spatial coder

$\tilde{e}(n_1, n_2, n)$

MC residual

spatial decoder

$\hat{x}(n_1, n_2, n)$

$\tilde{x}(n_1, n_2, n)$

MC

Z^{-1}

motion vectors

motion estimator

FIGURE 11.9   *Illustrative system diagram for forward motion-compensated DPCM*

11.2.2 MC SPATIOTEMPORAL PREDICTION

There are two types of motion-compensated hybrid coders. They differ in the kind of motion estimation they employ: *forward* motion estimation or *backward* motion estimation. The MC block in Figure 11.9 performs the motion compensation (warping) after the motion estimation block computes forward motion. The quantity $\tilde{x}(n_1, n_2, n)$ seen in the figure is also the decoded output, i.e., the encoder contains a decoder.

The actual decoder is shown in Figure 11.10, and consists of first a spatial decoder followed by the familiar DPCM temporal decoding loop as modified by the motion compensation warping operation MC that is controlled by the received motion vectors. In *forward MC*, motion vectors are estimated between the current input frame and the prior decoded frame at the encoder. Then the motion vectors must be coded and sent along with the MC residual data as side information, since the decoder does not have access to the input frame $x(n_1, n_2, n)$. In *backward MC*, we base the motion estimation on the previous two decoded frames $\tilde{x}(n_1, n_2, n-1)$ and $\tilde{x}(n_1, n_2, n-2)$, as shown in Figure 11.11. Then there is no need to transmit motion vectors as side information, since in the absence of channel errors, the decoder can perform the same calculation as the encoder, and come up with the same motion vectors. Of course, from the viewpoint of total computation, backward MC means doing the computationally intensive motion estimation and compensation work twice, once at the coder and once at the decoder. (You are asked to write the decoder block diagram for backward motion-compensated hybrid coding as an end-of-chapter problem.)

One problem with interframe coding is the required accuracy in representing the motion vectors. There has been some theoretical work on this subject [12] that uses a very simple motion model. Even so, insight is gained on required motion vector accuracy, especially as regards the level of noise in the input frames.

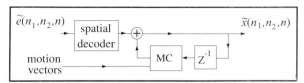

FIGURE 11.10   *Hybrid decoder for forward motion compensation*

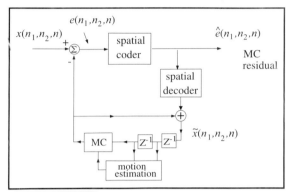

FIGURE 11.11   *Illustration of video encoder using backward motion compensation*

11.3 INTERFRAME CODING STANDARDS

Regarding interframe coding standards the most well known and influential is the MPEG family. The MPEG committee is a part of the *International Standards Organization* (ISO) and introduced its first video compression standard for CD-based video in 1988. It was called MPEG 1. A few years later they released a standard for SD video called MPEG 2. This standard has been widely used for distribution of digital video over satellite, DVD, terrestrial television broadcast, and cable. It was later extended to HD video. These are really standards for decoders. Manufacturers are free to create encoders with proprietary features that they feel improve picture quality, reduce needed bitrate, or improve functionality. The only requirement is that these coders create a bitstream that can be decoded by the standard decoder, thus ensuring interoperability. Since you cannot really create a decoder without an encoder, MPEG developed various test coders, or *verification models*, some of which have become widely available on the web. However, the performance of these "MPEG coders" varies and should not be considered to indicate the full capabilities of the standard. In the parlance of MPEG, these verification model coders are *nonnormative*, i.e., not part of the standard. The best available coders for an MPEG standard are usually proprietary.

11.3.1 MPEG 1

The first MPEG standard, MPEG 1, was intended for the SIF resolution, about half NTSC television resolution in each dimension. Exactly, this is 352×240 luma pixels. The chroma resolution is 2×2 reduced by half again to 176×120 chroma pixels. Temporally, SIF has a 30 Hz frame rate that is progressive, or non-interlaced. The aspect ratio is 4:3. The MPEG 1 standard was meant for bitrates between 1 and 2 Mbit/sec.

The overall coding method was intended for digital video and audio on 1x CDs in the late 1980s with a maximum data rate of 1.5 Mbps with 1.2 Mbps used for video. The intent was to provide for fast search, fast backward search, easy still-frame display, limited error propagation, and support for fast image acquisition starting at an arbitrary point in the bitstream. These needed functionalities make it difficult to use conventional frame-difference coding, i.e., to use temporal DPCM directly on all the frames. The MPEG solution is to block a number of frames into a *group of pictures* (GOP), and then to code these GOFs separately.

Each GOP must start with a frame that is intracoded, the *I* frame. This will limit efficiency, but facilitate random search, which can be conducted on the *I* frames only. An illustration of GOP structure is shown in Figure 11.12. If the GOP is not too long, say half a second, then the rapid image acquisition functionality can be achieved too. After the initial *I* frame, the remaining frames in the GOP are interframe coded. MPEG has two types of intercoded frames—*B* bidirectional MC-interpolated and *P* frames that are MC-predicted. The typical GOP size in video is $N = 15$, with $I = 1$ intraframe, $U = 4$ predicted frames, $B = 10$ interpolated frames. A parameter $M = 2$ here denotes the number of *B* frames between each *P* frame. After the *I* frame, we next predictively code the *P* frames, one by one, as seen in Figure 11.13. After its neighboring *P* frames are coded, then the bidirectional interpolation error of the *B* frames is calculated (and quantized) as illustrated in Figure 11.14. In MPEG parlance, with reference to this figure, the neighboring *I* and *P* frames serve as references for the coding of the *B* frames.

In MPEG 1, motion compensation is conducted by block matching on *macroblocks* of size 16×16. Exactly how to do the block matching is left open in the standard, as well as the size of the search area. Accuracy of the motion vectors was specified at the integer pixel level, sometimes called "full pixel accuracy," the accuracy naturally arising from block matching motion estimation. In fact, most MPEG 1 coders use a fast search method in their block matching.

We can see that MPEG 1 uses a layered coding approach. At the highest level is the GOP layer. The GOP in turn is made up of *pictures I, P, and B*. Each picture is composed of an array of macroblocks. Each macroblock consists of a 2×2 array of 8×8 blocks. The DCT is performed at this block level. There is one additional level, the *slice* level, that occurs within a picture. A slice is just a row or contiguous set of rows of macroblocks. Its main purpose is synchronization in the face of channel errors or lost packets in transmission. Variable-length coding (VLC) is performed only within a slice, and every slice is terminated with a unique *end-of-slice* (EOS) codeword. When this EOS codeword is detected in the sequence, end-of-slice is declared, and any remnant effects of past channel errors terminate. Another advantage of partitioning the coded data into slices is to provide more decoder access points, since there is usually a slice header that specifies frame number and position information. Usually cross-references across slices are disallowed, thus incurring at least a slight bitrate efficiency penalty for using small slices.

MPEG 1 achieves CBR by employing a so-called *video buffer verifier* (VBV) in the coder. It serves as a stand-in for the buffer that must be used at a decoder, since the decoder is fed by a constant bitrate by the channel. By keeping the VBV from underflow or overflow during coding, one can guarantee that there will be no buffer problems during decoding if a sufficiently large decoder buffer is employed.

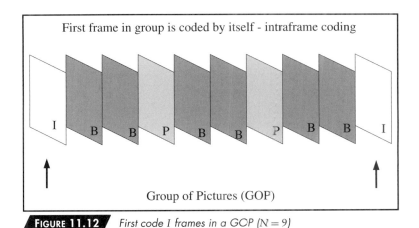

FIGURE 11.12   *First code I frames in a GOP (N = 9)*

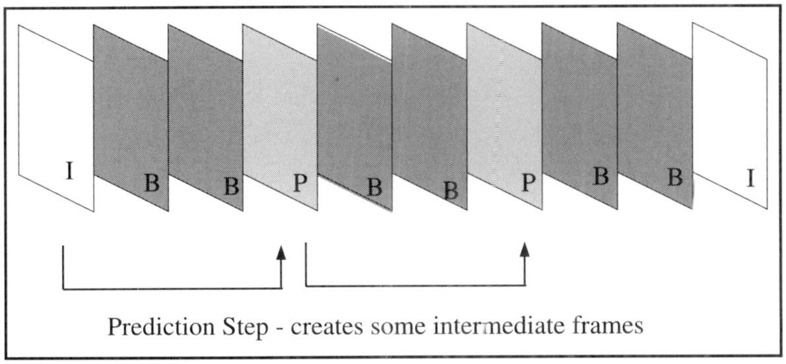

FIGURE 11.13   *Illustration of predictive coding of P frames in MPEG 1*

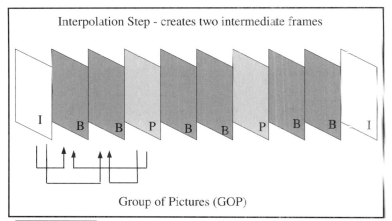

Interpolation Step - creates two intermediate frames

Group of Pictures (GOP)

FIGURE 11.14   *Illustration of coding the B frames based on bidirectional references*

11.3.2 MPEG 2—"A GENERIC STANDARD"

MPEG 2 (aka H.262) [19] includes MPEG 1 as subset, and does this by introducing *profiles* and *levels*. The level refers to the image size, i.e., CIF, SIF, or SD; frame rate; and whether the data is interlaced or progressive. The profile refers to the MPEG 2 features that are used in the coding. MPEG 2 also established new prediction modes for interlaced SD, permitting prediction to be *field based* or *frame based*, or a combination of the two. The predictions and interpolations have to consider the interlaced nature of common SD NTSC and PAL data. As a result, various combinations of field-based and frame-based prediction and interpolation modes are used in MPEG 2. For example, when bidirectionally predicting (interpolating) the *B* frames in Figure 11.15, there are both field-based and frame-based prediction modes. In field-based mode, which usually works best for fast motion, each field is predicted and each difference field is coded separately. In frame-based mode, which works best under conditions of slow or no motion, the frame is treated as a whole and coded as a unit. In field-based mode there is the question of what data to use as the reference field, i.e., which field, *upper* or *lower*.[2] The additional overhead information of the prediction mode, field or frame, has to be coded and sent along as overhead. This provision for interlaced video makes MPEG 2 more complicated than MPEG 1, which only handled progressive video, but reportedly gives it about a 0.5–2.5 dB advantage [16] versus frame-based mode alone. This is due to the combined use of frame/field prediction and frame/field DCT.

2. In MPEG 2, all data is stored in frames, and the field and frame numbers in a GOP are numbered starting with 0. Each frame consists of an upper or *even field*, which comes first, and a lower or *odd field*, which comes second.

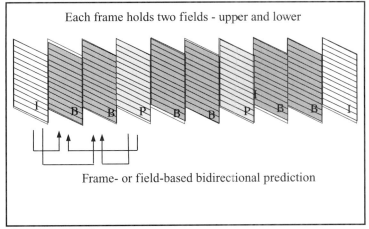

FIGURE 11.15   *New in MPEG 2 was the need to process interlaced frame data*

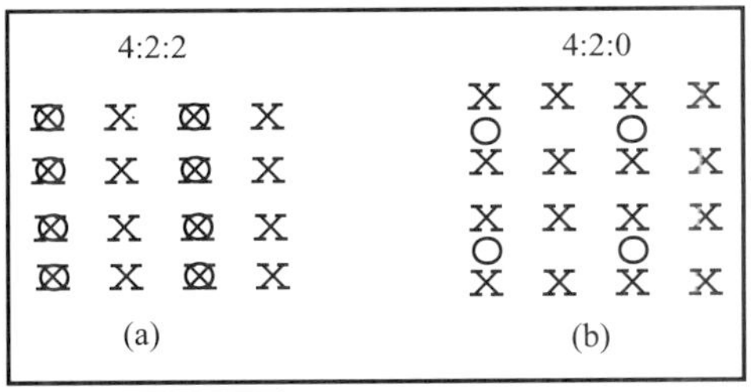

FIGURE 11.16   *MPEG 2 4:2:2 and 4:2:0 luma and chroma subsampling sites*

Commonly used color spaces in MPEG 2 are 4:2:2 for professional applications and 4:2:0 for consumer delivery, e.g., on DVD or digital satellite television. These subsampling structures are shown in Figure 11.16 [19].

EXAMPLE 11.3-1 (*MPEG 2 example*)
We coded the ITU 601 or SD test sequence *Susie* with a standard MPEG 2 coder available on the web. The bitrate was 3 Mbps at a frame rate of 30 fps. We first used all *I* frames, i.e., GOP = 1, then we set GOP = 15, or 1/2 second, and coded with *P* frames, and then with *P* and *B* frames. The PSNR results are shown in the Table 11.2. We see a jump in PSNR going from *I* frame only to *I* and *P*, showing the efficiency of the *P* frame in a hybrid coder. Not much objective improvement, though, for adding on the *B* frame is seen in this visual conferencing test clip. See video results on enclosed CD-ROM.

Table 11.2. Some MPEG 2 coding results at 3 Mbps for SD test clip *Susie*

PSNR	Y (dB)	U (dB)	V (dB)
I	33.8	44.7	44.6
I, P	39.0	45.7	45.5
I, B, P	39.1	46.1	45.8

11.3.3 THE MISSING MPEG 3—HIGH-DEFINITION TELEVISION

As can easily be seen, video spatial resolution is generally quite poor, with the so-called SD resolution not really adequate even for digital photos. Researchers for many years have been working on an increased resolution system. Reading from report 801-4 of the Consultative Committee for International Radio (CCIR; now ITU), in 1990, they defined HDTV as "a system designed to allow viewing at about three times picture height, so that the system is virtually transparent to the quality of portrayal that would have been perceived in the original scene or performance by a discerning viewer with normal visual acuity."

Researchers at NHK in Japan had started modern research on HDTV in the early 1970s [8]. They came up with an analog system called *Hi-Vision* that enjoyed some deployment, mainly in Japan, but the equipment was very bulky, heavy, and expensive. This spurred U.S. companies and researchers to try to invent an all-digital approach in the mid to late 1980s. A set of competing digital HDTV proposals then emerged in the United States. The Grand Alliance was formed in May 1993 from the these proponents and offered a solution, eventually based on extending MPEG 2. It covered both interlaced and progressive scanning formats in a 16:9 aspect ratio (near to the 2:1 and higher aspect ratio of motion pictures). The resulting digital television standard from the group now known as the Advanced Television Systems Committee (ATSC) offers 1080i and 720p digital video in cable and satellite systems in North America. (More information on these formats can be found in the appendix at the end of Chapter 10.) Since ATSC essentially voted in favor of MPEG 2, the HD and SD resolutions have essentially been folded together now into one coding family, just extending the profiles and levels of MPEG 2. Thus HD is coded with MPEG 2—Main Profile, High Level (MP@HL). The consumer/prosumer HDV format also uses MPEG 2 at High Level.

11.3.4 MPEG 4—NATURAL AND SYNTHETIC COMBINED

The original concept of MPEG 4 was to find a very much more efficient coder at low bitrates, e.g., 64 Kbps. When this target proved elusive, the effort turned to object-based video coding, where it remained till completion. In MPEG 4, so-called video objects (VOs) are the basic elements to be coded. Their boundaries are coded by *shape coders* and their interiors are coded by so-called *texture coders*.

The method was eventually extended up to SD and HD and is backward compatible with MPEG 2. A *synthetic natural hybrid coder* (SNHC) is capable of combining natural and synthetic (cartoon, graphics, text) together in a common frame. Nevertheless, most currently available embodiments of MPEG 4 work only with rectangular objects and one object per frame. Of course, this may change, but segmenting natural video into reasonable objects remains a hard problem.

Also starting with MPEG 4, there was an effort to separate the video coding from the actual transport bitstreams. Researchers wanted MPEG 4 video to be carried on many different networks with different packet or cell sizes and qualities of service. So MPEG 4 simply provides a *delivery multimedia integration framework* (DMIF) for adapting the output of the MPEG 4 video coder, the so-called *elementary streams* (ES), to any of several already existing transport formats, such as MPEG 2 *transport stream* (TS), asynchronous transfer mode (ATM), and the real-time transport protocol/internet protocol (RTP/IP). Complete information on MPEG 4 is available in the resource book by Pereira and Ebrahimi [35].

11.3.5 VIDEO PROCESSING OF MPEG-CODED BITSTREAMS

Various video processing tasks have been investigated for coded data and MPEG 2 in particular. In *video editing* of MPEG bitstreams, the question is how to take two or more MPEG input bitstreams and make one composite MPEG output bitstream. The problem with decoding/recoding is that it introduces artifacts and is computationally demanding. Staying as much in the MPEG 2 compressed domain as possible and reusing the editing mode decisions and motion vectors have been found essential. Since the GOP boundaries may not align, it is necessary to recode the one GOP where the edit point lies. Even then original bitrates may make the output *video buffer verifier* (VBV) overflow. The solution is to requantize this GOP and perhaps neighboring GOPs to reduce the likelihood of such buffer overflow. The recent introduction of the HDV format, featuring MPEG 2 compression of HD video in camera, brings the MPEG bitstream editing problem to the forefront. Many software products are emerging and promise to edit HDV video in its so-called *native mode*.

The *transcoding of MPEG* question is how to go from a high-quality level of MPEG to a lower one, without decoding and recoding at the desired output rate. Transcoding is of interest for video-on-demand (VoD) applications. Transcoding is also of interest in video production work where short GOP (IPIP) may be used internally for editing and program creation, while the longer *IBBPBBP* ··· long GOP structure is necessary for program distribution. Finally, there is the worldwide distribution problem, where 525 and 625 line systems[3] continue to coexist.

---

3. The reader should note that the 525 line system only has 486 visible lines, i.e., it is D1, which is 720×486. A similar statement is true for the 625 line system. The remaining lines are hidden!

Here a motion-compensated transcoding of the MPEG bitstream is required. For more on transcoding, see Chapter 6.3 in Bovik's handbook [48].

11.3.6 H.263 CODER FOR VISUAL CONFERENCING

The H.263 coder from the ITU evolved from their earlier H.261, or px64 coder. As the original name implies, it is targeted at rates that are a multiple of 64 Kbps. To achieve such low bitrates, they resort to a small QCIF frame size, and a variable and low frame rate, with bitrate control based on buffer fullness. If there is a lot of motion in detailed areas that generate a lot of bits, then the buffer fills and the frame rate is reduced, i.e., frames are dropped at the encoder. The user can specify a target frame rate, but often the H.263 coder at, say 64 Kbps, will not achieve a target frame rate of 10 fps. The H.263 coder features a *group of blocks* (GOB) structure, rather than a GOP, with *I* blocks being inserted randomly for refresh. While there are no *B* frames, there is the option for so-called *PB* frames. The coder has half-pixel accurate motion vectors like MPEG 2, and can use overlapped motion vectors from neighboring blocks to achieve a smoother motion field and reduced blocking artifacts. Also, there is an advanced prediction mode option and an arithmetic coder option.

The reason for targeting the GOB structure versus the GOP structure is the need to avoid the *I* frames in the GOP structure, because they require a lot of bits to transmit, a difficulty in videophone, which H.263 targets as a main application. In videophone, low bitrates and short latency requirement ($\leqslant 200$ msec) mitigate against the bit-hungry *I* frames. As a result, in H.263, slices or GOBs are updated by *I* slices randomly, thus reducing the variance on coded frame sizes that occurs with the GOP structure. High variance of bits/frame is not a problem in MPEG 2 because of its targeted entertainment applications, such as video streaming, including multicasting, digital broadcasting, and DVD. Some low bitrate H.263 coding results are contained on the enclosed CD-ROM.

11.3.7 H.264/AVC

Research on increasing coding efficiency continued through the late 1990s and it was found that up to a 50% increase in coding efficiency could be obtained by various improvements to the basic hybrid coding approach of MPEG 2. Instead of using one hypothesis for the motion estimation, multiple hypotheses from multiple locations in past frames could be used [39] together with an optimization approach to allocate the bits. By 2001, the video standards groups at ITU, Video Coding Experts Group (VCEG), and ISO MPEG convened a joint video team (JVT) to work on the new standard, to be called H.264 by ITU and MPEG 4, part 10 by the ISO. The common name is Advanced Video Coder (AVC). With reference to Figure 11.17, we see that more frame memory to store past frames

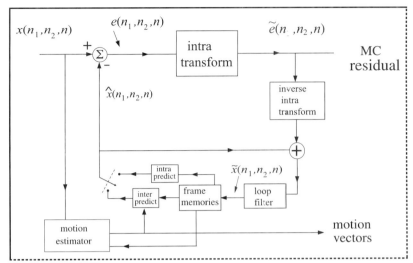

FIGURE 11.17   *System diagram of the H.264/AVC coder*

has been added to the basic hybrid coder of Figure 11.9. We also see the addition of a loop filter that serves to smooth out blocking artifacts. Further, before the intra transform, there is intra, or directional spatial prediction (explained below), hence the need to switch between intra and inter prediction modes as seen in Figure 11.17.

There are many new ideas in H.264/AVC that allow it to perform at almost twice the efficiency of the MPEG 2 standard, also known as H.262. There is a new variable blocksize motion estimation, with blocksizes ranging from 16×16 down to 4×4, and motion vector accuracy raised to one-quarter pixel from the half-pixel accuracy of MPEG 2. The permitted blocksize choices are shown in Figure 11.18. The 16×16 macroblock can be split in three ways to get 16×8, 8×16, or 8×8, as shown. If the accuracy of 8×8 blocks is not enough, one more round of such splitting finally results in the sub-macroblocks 8×4, 4×8, or 4×4. Note that in addition to what we would get by quadtree splitting (cf. Chapter 10), we get the possible rectangular blocks, which can be thought of as a level inserted between two quadtree square block levels.

To match this smallest blocksize, a 4×4 integer-based transform is introduced that is DCT-like, and separable using the 1-D four-point transform

$$\mathbf{H} = \begin{bmatrix} 1 & 1 & 1 & 1 \\ 2 & 1 & -1 & -2 \\ 1 & -1 & -1 & 1 \\ 1 & -2 & 2 & -1 \end{bmatrix}. \tag{11.3-1}$$

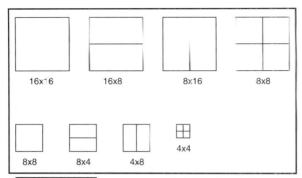

Allowed motion vector block sizes in H.264/AVC

The H.264/AVC coder is based on *slices*, with I, B, and P slices, as well as two new *switching slices* SP and SI. The slices are in turn made up of 16×16 macroblocks. The P slice can have I or P macroblocks. The B slice can have I, B, or P macroblocks. There is no mention of group of pictures, but there are I pictures, needed at the start to initialize this hybrid coder. There is nothing to prohibit a slice from being the size of a whole frame, so that there can effectively be P and B pictures as well.

In H.264/AVC, an I slice is defined as one whose macroblocks are all intracoded. A P slice has macroblocks that can be predictively coded with up to *one* motion vector per block. A B slice has macroblocks predictively (interpolatively) coded using up to *two* motion vectors per block. Additionally, there are new switching slices SP and SI that permit efficient jumping from place to place within or across bitstreams (cf., Section 12.3, Chapter 12).

Within an I slice, there is *intrapicture prediction*, done blockwise based on previously coded blocks. The intra prediction block error is then subject to the 4×4 integer-based transform and then quantization. The intra prediction is adaptive and directional based for 4×4 blocks, as indicated in Figure 11.19. A prespecified fixed blending of the available boundary values is tried in each of the eight prediction directions. Also available is a so-called DC option that predicts the whole block as a constant. For 16×16 blocks, only four intra prediction modes are possible: vertical, horizontal, DC, and *plane*, the last one coding the values of a best-fit plane for the macroblock. The motion compensation can use multiple references, so, for example, a block in a P slice can be predicted by one to four reference blocks in earlier frames. The H.264/AVC standard specifies the amount of reference frame storage that must be available at the decoder to store these past pictures, and five past frames is common.

Figure 11.20 illustrates the comparative PSNR versus bitrate performance of the verification models of H.264/AVC on the 15-fps CIF test clip *Tempete* (HLP, High-Latency Profile; ASP, Advanced Simple Profile; MP, Main Profile). The figure [40] shows considerable improvement over MPEG 2, of about a factor of 2 in-

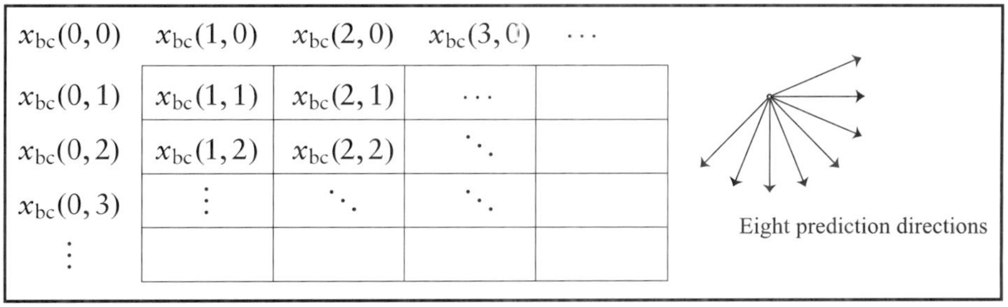

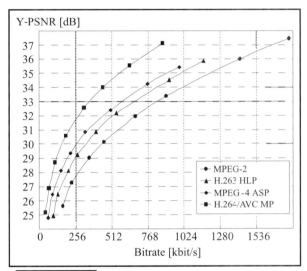

FIGURE 11.19 *Illustration of directional prediction modes of H.264/AVC in the case of 4 × 4 blocks*

FIGURE 11.20 *PSNR vs. bitrate for 15-fps CIF test clip Tempete. Reprinted with permission form Sullivan and Wiegand [40]*

creased compression at fixed PSNR. This improvement in compression efficiency is largely due to the greater exploitation of motion information made possible by the revolutionary increases in affordable computational power of the past 10 years. More information on the new H.264/AVC standard is available in the review article by Wiegand *et al.* [43], which introduces a special issue on this topic [38].

11.3.8 VIDEO CODER MODE CONTROL

A video coder like H.264/AVC has many coder *modes* to be determined. For each macroblock, there is the decision of inter or intra, and quantizer step-size, as well as motion vector blocksize. Each choice (mode) translates into a different number of coded bits for the macroblock. We can write the relevant Lagrangian form as

$$J_k \triangleq D_k(\mathbf{b}_k, \mathbf{m}_k, Q) + \lambda_{mode} R_k(\mathbf{b}_k, \mathbf{m}_k, Q), \qquad (11.3\text{-}2)$$

where D_k and R_k are the corresponding mean-square distortion and rate for block $\mathbf{b}_k$ when coding in mode $\mathbf{m}_k$, and Q is the quantizer step-size parameter. Here R_k must include the bits for the transformed block plus bits for the motion vector(s) and mode decision, the latter usually being negligible for 16×16 macroblocks. For the moment assume that the motion vectors have been determined prior to this optimization. Then we sum over all the blocks in the frame to get the total distortion and rate for that frame,

$$
\begin{aligned}
J &= D + \lambda_{mode} R \\
&= \sum_k \left\{ D_k(\mathbf{b}_k, \mathbf{m}_k, Q) + \lambda_{mode} R_k(\mathbf{b}_k, \mathbf{m}_k, Q) \right\} \\
&= \sum_k J_k,
\end{aligned}
$$

where D and R are the frame's total distortion and rate. Normally the blocks in a frame are scanned in the 2-D raster manner, i.e., left to right and then top to bottom. The joint optimization of all the modes and Q values for these blocks is a daunting problem. In practical coders, often what optimization there is, is done macroblock by macroblock, wherein the best choice of mode $\mathbf{m}_k$ and Q is done for block $\mathbf{b}_k$ conditional on the past of the frame in the NSHP sense, resulting in at most a stepwise optimal result. We can achieve this stepwise or greedy optimal result by evaluating J_k in (11.3-2) for all the modes and Q and then choosing the lowest value. This point will then be on the optimized D–R curve for some rate. To generate the entire curve, we sweep through the parameter λ_{mode}. Now, in the test model for H.263, an experimentally observed relation is used, that has been theoretically motivated in the high-rate Gaussian case [39],

$$\lambda_{mode} = cQ^2,$$

with the value $c = 0.85$, approximately on the experimental D–R curve. Therefore, the blockwise optimization in the test model for H.263 then becomes: for each value of quantization parameter Q, choose the mode m_k such that

$$\mathbf{m}_k = \arg\min_k \left\{ D_k(\mathbf{b}_k, \mathbf{m}_k, Q) + cQ^2 R_k(\mathbf{b}_k, \mathbf{m}_k, Q) \right\}.$$

A somewhat different approximation for $\lambda_{mode} = f(Q)$ is used in the H.264/AVC test model. In both cases, this then results in a sequence of macroblocks that is optimized in the so-called constant-slope sense. To actually get a CBR coded sequence, some kind of further rate control has to be applied to decide what value of Q to use for each macroblock in each frame. An easy VBR case results from the choice of constant Q, but then the total bitrate is unconstrained. Also, in practice, different Q values are used for I, P, and B slices or frames, with step-size increasing in a fixed manner. This choice also is usually fixed and not optimized over. Extension of this method to include the needed optimization of the motion vector bitrate in a variable blocksize motion field is contained in Sullivan and Wiegand [39].

11.3.9 NETWORK ADAPTATION

In H.264/AVC, there is a separation of the *video coding layer* (VCL), which we have been discussing, from the actual transport bitstream. As in MPEG 4 video, H.264/AVC is intended to be carried on many different networks with different packet or cell sizes and qualities of service. So MPEG simply provides a *network abstraction layer* (NAL) specification about how to adapt the output of the VCL to any of several transport streams, including transport on MPEG 2, ATM, and IP networks. (In Chapter 12, we discuss further some basic issues in network video.)

11.4 INTERFRAME SWT CODERS

The first 3-D subband coding was done by Karlson [22] in 1989. This first 3-D subband coding (Figure 11.21) used Haar or 2-tap subband/wavelet filters in the temporal direction and separable LeGall/Tabatabai (LGT) 5/3 spatial filters [25], whose 1-D component filters are as follows:

$$H_0(z) = \frac{1}{4}\left(-1 + 2z^{-1} + 6z^{-2} + 2z^{-3} - z^{-4}\right),$$

$$H_1(z) = \frac{1}{4}\left(1 - 2z^{-1} + z^{-2}\right).$$

$$G_0(z) = \frac{1}{4}\left(1 + 2z^{-1} + z^{-2}\right),$$

and

$$G_1(z) = \frac{1}{4}\left(1 + 2z^{-1} - 6z^{-2} + 2z^{-3} + z^{-4}\right).$$

The system was subjectively optimized using scalar quantization, and DPCM to increase coding efficiency for the base subband. The 3-D frequency domain subband division of Karlson, as shown in Figure 11.21, achieves the frequency decomposition shown in Figure 11.22.

A simple 3-D subband coder was also designed by Glenn *et al.* [13]; which used perceptual coding based on the human contrast sensitivity function, but like

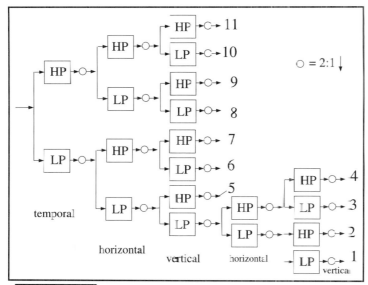

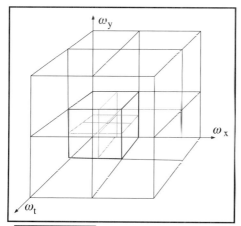

FIGURE 11.21 First 3-D subband filter, due to Karlson

FIGURE 11.22 Frequency decomposition of Karlson's filter tree

the Karlson coder, no motion compensation. A real-time demonstration hardware of Glenn's offered 30:1 compression of SD video up to a bitrate of to 6 Mbits/sec. The visual performance was said to be comparable to MPEG 2, in that it traded slight MPEG artifacts for this coder's slight softening of moving scenes. Such softening of moving objects is typical of 3-D transform coders that do not use motion compensation.

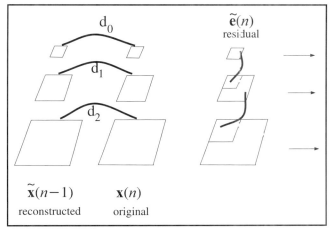

FIGURE 11.23   *Illustration of multiresolution coder*

11.4.1 MOTION-COMPENSATED SWT HYBRID CODING

Naveen and Woods [29,46] produced a hybrid SWT coder with the spatial scalability feature. Both pel-recursion, in the context of backward motion estimation, and block matching, for forward motion estimation, were considered. They used a subband pyramid for hierarchical motion estimation for the multiresolution transmission of HD and SD video. Their basic motion compensation approach placed the motion compensation inside the hierarchical coding loop to avoid drift.

An illustration of the motion compensation pyramid and coding pyramid is given in Figure 11.23, which shows two subband/wavelet pyramids on the left, for the previous reconstructed frame $\tilde{x}(n-1)$ and the current input frame $x(n)$. The prediction error residual, to be coded and transmitted, $e(n)$, is shown on the right pyramid. First, hierarchical block matching determines a set of displacement vector fields d_0, d_1, and d_2 of increasing resolution. A prediction residual is then created for the lowest resolution level, coded, and transmitted. The corresponding low-resolution reconstructed value is then used to predict the LL subband of the next higher resolution level. For this level, the prediction error is coded and transmitted for the LL subband, while the temporal prediction residuals are coded and transmitted for the higher resolutions. The process is repeated for the third spatial resolution. In this scheme there arises the need to make a bit assignment for the various levels to try to meet given quality requirements.

In the context of block matching motion estimation with SWT coding of the residual, the motion vector discontinuities at the block boundaries can increase the energy in the high-frequency coefficients. This situation does not arise in DCT coding because the DCT blocksize is kept at a subpartition size with respect to the motion vector blocks, e.g., common DCT size 8×8 with common motion vector blocksizes are 16×16 and 8×8 in MPEG 2. Thus the DCT transform does not see the artificial edge in the prediction residual frame caused by any motion-vector discontinuity at the motion-vector block boundaries. Quite the contrary is true of the SWT transform, where there are no block boundaries, so that any discontinuities caused by the block motion vector prediction are a serious problem. To counter this, a small amount of smoothing of motion vectors at the block boundaries was used in [29] for the case of forward motion compensation. The problem does not arise in the backward motion compensation case, where a smooth dense motion estimate can be used, e.g., the pel-recursive method, since the motion field does not have to be transmitted.

Gharavi [11] used block matching and achieved integer pixel accuracy in subbands, permitting the parallel processing of each subband at four times lower sample rate. However, Vandendorpe [42] pointed out that neighboring subbands must be used for *in-band* (in-subband) *motion compensation*. Bosveld [2] continued the investigation and found MSE prediction gains almost equal to those of the spatial domain case. The result is more embedded data sets that facilitate scalability. Usually just the nearest or bordering subbands contribute significantly to the in-band motion compensation.

11.4.2 3-D OR SPATIOTEMPORAL TRANSFORM CODING

Kronander [24] took the temporal DCT in blocks of 4 or 8 frames and motion-compensated toward a central frame in the block. A spatial subband/wavelet filter was then used to code the MC residual frames in the block. Decoding proceeded by first decoding the residual frames, and then inverting the MC process.

Consider a block of compressed images available at time k, and denote the lth such image

$$\mathbf{y}_{k,l} = \mathbf{x}_{k,l}\big(\mathbf{t} - \mathbf{v}_{k,l}(\mathbf{t})\,\triangle\, t\big); \quad l = 0, \ldots, L-1$$

where $\mathbf{t}$ is a spatial position with velocity $\mathbf{v}_{k,l}$ with respect to some central reference frame l'. We then perform spatial subband/wavelet coding/decoding of the temporal prediction residue,

$$\mathbf{y}_{k,l} \to \tilde{\mathbf{y}}_{k,l}.$$

Finally, we approximately invert the displacement information to obtain the decoded frames,

$$\tilde{\mathbf{x}}_{k,l}(\mathbf{t}) = \tilde{\mathbf{y}}_{k,l}\big(\mathbf{t} - \mathbf{u}_{k,l}(\mathbf{t})\,\triangle\, t\big),$$

where $\mathbf{u}_{k,l}$ is the inverse to $\mathbf{v}_{k,l}$, at least approximately, and then proceed to the next block, $k+l$.

Conditions for invertability, simply stated, are "if the transformation were painted on a rubber sheet ... a good compensation algorithm should distort the

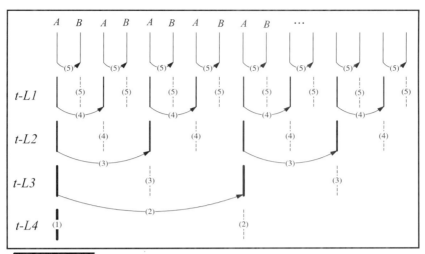

FIGURE 11.24   *Illustration of four-level Haar filter MCTF*

rubber but not fold it" [24]. This effectively precludes covered or occluded regions. The motion estimation methods used were full-search block matching—8×8 block, 16×16 search area—and hierarchical block matching with a 7×7 block and 3×3 search at each level. Motion was inverted by simple negation of displacement vectors.

Ohm's SWT coder [33] featured a *motion-compensated temporal filter* (MCTF), a type of 3-D subband/wavelet filter that attempts to filter along the local motion paths. For the temporal filtering, Ohm made use of the 2-tap Haar transform. Due to the fact that MCTF is nonrecursive in time, we can expect limited error propagation, and, importantly, no temporal drift in scalable applications. A four-temporal-level MCTF with GOP = 16 is shown in Figure 11.24. This MCTF uses the Haar filter, with motion fields between successive pairs of frames indicated with curved arcs. Proceeding down the diagram we go to lower temporal levels or frame rates, by a factor of two each time. The lowest frame rate here is one-sixteenth of the full frame rate. The numbers in parentheses denote the temporal levels, with 1 being the lowest and 5 (or full frame rate) being the highest. Note that the motion vectors are numbered with the temporal levels that they serve to reconstruct. With reference to Figure 11.24, we see motion vectors represented as rightward arrows, terminating at the location of t–H frames. Different from the now somewhat old forward and backward MC terminology used in hybrid coding, in MCTF parlance this is called *forward motion compensation*, with backward motion compensation being reserved for the case where the arrows point to the left. In both cases, the motion vectors must be transmitted.

A corresponding three-level diagram for a three-level LGT 5/3 filter is shown in Figure 11.25, where we notice that twice as much motion information is needed as in the Haar filter MCTF. Also note that the GOP concept is blurred by the

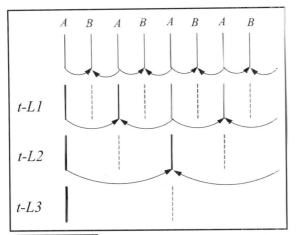

FIGURE 11.25   Illustration of three-level LGT 5/3 filter MCTF

references going outside the set of eight frames. So for 5/3 and longer SWT filter sets, it is better to talk about levels of the MCTF than about the GOP size.

In equations, we can write the lifting steps, at each temporal level, for each of these cases as follows:

Haar case:

$$h(\mathbf{n}) = x_A(\mathbf{n}) - x_B(\mathbf{n} - \mathbf{d}_f), \qquad (11.4\text{-}1)$$

$$l(n) = x_B(\mathbf{n}) + \frac{1}{2}h(\mathbf{n} - \mathbf{d}_f). \qquad (11.4\text{-}2)$$

LGT 5/3 case:

$$h(\mathbf{n}) = x_A(\mathbf{n}) - \frac{1}{2}\left[x_{B_f}(\mathbf{n} - \mathbf{d}_f) + x_{B_r}(\mathbf{n} - \mathbf{d}_r)\right], \qquad (11.4\text{-}3)$$

$$l(n) = x_B(\mathbf{n}) + \frac{1}{4}\left[h_f(\mathbf{n} + \mathbf{d}_f) + h_r(\mathbf{n} + \mathbf{d}_r)\right]. \qquad (11.4\text{-}4)$$

In these equations, we have labeled the frames in terms of A and B as shown in Figures 11.24 and 11.25 and also used $\mathbf{d}_f$ and $\mathbf{d}_r$ to refer to the forward and backward motion vectors, from frame A to reference frames B on either side, i.e., B_f and B_r, where f stands for *forward* and r stands for *reverse*. Note that the number of motion vectors is doubled from the Haar case. These operations are then repeated at the second and subsequent levels of decomposition. We can also notice that backward or reverse motion vectors, i.e., $\mathbf{d} = -\mathbf{d}_f$, are used in the update step for the LGT 5/3 lifting implementation. (This is also done in the Haar case.) Finally, in the subpixel accuracy case, several of the terms in these

Table 11.3. Choi and Woods results at 1.2 Mbps

Sequence	Y (dB)	U (dB)	V (dB)
Mobile	27.0	31.2	31.4
Table Tennis	33.8	39.4	39.8
Flower Garden	27.6	32.6	30.8

two equations must be computed by interpolation. The beauty of the lifting implementation, though, is that this is exactly invertible in the corresponding SWT synthesis operation, so long as the same interpolation method is used.

EXAMPLE 11.4-1 (*Haar MCTF 4 with four levels*)
With reference to Figure 11.24, t–L3 is an intraframe to be coded and sent every 16 frames. Hence there is a natural temporal blocksize of 16 here. The t–L4 frame subsequence has been filtered along the motion trajectory prior to being subsampled by 16. In addition to the t–L4 sequence, we must send the t–H subbands at levels 1 through 4, for a total of 16 frames per block. Of course, it is also necessary to transmit all of the motion vectors, the d's.

One issue with the MCTF approach is delay. A four-stage MCTF decomposition requires 16 frames, which may be too much delay for some applications, e.g., visual conferencing. However, one can combine the nonrecursive MCTF with the hybrid coder by limiting the number of MCTF stages to k and then applying *displaced frame difference* (DFD) prediction, i.e., a hybrid coder such as MPEGx, to the t–Lk subband. In a communications scenario, one may want to limit this kind of built-in algorithmic delay to 100 msec, which implies a frame rate greater than or equal to 10 fps. At this frame rate, no stages of MCTF could be allowed. If we raise the frame rate to 20 fps, then we can have one stage. At 50 fps, we can have two stages, etc., and still meet the overall goal of 100 msec algorithmic delay. So, as the frame rate goes up, the number of MCTF stages can grow too, even in the visual conferencing scenario. Of course, in streaming video, there is no problem with these small amounts of delay, and one can use five and six stages of MCTF with no significant penalty.

The Choi and Woods coder [6,7] had two levels of MCTF with 2-tap Haar filters and spatial SWT with Daubechies D4 filters, and variable blocksize motion compensation. This was followed by UTQs with adaptive arithmetic coding and prediction based on the subband-finite-state scalar quantization (FSSQ) coding technique [31]. An optimized rate allocation was performed across 3-D subbands using a Lagrangian method. The resulting coding performance in terms of PSNR at 1.2 Mbps is shown in Table 11.3 for various CIF clips.

11.5 SCALABLE VIDEO CODERS

In many applications of video compression, it is not known to the encoder what resolutions, frame rates, and/or qualities (bitrates) are needed at the receivers. Further, in a streaming or broadcast application, different values of these key video parameters may be needed across the communication paths or links. Scalable coders [47] have been advanced to solve this problem without the need for complicated and lossy transcoding.

DEFINITION 11.5-1 (*scalability*)
One coded video bitstream that can be efficiently decomposed for use at many spatial resolutions (image sizes), frame rates, regions of interest, and bitrates.

A scalable bitstream that has all four types of scalabilities will be referred to as *fully scalable*. We can think of region-of-interest capability as a restricted type of object-based scalability that can, together with resolution scalability, support zoom and pan functionality. This capability can support browsing, wherein a small version of a high-resolution image may be "zoomed and panned," to locate interesting parts for closer looks at higher resolution.

Scalability is needed in several areas, including digital television, heterogeneous computer networks, database browsing, to match various display formats (to adjust to window size on screen), to match various display frame rates, to deal with loading of a *video on demand* (VoD) server, etc. Scalability in database browsing facilitates efficient pyramid search of image and video databases. Scalability on heterogeneous networks can enable a network manager to do dynamic load control to match link capacities as well as terminal computer capabilities. A scalable encoder can also better match a variable *carrier-to-noise ratio* (CNR) on wireless channels.

The standard coders H.26x and MPEGx have a limited amount of scalability, usually just one or two levels. For spatial scalability, MPEG 2 uses a pyramid coding method, where the base layer is coded conventionally, and then an enhancement layer supplements this base layer for higher resolution, frame rate, or quality. Figure 11.26 shows a coder targeted for spatial scalability in HD and SD television.

Separate spatial and temporal scalability profiles, with two or three levels only, were standardized in MPEG 2, but are seldom used in practice. The limitations of these scalable profiles are lack of data conservation, limited range of scalability, coding errors not limited to the baseband, and drift for frequency scalable coder (reportedly building up to 7 dB over 12 frames [20]).

In the research area, Burt and Adelson [3] introduced a scalable pyramid coder with the desirable quantizer-in-loop property that can frequency shape (rolloff) lower resolution layers. Unfortunately, due to use of a Gaussian–Laplacian

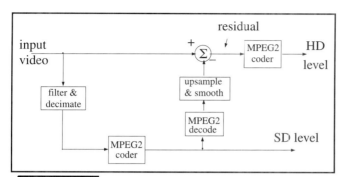

FIGURE 11.26   *Illustration of resolution scalability for SD and HD video using MPEG 2*

Table 11.4. PSNR results—forward motion compensation

PSNR (dB)	*MIT* (0.5 bpp)	*Surfside* (0.25 bpp)
Low res	38.0	42.2
Medium	33.5	41.6
High res	34.3	42.5
High (not scalable)	35.0	42.8

Table 11.5. Motion compensation via global pan vector at 1.5 Mbps

PSNR (dB)	MPEG 1	Highly scalable
Football	33.6	34.6
Table Tennis	32.8	34.6

lowpass pyramid, there is lack of data conservation. Also due to the pyramid-based coding-and-recoding structure, the coding noise and artifacts spread from low (baseband) to higher spatial frequencies. Naveen's multiresolution SWT coder [28,29] as sketched in Figure 11.23, is scalable in resolution with three spatial levels. There is no drift at lower resolution because the same motion information is used at the encoder as at the decoder. It uses efficient hierarchical MC that can incorporate rate constraints and frequency roll-off to make the lower resolution frames more video-like, i.e., reduce their spatial high frequencies. Table 11.4 gives some average luma PSNR values for HD test clips *MIT* and *Surfside*, obtained using forward MC.

The Taubman and Zakhor [41] MCTF algorithm featured global motion compensation, layered (embedded) quantization, and conditional zero coding to facilitate SNR scalability. The published PSNR results for two common CIF test clips at 1.5 Mbps, coding only the luma (*Y*) component, are shown in Table 11.5. Results were good on panning motion clips, but less efficient on clips with detailed local motion. Also provided in the table is an MPEG 1 result for comparison.

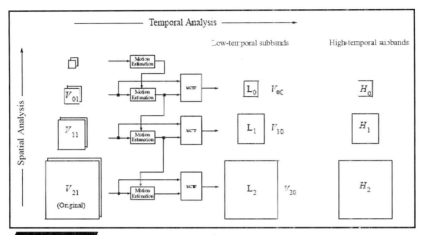

FIGURE 11.27 *RSC Demonstration Software: Three Resolutions, Two Frame Rates. Reproduced with permission [45]*

Table 11.6. RSC average PSNR (dB) on HD test clips

PSNR (dB) video level		Low frame rate			High frame rate		
		V_{00}	V_{10}	V_{20}	V_{01}	V_{11}	V_{21}
Mali	w/o roll-off	33.8	37.3	42.3	34.5	37.0	40.7
	with roll-off	36.0	39.4	42.2	37.2	39.5	40.6
MIT	w/o roll-off	30.4	28.7	32.1	31.9	31.1	33.2
	with roll-off	33.1	31.7	32.0	34.8	34.1	33.2

Our [45] *resolution scalable coder* (RSC), performed spatial subband analysis first, and then with one stage of MCTF at each resolution. RSC then applied the hybrid technique of DFD on the lowest spatiotemporal level to yield three spatial resolutions and two frame rates. This nonembedded approach used UTQs based on the generalized Gaussian distribution and multistage quantizers to achieve the resolution and frame-rate scalability. As seen in Figure 11.27, the system is of the type "2D + t" since it starts out with a spatial subband/wavelet pyramid on the incoming data. The various spatial levels are then fed into stages of motion estimation and MCTF, with outputs shown on the right, consisting of temporal low L_i and high H_i frames.

The low-frame-rate video hierarchy is given as V_{i0} and the high-rate hierarchy, denoted V_{i1}, is obtained by the addition of the corresponding H_i frame. Average PSNR results for the luma channel are given in Table 11.6, where each channel or spatiotemporal resolution was coded at 0.35 bpp. We see that high-frequency roll-off is moderately successful at keeping the PSNR relatively constant across the levels.

Hsiang and Woods [18] presented an MCTF scalable video coder, wherein an invertible motion-compensated 3-D SWT filter bank was utilized for video analysis/synthesis. The efficient embedded zero-block (image) coding (EZBC) scheme

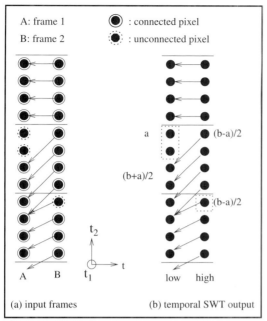

FIGURE 11.28   *Illustration of MCTF using Haar filters*

[17] was extended for use in coding the video subbands. Comparison on *Mobile* in CIF format to a commonly available MPEG 2 coder showed PSNR improvements ranging from 1 to 3 dB over a broad range of bitrates for this scalable coder.

11.5.1 MORE ON MCTF

MCTF plays a very important role in interframe SWT coding. In general, we try to concentrate the energy onto the temporal low subband, with the energy in the temporal high subbands thus as small as possible. But *covered* and *uncovered* pixels must be considered in acheiving this goal. With reference to the Haar filter MCTF shown in Figure 11.28a, let frame *A* be the reference frame, then some pixels in frame *B* will be uncovered pixels, i.e., they will have no reference in frame *A*. In Figure 11.28b we show the resulting t–L and t–H frames, as computed in-place, with the horizontal lines indicating a block matching size of 4×4. We see many connected pixels, but some in frame *A* are unconnected. They are either covered in frame *B* or else are just not the best match. The latter situation can happen due to expansion and contraction of motion fields as objects move toward or away from the camera.

We can see pixels in Figure 11.28 classified as *connected* and *unconnected*. If there is a one-to-one connection between two pixels, they are classified as connected pixels. If several pixels in frame B connect to the same pixel in frame A, only one of them can be classified as the connected pixel; the others are declared unconnected. These unconnected pixels in frame A are indicated by no reference between frames A and B. After the classification, we can perform the actual MC filtering on the connected pixels. For the unconnected pixels in frame A, their scaled original values are inserted into the temporal low subband without creating any problem. For the unconnected pixels in frame B, a scaled DFD can be substituted into the temporal high subband [7], if there is a pixel in frame A or in some earlier or later frame pair,[4] that has a good (but maybe not best) match. Otherwise some spatial or intraframe prediction/interpolation can be used, analogous to an I block in MPEG-coded video.

11.5.2 Detection of Covered Pixels

We conclude that, in order to make the covered pixels correspond to the real occurrence of the occlusion effect, we should employ a detection phase before the actual MC filtering. Following is a simple pixel-based algorithm to find the true covered pixels illustrated for the Haar MCTF case.

Covered Pixel Detection Algorithm with Reference to Figure 11.29

Step 1. Do backward motion estimation from frame B to frame A, in a frame pair.

Step 2. Get the state of connection of each pixel in frame A. There are three states:

Unreferred: a pixel that is not used as reference.

Uniconnected: a pixel that is used as reference by only one pixel in frame B.

Multiconnected: a pixel that is used as reference by more than one pixel in the current frame.

A multiconnected pixel in frame A has several corresponding pixels in frame B, so we compute the absolute DFD value with each of them, and just keep the minimum as uniconnected.

Step 3. Get the state of connection of each pixel in frame B. There are just two states:

Uniconnected: Here we have three cases:

Case 1: a pixel whose reference in frame A is uniconnected.

Case 2: a pixel whose reference in frame A is multiconnected, except if its absolute DFD value with the reference pixel is the only minimum, we declare it uniconnected.

---

4. Remember this is all feedforward, so any future reference only incurs a small additional delay at the transmitter and receiver. Of course, this could be a problem in some applications, e.g., visual conferencing.

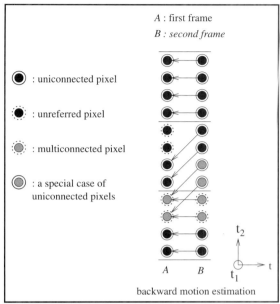

FIGURE 11.29 / *Diagram to illustrate covered pixel detection algorithm*

Case 3:   if there are several pixels in frame B pointing to the same reference pixel, and having the same minimum absolute DFD value, we settle this tie using the scan order.

Multiconnected: remaining pixels in frame B.

Step 4.   If more than half of the pixels in a 4×4 block of frame B are multiconnected, we try forward motion estimation. If motion estimation in this direction has smaller MCP error, we call this block a covered block, and pixels in this block are said to be covered pixels. We then use forward MCP, from the next frame, to code this block.

In Figure 11.30, we see the detections of this algorithm of the covered and uncovered regions, mainly on the side of the tree, but also at some image boundaries. With reference to Figure 11.31, we see quite strange motion vectors obtained around the sides of the tree as it "moves" due to the camera panning motion in *Flower Garden*. The motion field for the bidirectional MCTF is quite different, as seen in Figure 11.32, where the motion vector arrows generally look much more like the real panning motion in this scene. These figures were obtained from the MCTF operating two temporal levels down, i.e., with a temporal frame separation of four times that at the full frame rate.

FIGURE 11.30   *Example of unconnected blocks detected in Flower Garden*

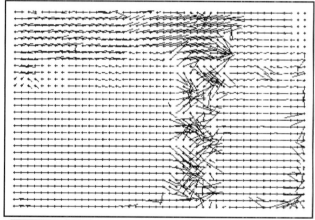

FIGURE 11.31   *Example of motion field of unidirectional MCTF*

11.5.3 BIDIRECTIONAL MCTF

Using the covered pixel detection algorithm, the detected covered and uncovered pixels are more consistent with the true occurrence of the occlusion effect. We can now describe a bidirectional MCTF process. We first get backward motion vectors for all 4×4 blocks via HVSBM. The motion vectors typically have quarter- or eighth-pixel accuracy and we use the MAD block matching criterion. Then we find covered blocks in frame B using the detection algorithm just introduced. The forward motion estimation is a by-product of the process of finding the covered

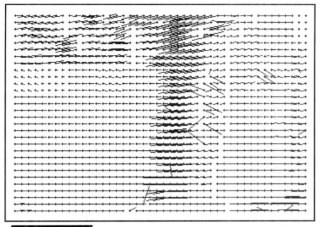

FIGURE 11.32 Example of motion field of bidirectional MCTF

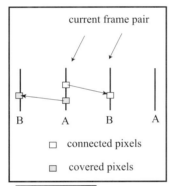

FIGURE 11.33 Bidirectional interpolation in forward direction for Haar MCTF

blocks. Then we do optimal tree pruning [7]. Thus, we realize a bidirectional variable blocksize motion estimation.

One can also perform MCTF in the *forward direction*, with reference to Figure 11.33. There are two kinds of blocks in frame *A*: connected blocks using forward motion estimation and covered blocks that use backward motion-compensated prediction. We can thus realize bidirectional MCTF. The shaded block in frame *A* is covered. For those covered pixels, we perform backward MCP. For the other regions, the MCTF is similar to that of Choi [7]. At GOP boundaries, we have to decide whether to use open or closed GOP, depending on delay and complexity issues.

We show in Figure 11.34 a t–L4 output from a unidirectional MCTF at temporal level four, i.e., one-sixteenth of the original frame rate. There are a lot of artifacts evident, especially in the tree. Figure 11.35 shows the corresponding t–L4

FIGURE 11.34 *Frame output from unidirectional MCTF at four temporal levels down*

FIGURE 11.35 *Four-temporal-level-down output of bidirectional MCTF (see the color insert)*

output from the bidirectional MCTF; we can detect a small amount of blurring, but no large artifact areas as were produced by the unidirectional MCTF.

Bidirectional MCTF along with unconnected pixel detection was implemented in a motion compensated EZBC video coder (MC-EZBC) in [5], where experimental coding results using the bidirectional Haar MCTF were obtained for the CIF test clips *Mobile, Flower Garden,* and *Coastguard.* Compared to the nonscalable ITU coder H.26L, a forerunner to H.264, MC-EZBC showed an approximate parity over the wide bit range 500–2500 Kbps. This is remarkable since MC-EZBC only codes the data once, and at the highest bitrate, with all the lower bitrate results being extracted out of the full bitstream. A fourth example showed results on *Foreman*, with about a 1 dB gap in favor of H.26L. Some MC-EZBC results in coding of 2K digital cinema data are contained in the MCTF video coding review article by Ohm *et al.* [34]. Video comparisons of MC-EZBC to both AVC and MPEG 2 results are contained in the enclosed CD-ROM.

11.6 OBJECT-BASED VIDEO CODING

Visual scenes are made up of objects and a background. If we apply image and video analysis to detect or pick out these objects and estimate their motion, this can be the basis of a potentially very efficient type of video compression [27]. This was a goal early on in the MPEG 4 research and development effort, but since reliable object detection is a difficult and largely unsolved problem, the object coding capability in MPEG 4 is largely reserved for artificially composed scenes where the objects are given and their locations are already known. In object-based coding, there is also the new problem of coding the outline or shape of the object, known as *shape coding*. Further, for a high-quality representation, it is necessary to allow for a slight overlap of the natural objects, or soft edge. Finally, there is the bit allocation problem, first to the object shape, and then to their inside, called *texture*, and their motion fields.

In Chapter 10 we looked at joint estimates of motion and segmentation. From this operation, we get both objects and motion fields for these objects. Each data can then be subject to either hybrid MCP or MCTF-type video coding. Here we show an application of Bayesian motion/segmentation to an example object-based research coder [14]. The various blocks of the object-based coder are shown in Figure 11.36, where we see the video input to the joint motion estimation and segmentation block of Chapter 10; the output here is a dense motion field **d** for each object with segmentation label *s*. This output is the input to the *motion/contour coder*, whose role is to code the motion field of each object and also to code its contour. The input video also goes into an *object classifier* and a *texture coder*. The object classifier decides whether the object can be predicted from the previous frame, a *P* object, or it is a new object occuring in this frame, an *I* object. Of course, in the first frame, all objects are *I* objects. The texture coder module computes the MCP of each object and codes the prediction residual.

The motion coder approximates the dense motion field of the Bayes estimate over each object with an *affine motion model* in a least-squares sense. An affine model has six parameters and represents rotation and translation, projected onto the image plane, as

$$d_1(t_1, t_2) = a_{11}t_1 + a_{12}t_2 + a_{13},$$
$$d_2(t_1, t_2) = a_{21}t_1 + a_{22}t_2 + a_{23},$$

where the position vector $\mathbf{t} = (t_1, t_2)^T$ on the object and $\mathbf{d}(t)$ is the corresponding displacement vector. We can see the translational motion discussed up to now as

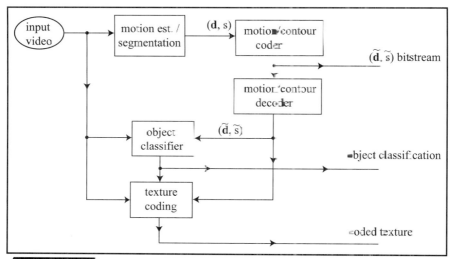

FIGURE 11.36   *System diagram of the hybrid MCP object video coder given in Han [14]*

the special case where only a_{13} and a_{23} are nonzero. The motion warping effect of an affine motion model has been found to well approximate the apparent motion of the pixels of rigid objects. The motion output of this coder is denoted $\tilde{d}$ in Figure 11.36. The other output of the motion/contour coder is the shape information $\tilde{s}$. Since the Bayesian model enforces a continuity along the object track as well as a spatial smoothness, this information can be well coded in a predictive sense. The details are contained in Han [15].

The texture coding proceeds for each object separately. Note that these are not really physical objects, but are the objects selected by the joint Bayesian MAP motion/segmentation estimate. Still, they correspond to at least large parts of physical objects. Note also that it is the motion field that is segmented, not the imaged objects themselves. As a result, when an object stops moving, it gets merged into the background segment. There are now a number of nice solutions to the coding of these irregular-shaped objects. There is the shape-adaptive DCT [37] that is used in MPEG 4. There are also various SWT extensions to nonrectangular objects [1, 26].

Here, we present an example from Han's thesis [14], using the SWT method of Bernard [1]. The *Carphone* QCIF test clip was coded at the frame rate of 7.5 fps and CBR bitrate of 24 Kbps, and the results were compared against an H.263 implementation from Telenor Research.

Figure 11.37 shows a decoded frame from the test clip coded at 24 Kbps via the object-based SWT coder. Figure 11.38 shows the same frame from the decoded output of the H.263 coder. The average PSNRs are 30.0 and 30.1 dB, respectively, so there is a very slight advantage of 0.1 dB to the H.263 coder. However, we can see that the image out of the object-based coder is sharper. (For

FIGURE 11.37   *QCIF Carphone coded via object-based SWT at 24 Kbps*

FIGURE 11.38   *QCIF Carphone coded at 24 Kbps by H.263*

typical frame segmentation, please refer to Figure 10.35b in Chapter 10.) Since the object-based coder was coded at a constant number of bits per frame, while the H.263 coder had use of a buffer, the latter was able to put more bits into frames with a lot of motion, and fewer bits into more quiet frames, so there was some measure of further improvement possible for the object-based SWT coder. See the enclosed CD-ROM for the coded video.

A comprehensive review article on object-based video coding is contained in Ebrahimi and Kunt [10].

11.7 COMMENTS ON THE SENSITIVITY OF COMPRESSED VIDEO

Just as with image coding, the efficient variable-length codewords used in common video coding algorithms render the data sensitive to errors in transmission. So, any usable codec must include general and repeated header information giving positions and length of coded blocks that are separated by synch words such as EOB. Of course, additionally the decoder has to know what to do upon encountering an error, i.e., an invalid result, such as EOB in the wrong place or an illegal codeword resulting from a noncomplete VLC code tree. It must do something reasonable to recover from such a detected error. Thus at least some redundancy must be contained in the bitstream to enable a video codec to be used in a practical environment, where even one bit may come through in error, say for example, in a storage application, such as a DVD.

Additionally, when compressed video is sent over a channel or network, it must be made robust with respect to much more frequent errors and data loss. Again, this is due to VLCs and spatiotemporal predictions used in the compression algorithms. As we have seen, the various coding standards place data into slices or groups of blocks that terminate with a synch word, which prevents the error propagation across this boundary. For packet-based wired networks such as the Internet, the main source of error is lost packets, as errors are typically corrected in lower network layers before the packet is passed up to the application layer and given to the decoder. Because of VLC, lost packets may make further packets useless until the end of the current slice or group of blocks. Forward error correction codes can be used to protect the compressed data. Slice lengths are designed based on efficiency, burst length, and preferred network packet or cell size. Further, any remaining redundancy, after compression, can be used at the receiver for a postdecoding cleanup phase, called *error concealment* in the literature. Combinations of these two methods can be effective too. Also useful is a request for retransmission or ACK strategy, wherein a lost packet is detected at the coder by not receiving an acknowledgment from the receiver, and based on its importance, may be resent. This strategy generally will require a larger output buffer and may not be suitable for real-time communication.

Scalable coding can be very effective to combat the uncertainty of a variable transmission medium such as the Internet. The ability to respond to variations in the *available bandwidth* (i.e., the bitrate that can be sustained without too much packet loss) gives scalable coders a big advantage in such situations. The next chapter concentrates on video for networks.

11.8 CONCLUSIONS

As of this writing, block transform methods dominate current video coding standards. For the past 10 years, though, research has concentrated on SWT methods. This is largely due to the high degree of scalability that is inherent in this approach. With the expected convergence of television and the Internet, the need for a relatively resolution-free approach is now bringing SWT compression to the forefront in emerging standards for digital video in the future. For super HD (SHD), to match film and graphics arts resolution, an 8×8 DCT block is quite small! Conversely, a 16×16 DCT means a lot more computation. Subband/wavelet schemes computationally scale nicely to SHD, close to digital cinema, where already the DCI has chosen M-JPEG 2000 as their recommended intraframe standard.

FIGURE 11.39   *4:2:0 color space structure of MPEG 2*

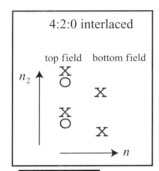

FIGURE 11.40   *4:2:0 chrominance samples (O) are part of top field only*

An overview of the state-of-the-art in video compression is contained in Section 6 "Video Compression" in A. Bovik's handbook [49]. Currently, the MPEG/VCEG joint video team (JVT) is developing a scalable video code (SVC) based on the H.264/AVC tools. It is layered 2D + t coder that makes use of the MCTF technique.

11.9 PROBLEMS

1   How many gigabytes are necessary to store 1 hour of SD 720×486 progressive video at 8 bits/pixel and 4:2:2 color space? How much for 4:4:4? How much for 4:2:0?

2   A certain analog video monitor has a 30 MHz video bandwidth. Can it resolve the pixels in a 1080p video stream? Remember, 1080p is the ATSC format with 1920×1080 pixels and 60 frames per second.

3   The NTSC DV color space 4:1:1 is specified in Figure 11.5 and the NTSC MPEG 2 color space 4:2:0 is given in Figure 11.39.

   (a)   Assuming that each format is progressive, what is the specification for the chroma sample rate change required? Specify the upsampler, the ideal filter specifications, and the downsampler.

   (b)   Do the same for the actual interlace format that is commonly used, but note that the chroma samples are part of the top field only, as shown in Figure 11.40.

4 Use a Lagrangian formulation to optimize the intraframe coder with distortion $D(n)$ as in (11.1-4), and average bitrate $R(n)$ in (11.1-5). Consider frames 1 to N and find an expression for the resulting average mean-square distortion as a function of the average bitrate.

5 Write the decoder diagram for the backward motion-compensated hybrid encoder shown in Figure 11.11.

6 Discuss the inherent or built-in delay of an MPEG hybrid coder based on the number of B frames that it uses. What is the delay when the successive number of B frames is M?

7 In an MPEG 2 coder, let there be two B frames between each non B frame, and let the GOP size be 15. At the source we denote the frames in one GOP as follows:

$$I_1 \ B_2 \ B_3 \ P_4 \ B_5 \ B_6 \ P_7 \ B_8 \ B_9 \ P_{10} \ B_{11} \ B_{12} \ P_{13} \ B_{14} \ B_{15}$$
$$1 \quad 2 \quad 3 \quad 4 \quad 5 \quad 6 \quad 7 \quad 8 \quad 9 \quad 10 \quad 11 \quad 12 \quad 13 \quad 14 \quad 15$$

(a) What is the order in which the frames must be coded?

(b) What is the order in which the frames must be decoded?

(c) What is the required display order?

(d) Based on answers a–c, what is the inherent delay in this form of MPEG 2? Can you generalize your result from two successive B frames to M successive B frames?

8 In MPEG 1 and 2, there are *open* and *closed* GOPs. In the closed GOP case, the bidirectional predictor cannot use any frames outside the current GOP in calculating B frames. In the open GOP case, they can be used. Consider a GOP of 15 frames, using pairs of B frames between each P frame. Which frames would be affected by the difference? How would the open GOP concept affect decodability?

9 Find the inverse transform for the 4×4 DCT-like matrix (11.3-1) used in H.264/AVC. Show that 8-bit data may be transformed and inverse transformed without any error-using 16-bit arithmetic.

10 Show the synthesis filter corresponding to the lifting analysis equations, (11.4-1) and (11.4-2). Is the reconstruction exact even in the case of sub-pixel accuracy? Why? Do the same for the LGT 5/3 SWT as in (11.4-3) and (11.4-4).

11 In an MCTF-based video coder, unconnected pixels can arise due to expansion and contraction of the motion field, but also, and more importantly, due to occlusion. State the difference between these two cases and discuss how they should be handled in the coding stage that comes after the MCTF.

REFERENCES

[1] H. J. Bernard, *Image and Video Coding Using a Wavelet Decomposition*, Ph.D. thesis, Delft Univ. of Technology, The Netherlands, 1994.

[2] F. Bosveld, *Hierarchical Video Compression Using SBC*, Ph.D. thesis, Delft Univ. of Technology, The Netherlands, 1996.

[3] P. J. Burt and E. H. Adelson, "The Laplacian Pyramid as a Compact Image Code," *IEEE Trans. Commun.*, 31, 532–540, April 1983.

[4] J. C. Candy *et al.*, "Transmitting Television as Clusters of Frame-to-Frame Differences," *Bell Syst. Tech. J.*, 50, 1889–1917, July–Aug. 1971.

[5] P. Chen and J. W. Woods, "Bidirectional MC-EZBC with Lifting Implementation," *IEEE Trans. Video Technol.*, 14, 1183–1194, Oct. 2004.

[6] S.-J. Choi, *Three-Dimensional Subband/Wavelet Coding of Video*, Ph.D. thesis, ECSE Dept., Rensselaer Polytechnic, Troy, NY, Aug. 1996.

[7] S. Choi and J. W. Woods, "Motion-compensated 3-D Subband Coding of Video," *IEEE Trans. Image Process.*, 8, 155–167, Feb. 1999.

[8] K. P. Davies, "HDTV Evolves for the Digital Era," *IEEE Commun. Magazine*, 34, 110–112, June 1996 (HDTV mini issue).

[9] P. H. N. de With and A. M. A. Rijckaert, "Design Considerations of the Video Compression System of the New DV Camcorder Standard," *IEEE Trans. Consumer Electr.*, 43, 1160–1179, Nov. 1997.

[10] T. Ebrahimi and M. Kunt, "Object-Based Video Coding," in *Handbook of Image and Video Processing*, 1st edn., A. Bovik, ed., Academic Press, San Diego, CA, 2000, Chapter 6.3.

[11] H. Gharavi, "Motion Estimation within Subbands," in *Subband Image Coding*. J. W. Woods, ed., Kluwer Academic Publ., Boston, MA, 1991. Chapter 6.

[12] B. Girod, "The Efficiency of Motion-Compensating Prediction for Hybrid Coding of Video Sequences," *IEEE J. Select. Areas Comm.*, SAC-5, 1140–1154, Aug. 1987.

[13] W. Glenn *et al.*, "Simple Scalable Video Compression Using 3-D Subband Coding," *SMPTE J.*, March 1996.

[14] S.-C. Han, *Object-Based Representation and Compression of Image Sequences*, Ph.D. thesis, ECSE Dept., Rensselaer Polytechnic Institute, Troy, NY, 1997.

[15] S.-C. Han and J. W. Woods, "Adaptive Coding of Moving Objects for Very Low Bitrates," *IEEE J. Select. Areas Commun.*, 16, 56–70, Jan. 1998.

[16] B. G. Haskell, A. Puri, and A. N. Netravali, *Digital Video: An Introduction to MPEG-2*, Chapman and Hall, New York, NY, 1997.

[17] S.-T. Hsiang and J. W. Woods, "Embedded Image Coding using Zeroblocks of Subband/Wavelet Coefficients and Context Modeling," *MPEG-4 Workshop and Exhibition at ISCAS 2000*. IEEE, Geneva, Switzerland, May 2000.

[18] S.-T. Hsiang and J. W. Woods, "Embedded Video Coding Using Invertible Motion Compensated 3-D Subband/Wavelet Filter Bank," *Signal Process. Image Commun.*, 16, 705–724, May 2001.

[19] International Standards Organization. *Generic Coding of Moving Pictures and Associated Audio Information*, Recommendation H.262 (MPEG 2), ISO/IEC 13818-2, Mar. 1994.

[20] A. W. Johnson, T. Sikora, T. K. Tan, and K. N. Ngan, "Filters for Drift Reduction in Frequency Scalable Video Coding Schemes," *Electron. Lett.*, 30, 471–472, March 1994.

[21] C. B. Jones, "An Efficient Coding System for Long Source Sequences," *IEEE Trans. Inform. Theory*, IT-27, 280–291, May 1981.

[22] G. D. Karlson, *Subband Coding for Packet Video*, MS thesis, Center for Telecommunications Research, Columbia University, New York, NY, 1989.

[23] G. D. Karlson and M. Vetterli, "Packet Video and its Integration into the Network Architecture," *J. Select. Areas Commun.*, 7, 739–751, June 1989.

[24] T. Kronander, "Motion Compensated 3-Dimensional Waveform Image Coding," *Proc. ICASSP*. IEEE, Glasgow, Scotland, 1989.

[25] D. LeGall and A. Tabatabai, "Sub-band Coding of Digital Images Using Symmetric Short Kernel Filters and Arithmetic Coding Techniques," *Proc. ICASSP 1988*. IEEE, New York, April 1988. pp. 761–764.

[26] J. Li and S. Lei, "Arbitrary Shape Wavelet Transform with Phase Alignment," *Proc. ICIP*, 3, 683–687, 1998.

[27] H. Musmann, M. Hotter, and J. Ostermann, "Object-Oriented Analysis-Synthesis Coding of Moving Images," *Signal Process.: Image Commun.*, 1, 117–138, Oct. 1989.

[28] T. Naveen, F. Bosveld, R. Lagendijk, and J. W. Woods, "Rate Constrained Multiresolution Transmission of Video," *IEEE Trans. Video Technol.*, 5, 193–206, June 1995.

[29] T. Naveen and J. W. Woods, "Motion Compensated Multiresolution Transmission of Video," *IEEE Trans. Video Technol.*, 4, 29–43 Feb. 1994.

[30] T. Naveen and J. W. Woods, "Subband and Wavelet Filters for High-Definition Video Compression," in *Handbook of Visual Communications*. H.-M. Hang and J. W. Woods, eds., Academic Press, San Diego, CA, 1995. Chapter 8.

[31] T. Naveen and J. W. Woods, "Subband Finite-State Scalar Quantization," *IEEE Trans. Image Process.*, 5, 150–155, Jan. 1996.

[32] A. Netravali and J. Limb, "Picture Coding: A Review," *Proc. IEEE*, March 1980.

[33] J.-R. Ohm, "Three-Dimensional Subband Coding with Motion Compensation," *IEEE Trans. Image Process.*, 3, 559–571, Sept. 1994.

[34] J. Ohm, M. van der Schaar, and J. W. Woods, "Interframe Wavelet Coding— Motion Picture Representation for Universal Scalability," *Signal Process.: Image Commun.*, 19, 877–908, Oct. 2004.

[35] F. Pereira and T. Ebrahimi, eds., *The MPEG-4 Book*, Prentice-Hall, Upper Saddle River, NJ, 2002.

[36] A. Secker and D. Taubman, "Highly Scalable Video Compression with Scalable Motion Coding," *IEEE Trans. Image Process.*, **13**, 1029–1041, Aug. 2004.

[37] T. Sikora, "Low-Complexity Shape-Adaptive DCT for Coding of Arbitrarily Shaped Image Segments," *Signal Process.: Image Comm.*, 7, 381–395, 1995.

[38] Special Issue on H.264/AVC, *IEEE Trans. Circ. Syst. Video Technol.*, **13**, July 2003.

[39] G. J. Sullivan and T. Wiegand, "Rate-Distortion Optimization for Video Compression," *IEEE Signal Process. Magazine*, **15**, 74–90, Nov. 1998.

[40] G. J. Sullivan and T. Wiegand, "Video Compression—From Concepts to the H.264/AVC Standard," *Proc. IEEE*, **93**, 18–31, Jan. 2005.

[41] D. Taubman and A. Zakhor, "Multirate 3-D Subband Coding of Video," *IEEE Trans. Image Process.*, **3**, 572–588, Sept. 1994.

[42] L. Vandendorpe, *Hierarchical Coding of Digital Moving Pictures*, Ph.D. thesis, Univ. of Louvain, Belgium, 1991.

[43] T. Wiegand, G. J. Sullivan, G. Bjontegaard, and A. Luthra, "Overview of the H.264/AVC Video Coding Standard," *IEEE Trans. Circ. Syst. Video Technol.*, **13**, 560–576, July 2003.

[44] J. W. Woods, S.-C. Han, S.-T. Hsiang, and T. Naveen, "Spatiotemporal Subband/Wavelet Video Compression," in *Handbook of Image and Video Processing*, 1st edn. A. Bovik, ed., Academic Press, San Diego, CA, 2000. Chapter 6.2, pp. 575–595.

[45] J. W. Woods and G. Lilienfield, "Resolution and Frame-rate Scalable Video Coding," *IEEE Trans. Video Technol.*, **11**, 1035–1044, Sept. 2001.

[46] J. W. Woods and T. Naveen, "Subband Coding of Video Sequences," *Proc. VCIP*. SPIE, Bellingham, WA, Nov. 1989. pp. 724–732.

[47] D. Wu, Y. T. Hou, and Y.-Q. Zhang, "Scalable Video Coding and Transport Over Broad-Band Wireless Networks," *Proc. IEEE*, **89**, 6–20, Jan. 2001.

[48] S. Liu and A. Bovik, "Digital Video Transcoding," Chapter 6.3, in *Handbook of Imaging and Video Processing*, 2nd edn., A. Bovik, ed., Elsevier/Academic Press, Burlington, MA. 2005.

[49] Chapter 6, in *Handbook of Imaging and Video Processing*, 2nd edn., A. Bovik, ed., Elsevier/Academic Press, Burlington, MA, 2005.

VIDEO TRANSMISSION OVER NETWORKS 12

Many video applications involve transmission over networks, e.g., visual conversation, video streaming, video on demand, and video downloading. Various kinds of networks are involved, e.g., integrated services digital network (ISDN), wireless local area networks (WLANs) such as IEEE 802.11, storage area networks (SANs), private internet protocol (IP) networks, public Internet, and wireless networks. If the network lacks congestion and has no bit errors, then the source coding bitstreams described in Chapter 11 could be sent directly over the network. Otherwise some provision has to be made to protect the rather fragile compressed data kept from the source coder. In typical cases, even a single bit error can cause the decoder to terminate upon encountering illegal parameter values.

In this chapter we will encounter some of the new aspects that arise with video transmission over networks. We review the error-resilient features of some standard video coders as well as the transport error protection features of networks, and show how they can work together to protect the video data. We present some robust transmission methods for scalable SWT-based coders such as MC-EZBC as well as for standard H.264/AVC. The chapter ends with a section on joint source–network coding, wherein the source–channel coding is continued into the network on overlay nodes that also support digital item adaptation (transcoding). We start with an overview of IP networks. General introductions to networks can be found in the textbooks by Tenenbaum [47] and Kurose and Ross [28].

12.1 VIDEO ON IP NETWORKS

This section introduces the problem of transmitting video on *internet protocol* (IP) networks such as the public Internet. We first provide an overview of such networks. Then we briefly introduce the error-resilience features found on common video coders. Finally, we consider the so-called *transport-level* coding done at the network level to protect the resulting error-resilient coded video data.

A basic system diagram for networked transmission of video on a single path is shown in Figure 12.1. In this case the *source coder* will usually have its error-resilience features turned on and the *network coder* will exploit these features at the transport level. The network coder will generally also exploit feedback on congestion in the network in its assignment of data to network packets, and choice of forward error correction (FEC) or ACK, to be discussed later in this chapter. This diagram is for *unicast*, so there is just one receiver for the one video source shown here. *Multicast*, or general *broadcast*,[1] introduces additional issues, one of which is the need for scalability due to the expected heterogeneity of multicast receivers and link qualities, i.e., usable bandwidth, packet loss, delay and delay variation (jitter), and bit-error rate. While we do not address multicast here, we do

---

1. The term "broadcast" may mean different things in the video coding community and the networks community. Here, we simply mean a very large-scale multicast to perhaps millions of subscribers, not just a wholesale flooding of the network.

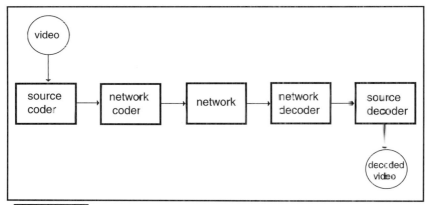

FIGURE 12.1 *System diagram of networked video transmission*

look at the scalability-related problem of digital item adaptation or transcoding later on in this chapter.

12.1.1 OVERVIEW OF IP NETWORKS

Our overview of IP networks starts with the basic concept of a *best-efforts* network. We briefly introduce the various network levels, followed by an introduction to the transport control protocol (TCP) and the *real-time protocol* (RTP) for video. Since the majority of network traffic is TCP, it is necessary that video transmissions over RTP be what is called TCP-friendly, meaning that, at least on the average, each RTP flow uses a fraction of network resources, similar to that of a TCP flow. We then discuss how these concepts can be used in packet loss prevention.

BEST-EFFORTS NETWORK

Most IP networks, including the public Internet, are best-efforts networks, meaning that there are no guarantees of specific performance, only that the network protocol is fair and makes a best effort to get the packets to their destination, perhaps in a somewhat reordered sequence with various delays. At intermediate nodes in the network, packets are stored in buffers of finite size that can overflow. When such overflow occurs, then the packet is lost, and this packet loss is the primary error that occurs in modern wired IP networks. So these networks cannot offer a guaranteed *quality of service* (QoS) such as is obtained on the private switched networks (e.g., T1, T2, or ATM). However, significant recent work of the Internet Engineering Task Force (IETF) [21] has concentrated on *differentiated services* (DiffServ), wherein various priorities can be given to classes of packet traffic.

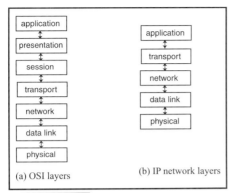

FIGURE 12.2   *Network reference models (stack)*

LAYERS AND PROTOCOL STACK

Figure 12.2a shows the Open System Interconnection (OSI) network layers, also called a protocol stack [47]. The lowest level is the *physical layer*, called layer 1, the level of bits. At level 2, the data link layer, the binary data is aggregated into *frames*, while the next higher layer 3, the *network layer*, deals only with packets. In the case of bit errors that cannot be corrected, a packet is rejected at the data link layer and so is not passed up to the network level. However, bit errors are significant only in wireless networks. For our purposes we can skip the presentation and session layers and go directly to the *application layer* (Figure 12.2b), where our video coders and decoders sit. On transmission, the application sends a message down the protocol stack, with each level requesting services from the one below. On reception, the information is passed back up the stack. Hopefully, the original message is then reconstructed.

The presentation and session layers are not mentioned in Kurose and Ross [28], who compress the stack to five layers, typical in Internet discussions. The *internet protocol* (IP) works in the network layer. The TCP works in the transport layer. The IP layer offers no reliability features. It just packetizes the data, adds a destination header, and sends it on its way. But the internet protocol at the network layer is responsible for all of the routing or path selection to get from one node to another all through the network to the final addressee node.

TCP AND UDP AT TRANSPORT LEVEL

The extremely reliable TCP controls the main load of the Internet traffic. It attains the reliability needed for general computer data by retransmitting any lost IP packets, thus incurring a possibly significant increased delay upon packet loss. Most media traffic does not require such high reliability, but conversely cannot tolerate excessive variation in time delay, so-called *time jitter*. The ubiquitous TCP includes congestion avoidance features that heavily modulate its sending rate to avoid packet loss, signaled to the sender by small backward-flowing control packets. TCP flows start up slowly, and upon encountering significant packet loss, the sending rate is drastically reduced (by half) through a process called *additive-increase multiplicative-decrease* (AIMD) [28]. This increase and decrease is conducted at network epochs as determined by the so-called *round-trip time* (RTT), the average time it takes a packet to make a round trip from source to receiver.

RTP AT THE APPLICATION LAYER

For real-time transmission, RTP does not employ TCP at all. The RTP operates at the application layer and uses the *user datagram protocol* (UDP), which, unlike TCP, simply employs single IP transmissions with no checking for ACKs and no retransmissions. Each UDP packet, or datagram, is sent separately, with no throttling down of rate as in TCP. Thus video sent via RTP will soon dominate the flows on a network unless some kind of rate control is applied. More information on these and other basic network concepts can be found in Kurose and Ross [28].

TCP FRIENDLY

The flow of video packets via RTP/UDP must be controlled so as not to dominate the bulk of network traffic, which flows via TCP. Various TCP-friendly rate equations have been devised [31,34] that have been shown to use on average a nearly equal amount of network bandwidth as TCP. One such equation [34] is

$$R_{TCP} = \frac{MTU}{RTT\sqrt{\frac{2p}{3}} + RTO\sqrt{\frac{27p}{8}}p(1 + 32p^2)}. \tag{12.1-1}$$

where RTT is the round-trip time, RTO is the receiver timeout time (i.e., the time that a sender can wait for the ACK before it declares the packet lost), MTU is the maximum transmission unit (packet) size, and p is the packet loss probability, also called *packet loss rate*. If the RTP video sender follows (12.1-1), then its sending rate will be TCP-friendly and should avoid being the immediate cause of network congestion.

As mentioned previously, the main loss mechanism in wired IP networks is packet overflow at the *first-in first-out* (FIFO) internal network buffers. If the source controls its sending rate according to (12.1-1), then its average loss should be approximately that of a TCP flow, which is a low 3–6% [35]. So, control of the video sending rate offers two benefits, one to the sender and one to the network. Bit errors are not considered a problem in wired networks. Also, any possible small number of bit errors is detected and not passed up the protocol stack, so from the application's viewpoint, the packet is lost. In the wireless case, bit errors are important equally with packet loss.

12.1.2 ERROR-RESILIENT CODING

Error-resilient features are often added to a video coder to make it more robust to channel errors. Here we discuss data partitioning and slices that are introduced for purposes of resynchronization. We also briefly consider reversible VLCs, which can be decoded starting from either the beginning or the end of the bitstream. We introduce the multiple description feature, whereby decoding at partial quality is still possible even if some of the descriptions (packets) are lost.

DATA PARTITIONING AND SLICES

If all of the coded data is put into one partition, then a problem in transmission can cause the loss of the entire remaining data, due first to the use of VLCs and then to the heavy use of spatial and temporal prediction in video coders. In fact, if synchronization is lost somewhere, it might never be regained without the insertion of *synch words*[2] into the bitstream. In MPEG 2, *data partitions* (DPs) consisted of just two partitions on a block basis: one for the motion vector and low-frequency DCT coefficients, and a second one for high-frequency coefficients [15]. Similar to the DP is the concept of *slice*, which breaks a coded frame up into separate coded units, wherein reference cannot be made to other slices in the current frame. Slices start (or end) with a synch word, and contain position, length of slice in pixels, and other needed information in a header, permitting the slice to be decoded independent of prior data. Because slices cannot refer to other slices in the same frame, compression efficiency suffers somewhat.

REVERSIBLE VARIABLE-LENGTH CODING (VLC)

Variable-length codes such as Huffman are constructed on a tree and thus satisfy a *prefix condition* that makes them uniquely decodable, i.e., no codeword is the prefix of another codeword. Also, since the Huffman code is optimal, the code tree is fully populated, i.e., all leaf nodes are valid messages. If we lose a middle segment in a string of VLC codewords, in the absence of other errors, we can only decode (forward) correctly from the beginning of the string up to the lost segment. However, at the error we lose track and will probably not be able to correctly decode the rest of the string, after the lost segment. On the other hand, if the VLC codewords satisfied a *suffix condition*, then we could start at the end of the string and decode back to the lost segment, and thus recover much of the lost data. Now, as mentioned in Chapter 8, most Huffman codes cannot satisfy both a suffix and a prefix condition, but methods that find *reversible VLCs* (RVLCs), starting from a conventional VLC, have been discovered [45]. Unfortunately, the resulting RVLCs lose some efficiency versus Huffman codes.

---

2. A synch word is a long and unique codeword that is guaranteed not to occur naturally in the coded bitstream (see end-of-chapter problem 1).

Regarding structural VLCs, the Golomb–Rice codes are near-optimal codes for the geometric distribution, and the Exp–Golomb code are near optimal for distributions more peaky than exponential, which can occur in the prediction residuals of video coding. RVLCs with the same length distribution as Golomb–Rice and Exp–Golomb were discovered by Wen and Villasenor [53], and Exp–Golomb codes are used in the video coding standard H.264/AVC. Finally, a new method is presented in Farber *at al.* [13] and Girod [16] that achieves bidirectional decoding with conventional VLCs. In this method, forward and reverse code streams are combined with a delay for the maximum-length codeword, and the compression efficiency is almost the same as for Huffman coding when the string of source messages is long.

MULTIPLE DESCRIPTION CODING (MDC)

The main concept in *multiple description coding* (MDC) is to code the video data into two or more multiple streams that each carry a near-equal amount of information. If all of the streams are received, then decoding proceeds normally. If fewer descriptions are received, then the video can still be decoded, but at a reduced quality. Also, the quality should be almost independent of which descriptions are received [48]. If we now think of the multiple descriptions as packets sent out on a best-efforts network, we can see the value of an MDC coding method for networked video.

EXAMPLE 12.1-1 (*scalar MDC quantizer*)
Consider an example where we have to quantize a continuous random variable X whose range is $[1, 10]$, with a quantizer having 10 equally spaced output values: $1, 2, \ldots, 9, 10$. Call this the original quantizer. We can also accomplish this with two separate quantizers, called *even quantizer* and *odd quantizer*, each with step-size 2, that output even and odd representation values, respectively. Clearly, if we receive either even or odd quantizer output, we can reconstruct the value of X up to an error of ± 1, and if we receive both even and odd quantizer outputs, then this is the same as receiving the original quantizer output having 10 levels, so that the maximum output error is now ± 0.5. Figure 12.3 illustrates this: the even quantizer output levels are denoted by the dashed lines and the odd quantizer output levels are denoted by the solid lines. We can see a redundancy in the two quantizers, because the dashed and solid horizontal lines overlap horizontally.

One simple consequence from Example 12.1-1 is that the MDC method introduces a redundancy into the two (or more) code streams or descriptions. It is this redundancy that protects the data against loss of one or another bitstream. The concept of MDC has been extended far beyond simple scalar quantizers, e.g., MD transform coding [51] and MD-FEC coding [38], which we will apply to a layered code stream later in this chapter.

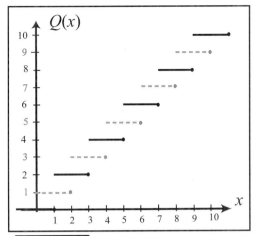

FIGURE 12.3 *MDC scalar quantizer of X uniform on* [1,10]

12.1.3 TRANSPORT-LEVEL ERROR CONTROL

The preceding error-resilient features can be applied in the video coding or application layer. Then the packets are passed down to the *transport level*, for transmission over the channel or network. We look at two powerful and somewhat complementary techniques, error control coding and acknowledgment/retransmission (ACK/NAK). Error control coding is sometimes called *forward error correction* (FEC) because only a forward channel is used. However, in a packet network there is usually a backward channel, so that acknowledgments can be fed back from receiver to transmitter, resulting in the familiar ACK/NAK signal. Using FEC we must know the present channel quality fairly well or risk wasting error control (parity) bits, on the one hand, or not having enough parity bits to correct the error, on the other hand. In the simpler ACK/NAK case, we do not compromise the forward channel bandwidth at all, and only transmit on the backward channel a very small ACK/NAK packet, but we do need the existence of a backward channel. In delay-sensitive applications such as visual conferencing, we generally cannot afford to wait for the ACK and the subsequent retransmission, due to stringent total delay requirements ($\leqslant$250 msec [44]). Some data "channels" (where there is no backward channel) are the CD, the DVD, and TV broadcast. There is a back channel in Internet unicast, multicast, and broadcast. However, in the latter two, multicast and broadcast, there is the need to consolidate the user feedback at overlay nodes to make the system scale to possibilities of large numbers of users.

FORWARD ERROR CONTROL CODING

The FEC technique is used in many communication systems. In the simplest case, it consists of a block coding wherein a number $n - k$ of parity bits are added to k binary information bits to create a binary channel codeword of length n. The Hamming codes are examples of binary linear codes, where the codewords exist in n-dimensional binary space. Each code is characterized by its minimum Hamming distance d_{min}, which is the minimum number of bit differences between any two codewords. It takes d_{min} number of bit errors to change one codeword into another, so the error detection capability of a code is $d_{min} - 1$, and the error correction capability of a code is $\lfloor d_{min}/2 \rfloor$, where $\lfloor \cdot \rfloor$ is the least integer function. This last is so because if fewer than $\lfloor d_{min}/2 \rfloor$ errors occur, the received string is still closer (in Hamming distance) to its error-free version than to any other codeword. Now, Reed–Solomon (RS) codes are also linear but operate on symbols in a so-called Galois field with 2^l elements. Codewords, parity words, and minimal distance are all computed using the arithmetic of this field. An example is $l = 4$, which corresponds to hexadecimal arithmetic with 16 symbols. The $(n, k) = (15, 9)$ RS code has hexadecimal symbols and can correct three symbol errors. It codes nine hexadecimal information symbols (36 bits) into 15 symbol codewords (60 bits) [2]. The RS codes are perfect codes, meaning that the minimum distance between codewords attains the maximum value $d_{min} = n - k + 1$ [14]. Thus an (n, k) RS code can detect up to $n - k$ symbol errors. The RS codes are very good for bursts of errors since a short symbol error burst translates into an l times longer binary error burst, when the symbols are written in terms of their l-bit binary code [36]. These RS codes are used in the CD and DVD standards to correct error bursts on decoding.

AUTOMATIC REPEAT REQUEST

A simple alternative to using error control coding is the *automatic repeat request* (ARQ) strategy of acknowledgment and retransmission. It is particularly attractive in the context of an IP network, and is used exclusively by TCP in the transport layer. There is no explicit expansion needed in available bandwidth, as would be the case with FEC, and no extra congestion, unless a packet is not acknowledged, i.e., no ACK is received. Now TCP has a certain *timeout* interval [28], at which time the sender acts by retransmitting the unacknowledged packet. Usually the timeout is set to the *round-trip time* (RTT) plus four standard deviations. (Note that this can lead to duplicate packets being received under some circumstances.) At the network layer, the IP protocol has a header checksum that can cause some packets to be discarded there. While ARQ techniques typically result in too much delay for visual conferencing, they are quite suitable for video streaming, where playout buffers are typically 5 seconds or more in length.

12.1.4 WIRELESS NETWORKS

The situation in video coding for wireless networks is less settled than that for the usually more benign wired case. In a well-functioning wired IP network, usually bit errors in the cells (packets) can almost be ignored, being of quite low probability, especially at the application layer. In the wireless case the *signal-to-noise ratio* (SNR) is highly variable, with bit errors occurring in bursts and at a much higher rate. Methods for transporting video over such networks must consider both packet loss and bursty bit errors.

The common protocol TCP is not especially suitable for the wireless network because it interprets packet loss as indicating congestion, while in a wireless network cell loss may be largely due to bit errors, in turn causing the packet to be deleted at the link layer [28]. Variations of TCP have been developed for wireless transmission [7], one of which is just to provide for link layer retransmissions. So-called *cross-layer* techniques have recently become popular [39] for exploiting those packets with bit errors. A simple cross-layer example is that packets with detected errors could be sent up to the application level instead of being discarded. There RVLCs could be used to decode the packet from each end till the location where the error occurred, thus recovering much of the loss.

12.1.5 JOINT SOURCE–CHANNEL CODING

A most powerful possibility is to design the video coder and channel coder together to optimize overall performance in the operational rate-distortion sense. If we add FEC to the source coder bitrate, then we have a higher channel bitrate. For example, a channel rate $R_C = 1/2$ Reed–Solomon code would double the channel bitrate. This channel bitrate translates into a probability of bit error. Depending on the statistics of the channel, i.e., independent bit errors or bursts, the bit error can be converted to a word error and frame error, where a frame would correspond to a packet, which may hold a slice or a number of slices. Typically these packets with bit errors will not be passed up to the application layer, and will be considered as lost packets. Finally, considering error concealment provisions in the decoder, this packet loss rate can translate into an average PSNR for the decoded video frames. Optimization of the overall system is called *joint source–channel coding* (JSCC), and essentially consists of trading off the source coder bitrate versus FEC parity bits for the purpose of increasing the average decoder PSNR. Modestino and Daut [32] published an early paper introducing JSCC for image transmission.

Bystrom and Modestino [8] introduced the idea of a universal distortion-rate characteristic for the JSCC problem. Writing the total distortion from source and channel coding as D_{S+C}, they express this total distortion as a function of inverse bit-error probability on the channel p_b for various source coding rates R_S. A sketch of such a function is shown in Figure 12.4. Such a function can be obtained either analytically or via simulation for the source coder in question. However, pre-computing such a curve can be computationally demanding. The characteristic is independent of the channel coding, modulation, or channel model. However,

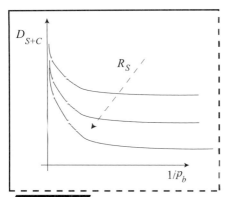

FIGURE 12.4   *Total distortion plotted versus reciprocal of channel bit error probability, with source coding rate as a parameter*

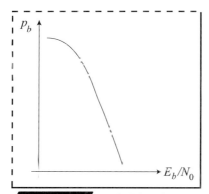

FIGURE 12.5   *Probability of bit error versus channel signal-to-noise ratio*

the method does assume that successive bit errors are independent, but Bystrom and Modestino [8] stated that this situation can be approximately attained by standard interleaving methods.

This universal distortion-rate characteristic of the source can be combined with a standard bit-error probability curve of the channel, as shown in Figure 12.5, to obtain a plot of D_{S+C} versus either source rate R_S (source bits/pixel) or what they [8] called $R_{S+C} \triangleq R_S/R_C$, where R_C is the channel coding rate, expressed in terms of source bits divided by total bits, i.e., source bits plus parity bits. The units of R_{S+C} are then bits/c.u., where c.u. = channel use. The resulting curves look something like that of Figure 12.6.

To obtain such a curve for a certain E_b/N_0 or SNR on the channel, we first find p_b from Figure 12.5. Then we plot the corresponding vertical line on Figure 12.4, from which we can generate the plot in Figure 12.6 for a particular R_C of the used FEC, assumed modulation, and channel model. Some applications

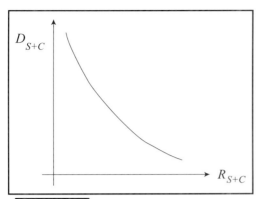

FIGURE 12.6   *Total distortion versus source bits per channel use*

to an ECSQ subband video coder and an MPEG 2 coder over an additive white Gaussian noise (AWGN) channel are contained in [8].

12.1.6 ERROR CONCEALMENT

Conventional source decoding does not make use of any *a priori* information about the video source, such as source model or power spectral density, in effect making it an unbiased or maximum-likelihood source decoder. However, better PSNR and visual performance can be expected by using model-based decoding, e.g., a postprocessing filter. For block-based intraframe coders, early postprocessing concealed blocking artifacts at low bitrates by smoothing nearest-neighbor columns and rows along DCT block boundaries.

Turning to channel errors and considering the typical losses that can occur in network transmission, various so-called error concealment measures have been introduced to counter the common situations where blocks and reference blocks in prior frames or motion vectors are missing, due to packet loss. If the motion vector for a block is missing, it is common to make use of an average of neighboring motion vectors to perform the needed prediction. If the coded residue or displaced frame difference (DFD) is missing for a predicted block (or region), often just the prediction is used without any update. If an I block (or region) is lost, then some kind of spatial prediction from the neighborhood is often used to fill in the gap. Such schemes can work quite well for missing data, especially in rather smooth regions of continuous motion. Directional interpolation based on local geometric structure has also been found useful for filling in missing blocks [55].

A particularly effective method to deal with lost or erroneous motion vectors was presented in Lam *et al.* [29]. A boundary matching algorithm is proposed to estimate a missing motion vector from a set of candidates. The candidates include the motion vector for the same block in the previous frame, the available neighboring motion vectors, an average of available neighboring motion vectors, and the median of the available neighboring motion vectors. The estimate is then chosen that minimizes a side-match error, i.e., the sum of squares of the first-order difference across the available top, left and bottom block boundaries. A summary of error concealment features in the H.26L test model is contained in Wang *et al.* [50].

Some error concealment features of H.264/AVC will be presented in Section 12.3. In connection with the standards, though, it should be noted that error concealment is *nonnormative*, meaning that error concealment is not a mandated part of the standard. Implementors are thus free to use any error concealment methods that they find appropriate for the standard encoded bitstreams. When we speak of error concealment features of H26L or H.264/AVC, we mean error concealment features of the test or verification models for these coders. An overview of early error concealment methods is contained in the Katsaggelos and Galatsanos book [25].

12.2 ROBUST SWT VIDEO CODING (BAJIĆ)

In this section we introduce dispersive packetization, which can help error concealment at the receiver to deal with packet losses. We also introduce a combination of multiple description coding with FEC that is especially suitable for matching scalable and embedded coders to a best-efforts network such as the current Internet.

12.2.1 DISPERSIVE PACKETIZATION

Error concealment is a popular strategy for reducing the effects of packet losses in image and video transmission. When some data from the compressed bitstream (such as a set of macroblocks, subband samples, and/or motion vectors) is lost, the decoder attempts to estimate the missing pieces of data from the available ones. The encoder can help improve estimation performance at the decoder side by packetizing the data in a manner that facilitates error concealment. One such approach is *dispersive packetization*.

The earliest work in this area seems to be the even–odd splitting of coded speech samples by Jayant [22]. Neighboring speech samples are more correlated than are samples that are far from each other. When two channels are available for speech transmission, sending even samples over one of the channels and odd samples over the other, will facilitate error concealment at the decoder if one of the channels fails. This is because each missing even sample will be surrounded by two available odd samples, which can serve as a basis for interpolation, and vice versa. If multiple channels (or packets) are available, a good strategy would be to group together samples that are maximally separated from each other. For example, if

N channels (packets) are available, the ith group would consist of samples with indices $\{i, i + N, i + 2N, \ldots\}$. In this case, if any one group of samples is lost, each sample from that group would have $N - 1$ available neighboring samples on either side. Even–odd splitting is a special case of this strategy for $N = 2$. The idea of splitting neighboring samples into groups that are transmitted independently (over different channels, or in different packets) also forms the basis of dispersive packetization of images and video.

DISPERSIVE PACKETIZATION OF IMAGES

In the 1-D example of a speech signal, splitting samples into N groups that contain maximally separated samples amounts to simple extraction of N possible phases of the speech signal subsampled by a factor of N. In the case of multidimensional signals, subsampling by a factor of N can be accomplished in many different ways (cf., Chapter 2 and also [12]). The particular subsampling pattern that maximizes the distance between the samples that are grouped together is the one that solves the sphere packing problem [10] in the signal domain. Digital images are usually defined on a subset of the 2-D integer lattice $\mathbb{Z}^2$. A fast algorithm for sphere packing onto $\mathbb{Z}^2$ was proposed in Bajić and Woods [4], and it can be used directly for dispersive packetization of raw images.

In practice, however, we are often interested in transmitting transform-coded images. In this case, there are two ways to accomplish dispersive packetization, as shown in Figure 12.7. Samples can be split either before (Figure 12.7a) or after (Figure 12.7b) the transform. The former approach was used in Wang and Chung [49], where the authors propose splitting the image into four subsampled versions and then encoding each version by the conventional JPEG coder. The latter approach was used in several works targeted at SWT-coded images [11,6]. Based on the results reported in these papers, splitting samples before the trans-

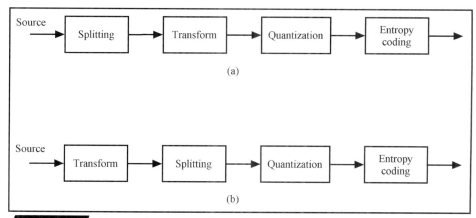

FIGURE 12.7   *Splitting of source samples can be done (a) before the transform or (b) after the transform*

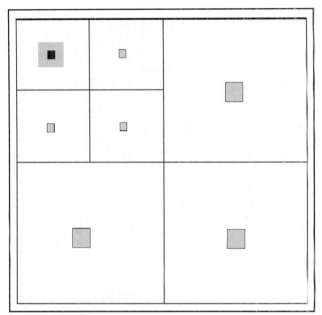

FIGURE 12.8   *One low-frequency SWT sample (black) and its SWT neighborhood (gray)*

form yields lower compression efficiency but higher error resilience, while splitting after the transform gives higher compression efficiency with lower error resilience. For example, the results in Wang and Chung [49] for the *Lena* image show the coding efficiency reduction of over 3 dB in comparison with conventional JPEG, while acceptable image quality is obtained with losses of up to 75%. On the other hand, the methods in Rogers and Cosman [40] and Bajić and Woods [6] are only about 1 dB less efficient than SPIHT [41] (which is a couple of decibels better than JPEG on the *Lena* image), but are targeted at lower packet losses, up to 10–20%.

We will now describe the method of [6], which we refer to as *dispersive packetization* (DP). Figure 12.8 shows a two-level SWT of an image. One sample from the lowest frequency LL subband is shown in black and its neighborhood in the SWT domain is shown in gray. In DP, we try to store neighboring samples in different packets. If the packet containing the sample shown in black is lost, many of its neighbors will still be available. Due to the intraband and interband relationships among subband/wavelet samples, the fact that many neighbors of a missing sample are available can facilitate error concealment.

Given the target image size and the maximum allowed packet size (both in bytes), we can determine the suitable number of packets, say N. Using the lattice partitioning method from [4], a suitable 2-D subsampling pattern $p : \mathbb{Z}^2 \rightarrow \{0, 1, \ldots, N-1\}$ can be constructed for the subsampling factor of N. This pattern is used in the LL subband—the sample at location (i, j) in the LL subband is stored in the packet $p(i, j)$. Subsampling patterns for the higher frequency pat-

0	1	2	3	1	2	3	0	0	1	2	3	0	1	2	3
2	3	0	1	3	0	1	2	2	3	0	1	2	3	0	1
0	1	2	3	1	2	3	0	0	1	2	3	0	1	2	3
2	3	0	1	3	0	1	2	2	3	0	1	2	3	0	1
2	3	0	1	3	0	1	2	0	1	2	3	0	1	2	3
0	1	2	3	1	2	3	0	2	3	0	1	2	3	0	1
2	3	0	1	3	0	1	2	0	1	2	3	0	1	2	3
0	1	2	3	1	2	3	0	2	3	0	1	2	3	0	1
1	2	3	0	1	2	3	0	2	3	0	1	2	3	0	1
3	0	1	2	3	0	1	2	0	1	2	3	0	1	2	3
1	2	3	0	1	2	3	0	2	3	0	1	2	3	0	1
3	0	1	2	3	0	1	2	0	1	2	3	0	1	2	3
1	2	3	0	1	2	3	0	2	3	0	1	2	3	0	1
3	0	1	2	3	0	1	2	0	1	2	3	0	1	2	3
1	2	3	0	1	2	3	0	2	3	0	1	2	3	0	1
3	0	1	2	3	0	1	2	0	1	2	3	0	1	2	3

FIGURE 12.9   *Example of a 16 × 16 image with two levels of SWT decomposition, packetized into four packets*

terns are obtained by modulo shifting in the following way. Starting from the LL subband and going toward higher frequencies in a Z-scan, label the subbands in increasing order from 0 to K. Then, the sample at location (i, j) in subband k is stored in packet $(p(i, j) + k) \mod N$. As an example, Figure 12.9 shows the resulting packetization into $N = 4$ packets of a 16 × 16 image with two levels of subband/wavelet decomposition. One LL sample from packet number 3 is shown in black. Note that out of a total of 23 SWT domain neighbors (shown in gray) of this LL sample, only three are stored in the same packet. (For details on entropy coding and quantization, the reader is referred to [6] and references therein.)

As mentioned before, DP facilitates easy error concealment. Even simple error concealment algorithms can produce good results when coupled with DP. For example, in [6], error concealment is carried out in the three subbands (LL, HL, and LH) at the lowest decomposition level, where most of the signal energy is usually concentrated. Missing samples in the LL subband are interpolated bilinearly from the four nearest available samples in the horizontal and vertical directions. Missing samples in the LH and HL subbands are interpolated linearly from nearest available samples in the direction in which the subband has been lowpass filtered. Missing samples in higher frequency subbands are set to zero. More sophisticated error concealment strategies may bring further improvements [3]. As an example, Figure 12.10 shows PSNR vs. packet loss comparison between the *packetized zero-tree wavelet* (PZW) method from Rogers and Cosman [40], and

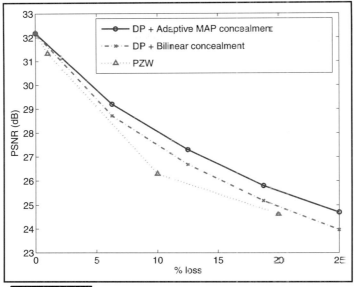

FIGURE 12.10   *PSNR versus packet loss on Lena*

two versions of DP—one paired with bilinear concealment [6] and the other with adaptive MAP concealment [3]. The example is for a 512 × 512 grayscale *Lena* image encoded at 0.21 bpp.

DISPERSIVE PACKETIZATION OF VIDEO

Two-dimensional concepts from the previous section can be extended to three dimensions for dispersive packetization of video. For example, Figure 12.11 shows how alternating the packetization pattern from frame to frame (by adding 1 modulo N) will ensure that samples from a common SWT neighborhood will be stored in different packets. This type of packetization can be used for intraframe-coded video. When motion compensation is employed, each level of the motion-compensated spatiotemporal pyramid consists of an alternating sequence of frames and motion vector fields. In this case, we want to ensure that motion vectors are not stored in the same packet as the samples they point to. Again, adding 1 modulo N to the packetization pattern used for samples will shuffle the pattern so that it can be used for motion vector packetization. This is illustrated in Figure 12.12.

   As an illustration of the performance of video DP, we present an example that involves the grayscale SIF *Football* sequence at 30 fps. The sequence was encoded at 1.34 Mbps with a GOP size of four frames using the DP video coder from [6]. The Gilbert model was used to simulate transmission over a packet-based network, with the "good" state representing packet reception and the "bad" state representing packet loss. Average packet loss was 10.4%.

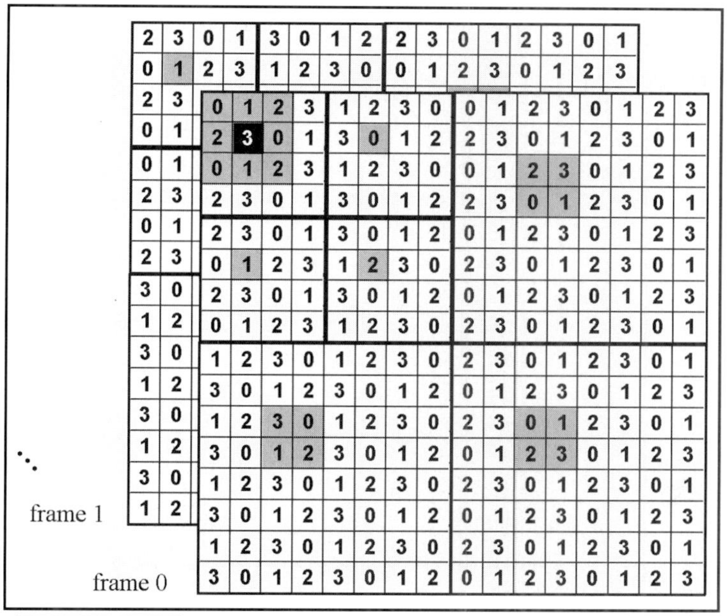

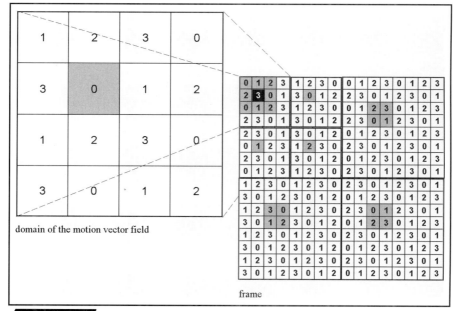

FIGURE 12.11   *Extension of dispersive packetization to three dimensions*

FIGURE 12.12   *Dispersive packetization of motion-compensated SWT video*

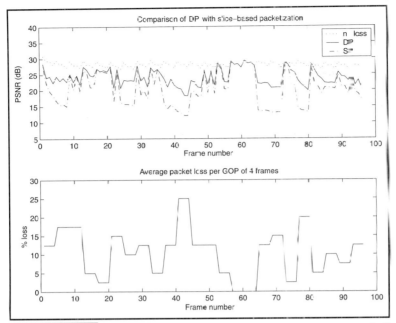

FIGURE 12.13   *Comparative performance of DP and SP schemes on Football*

Figure 12.13 shows the results of a sample run from the simulation. The top part shows the PSNR of the decoded video in three cases: no loss, DP, and *slice-based packetization* (SP). The SP video is similar to the H.26x and MPEGx recommendations for packet-lossy environments, where each slice (a row of macroblocks) is stored in a separate packet. The bottom part of the figure shows the corresponding packet loss profile. In this case, DP provides an average 3–4 dB advantage in PSNR over SP.

Visual comparison between DP and SP is shown in Figure 12.14, which corresponds to frame 87 of the *Football* sequence, with 10% loss. The figure shows (a) original frame, (b) coded frame with no loss (PSNR = 27.7 dB), (c) decoded SP frame (PSNR = 20.6 dB), and (d) decoded DP frame (PSNR = 22.2 dB); (e) and (f) are zoomed-in segments of (c) and (d), respectively, illustrating the performance of the two methods on small details. Observe that details are better preserved in the DP frame.

12.2.2 MULTIPLE DESCRIPTION FEC

The multiple description forward error correction (MD-FEC) technique is a way of combining information and parity bits into descriptions or packets in order to achieve robustness against packet loss through unequal error protection. Among

FIGURE 12.14 *Frame 87 of the* Football *sequence; SIF at 1.34 Mbps*

the first works on the topic was the *priority encoding transmission* (PET) in Albanese *et al.* [1]. In PET, a message is broken into smaller segments, and each segment is assigned a priority level between 0 and 1. Then, each segment is encoded into a set of packets using a version of the RS error-control codes, in a way that guarantees that the segment is decodable if a fraction of the number of pack-

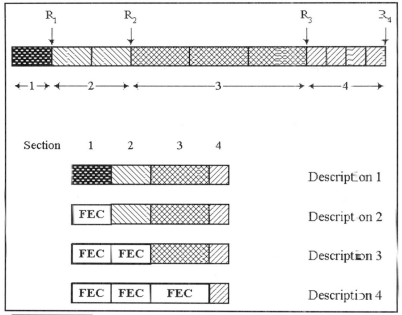

FIGURE 12.15 *Simple illustration of MD-FEC*

ets, at least equal to its priority, is received. Later MD-FEC [33,37] used a very similar strategy, but assumed that the message (image or video clip) is encoded into an embedded or SNR scalable bitstream. This also facilitates assignment of priorities based on distortion–rate $(D-R)$ characteristics. Our description of the method follows the work of Puri *et al.* in [37].

Consider a message encoded into an embedded bitstream. The bitstream is first divided into N sections: $(0, R_1], (R_1, R_2], \ldots (R_{N-1}, R_N]$. Section k is next split into k subsections, and encoded by an (N, k) RS code. Such a code can correct up to $N - k$ erasures. Individual symbols of RS codewords are concatenated to make descriptions. Each description is stored in a separate packet, so we will use the terms "description" and "packet" interchangeably. An illustration of the MD-FEC procedure for $N = 4$ is given in Figure 12.15. In this example, section 1 is protected by RS (4, 1) code, section 2 by RS (4, 2) code, section 3 by RS (4, 3) code, and section 4 is unprotected. If no packets are lost, the bitstream can be decoded up to R_4. If one packet is lost, we are guaranteed to be able to decode up to R_3, since RS (4, 1) can correct any single erasure. However, if the one packet that was lost was packet 4, then we can decode all the way up to $R_3 + 3(R_4 - R_3)/4$. In general, if j out of N packets are received, we can decode up to some rate in the range $[R_j, R_j + j(R_{j+1} - R_j)/(j + 1)]$.

Given the channel packet loss characteristics, the goal of FEC assignment in MD-FEC is to minimize the expected distortion subject to a total rate constraint.

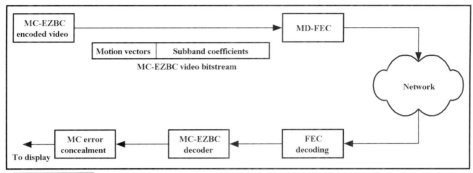

FIGURE 12.16   *Video streaming system based on MC-EZBC*

It is a bit assignment problem, and can be solved in several ways. The reader is referred to [37] for a possible fast solution. In an embedded bitstream, we have $D(R_i) \leqslant D(R_j)$ for $R_i > R_j$, and this fact can be used to reduce computational burden. Designing the MD-FEC amounts to a joint optimization over both the source and channel coders.

3D-SPIHT [26]

In [37], 3D-SPIHT, a spatiotemporal extension of the SPIHT image coder (cf., Section 8.6, Chapter 8), was used as the video coder to produce an embedded bitstream for MD-FEC. In a typical application, MD-FEC would be applied on a GOP-by-GOP basis. Channel conditions are monitored throughout the streaming session and, for each GOP, an FEC assignment is carried out using recent estimates of the channel loss statistics.

MC-EZBC

One can also use MC-EZBC (cf., Section 11.5, Chapter 11) to produce an embedded bitstream for MD-FEC. Since motion vectors form the most important part of interframe video data, they would be placed first in the bitstream, followed by SWT samples encoded in an embedded manner. A video streaming system based on MC-EZBC was proposed in [5]. The block diagram of the system is shown in Figure 12.16. As with 30-SPIHT, MD-FEC is applied on a GOP-by-GOP basis. A major difference from the system [37] based on 3D-SPIHT is the MC error-concealment module at the decoder. Since motion vectors receive the most protection, they are the most likely part of the bitstream to be received. MC error concealment is used to recover from short burst losses that are hard to predict by simply monitoring channel conditions, and therefore often are not compensated for by MD-FEC.

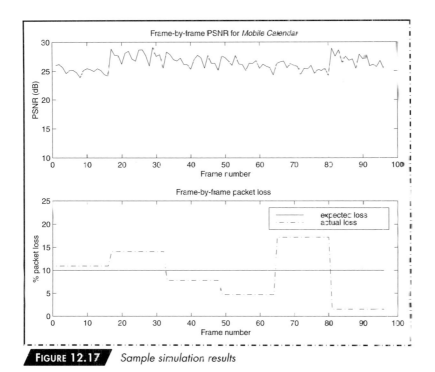

FIGURE 12.17   *Sample simulation results*

EXAMPLE 12.2-1 (*MD-FEC*)

This example illustrates the performance of MD-FEC for video transmission. The *Mobile Calendar* sequence (SIF resolution, 30 fps) was encoded into an SNR-scalable bitstream using the MC-EZBC video coder. The sequence was encoded with a GOP size of 16 frames. FEC assignment was computed for each GOP separately, with 64 packets per GOP and a total rate of 1.0 Mbps, assuming a random packet loss of 10%. Video transmission was then simulated over a two-state Markov (Gilbert) channel, which is often used to capture the bursty nature of packet loss in the Internet. This channel has two states: "good," corresponding to successful packet reception, and "bad," corresponding to packet loss. The channel was simulated for 10% average loss rate and an average burst length of 2 packets.

Results from a sample simulation run are shown in Figure 12.17. The top part of the figure shows PSNR for 96 frames of the sequence. The bottom part shows the expected loss rate of 10% (solid line) and the actual loss (dashed line). As long as the actual loss does not deviate much from the expected value used in FEC assignment, MD-FEC is able to provide consistent video quality at the receiver. Please see robust SWT video results on enclosed CD-ROM.

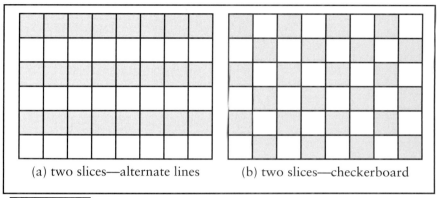

(a) two slices—alternate lines   (b) two slices—checkerboard

FIGURE 12.18   *Interleaved and FMO mapping of macroblocks onto two slices: the gray slice and the white slice*

12.3 ERROR-RESILIENCE FEATURES OF H.264/AVC

The current VCEG/MPEG coding standard H.264/AVC inherits several error-resilient features from earlier standards such as H.263 and H.26L, but also adds several new ones. These features of the source coder are intended to work together with the transport error protection features of the network to achieve a needed *quality of service* (QoS). The error-resilience features in H.264/AVC are syntax, data partitioning, slice interleaving, flexible macroblock ordering, SP/SI switching frames, reference frame selection, intrablock refreshing, and error concealment [27].

H.264/AVC is conceptually separated into a *video coding layer* (VCL) and a *network abstraction layer* (NAL). The output of the VCL is a slice, consisting of an integer number of macroblocks. These slices are independently decodable due to the fact that positional information is included in the header, and no spatial references are allowed outside the slice. Clearly, the shorter the slice, the lower the probability of loss or other compromise. For highest efficiency, there should be one slice per frame, but in poorer channels or lossy environments, slices consisting of a single row of macroblocks have been found useful. The role of the NAL is to map slices to transmission units of the various types of networks, e.g., map to packets for an IP-based network.

12.3.1 SYNTAX

The first line of defense from channel errors in video coding standards is the semantics or syntax [27]. If illegal syntatical elements are received, then it is known that an error has occurred. At this point, error correction can be attempted or the data declared lost, and error concealment can commence.

12.3.2 DATA PARTITIONING

By using the *data partitioning* (DP[3]) feature, more important data can be protected with stronger error control features, e.g., forward error correction and acknowledgement-retransmission, to realize unequal error protection (UEP). In H.264/AVC, coded data may be partitioned into three partitions, A, B, and C. In partition A is header information including slice position. Partition B contains the coded block patterns of the intrablocks, followed by their transform coefficients. Partition C contains coded block patterns for interblocks and their transform coefficients. Coded block patterns contain the position and size information for each type of block, and incur a small overhead penalty for choosing DP.

12.3.3 SLICE INTERLEAVING AND FLEXIBLE MACROBLOCK ORDERING

A new feature of H.264/AVC is *flexible macroblock ordering* (FMO) that generalizes slice interleaving from earlier standards. The FMO option permits macroblocks to be distributed across slices in such a way as to aid error concealment when a single slice is lost. For example, we can have two slices as indicated in Figure 12.18 (a or b). The indicated squares can be as small as macroblocks of 16×16 pixels, or they could be larger. Figure 12.18a shows two interleaved slices, the gray and the white. If one of the two interleaved slices is lost, in many cases it will be possible to interpolate the missing areas from the remaining slice, both above and below.

In Figure 12.18b, a gray macroblock can more easily be interpolated if its slice is lost, because it has four nearest-neighbor macroblocks from the white slice, presumed received. On the other hand, such extreme macroblock dispersion as shown in this checkerboard pattern would mean that the nearest-neighbor macroblocks could not be used in compression coding of a given macroblock, and hence there would be a loss in compression efficiency. Of course, we must trade off the benefits and penalties of an FMO pattern for a given channel error environment. We can think of FMO as akin to dispersive packetization.

12.3.4 SWITCHING FRAMES

This feature allows switching between two H.264/AVC bitstreams stored on a video server. They could correspond to different bitrates, in which case the switching permits adjusting the bitrate at the time of the so-called *SP* or switching frame, while not waiting for an *I* frame. While not as efficient as a *P* frame, the *SP* frame can be much more efficient than an *I* frame. Another use for a switching frame is

---

3. Please do not confuse with dispersive packetization; which meaning of DP should be clear in context.

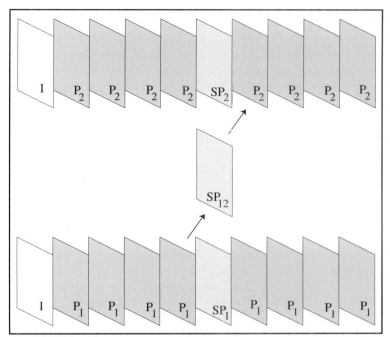

FIGURE 12.19   *Illustration of switching from bottom bitstream 1 to top bitstream 2 via switching frame SP_{12}*

to recover from a lost or corrupted frame, whose error would otherwise propagate till reset by an *I* frame or slice. This can be accomplished, given the existence of such switching frames, by storing two coded bitstreams in the video server, one with a conventional reference frame choice of the immediate past frame, and the second bitstream with a reference frame choice of a frame further back, say two or three frames back. Then if the receiver detects the loss of a frame, and this loss can be fed back to the server quickly enough, the switch can be made to the second bitstream in time for the computation and display of the next frame, possibly with a slight freeze-frame delay. The *SP* frame can be also be used for fast-forward searching through the video program.

With reference to Figure 12.19, we illustrate a switching situation where we want to switch from the bottom bitstream 1 to the top bitstream 2. Normally this could only be done at the time of an *I* frame, since the *P* frame from bitstream 2 needs the decoded reference frame from bitstream 2, in order to decode properly. If the decoded frame from bitstream 1 is used, as in our switching scenario, then the reference would not be as expected and an error would occur, and propagate via intra prediction within the frame and then via inter prediction to the following frames. Note that this switching can just as well be done on a slice basis as well as the frame basis we are considering here.

The question now is how can this switching be done, in such a way as to avoid error in the frames following the switch. First, the method involves special *primary* switching frames SP_1 and SP_2 inserted into the two bitstreams at the point of the expected switch. The so-called *secondary* switching frame SP_{12} then must have the same reconstructed value as SP_2, even though they have different reference frames and hence different predictions. The key to accomplishing this is to operate with the quantized coefficients in the block transform domain, instead of making the prediction in the spatial domain. The details are shown in Karczewicz and Kurceren [24] where a small efficiency penalty is noted for using switching frames.

12.3.5 REFERENCE FRAME SELECTION

In H.264/AVC we can use frames for reference other than the immediately last one. If the video is being coded now, rather than stored on a server, then feedback information from the receiver can indicate that one of the reference frames is lost; then the *reference frame selection* (RFS) feature can alter the choice of reference frames for encoding the current frame, in such a way as to exclude the lost frame, and therefore terminate any propagating error. Note that so long as the feedback arrives within a couple of frame times, it can still be useful in terminating propagation of an error in the output, which would normally only end at an *I* frame or perhaps at the end of the current slice. Note that RFS can make use of the multiple reference frame feature of H.264/AVC, thus this feature also has a role in increasing the error resilience.

12.3.6 INTRABLOCK REFRESHING

A basic error resilience method is to insert intraslices instead of intraframes. These intraslices can be inserted in a random manner, so as not to increase the local data rate the way that intraframes do. Intraslices have been found useful for keeping delay down in visual conferencing. Intraslices terminate error propagation within the slice, the same way that *I* frames do for the frame.

12.3.7 ERROR CONCEALMENT IN H.264/AVC

While error concealment is not part of the normative H.264/AVC standard, there are certain nonnormative (informative) features in the test model [44]. First, a lost macroblock is concealed at the decoder using a similar type of directional interpolation (prediction) as is used in the coding loop at the encoder. Second, an interframe error concealment is made from the received reference blocks, wherein the candidate with the smallest boundary matching error [29,50] is selected, with the zero motion block from the previous frame always a candidate.

FIGURE 12.20   *Slice consists of one row of macroblocks (see the color insert)*

FIGURE 12.21   *Slice consists of three contiguous rows of macroblocks (see the color insert)*

EXAMPLE 12.3-1 (*selection of slice size*) (Soyak and Katsoggelos)
In this example, baseline H.264/AVC is used to code slice sizes consisting of both one row and three rows of macroblocks. The *Foreman* CIF clip was used at 30 fps and 384 Kbps. Because of the increased permitted use of intra prediction by the larger slice size, the average luma PSNRs without any packet loss were 36.2 dB and 36.9 dB, respectively. Then the packets were subjected to a loss simulation as follows. The smaller packets from the one-row slices were subjected to a 1% packet loss rate, while the approximately three-times longer packets from the three-row slices were subjected to a comparable 2.9% packet loss rate. Simple error concealment was used to estimate any missing macroblocks using a median of neighboring motion vectors.

The resulting average luma PSNRs became 30.8 and 25.6 dB, respectively. Typical frames grabbed from the video are shown in Figures 12.20 and 12.21.

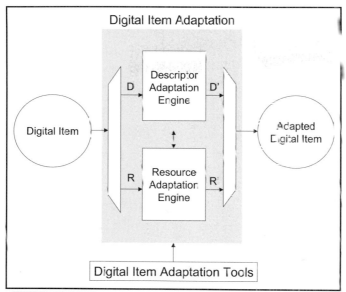

FIGURE 12.22 *Concept DIA model (from International Standardization Organization [20])*

We can see the greater effect of data loss in Figure 12.21. Video results are contained in the enclosed CD-ROM in the Network Video folder.

12.4 JOINT SOURCE–NETWORK CODING

Here we talk about methods for network video coding that make use of overlay networks. These are networks that exist on top of the public Internet at the application level. We will use them for transcoding or digital item adaptation and an extension to MD-FEC. By *joint source–network coding* (JSNC), we mean that source coding and network coding are performed jointly in an effort to optimize the performance of the combined system. Digital item adaptation can be considered as an example of JSNC.

12.4.1 DIGITAL ITEM ADAPTATION (DIA) IN MPEG 21

MPEG 21 has the goal of making media, including video, available everywhere [19,20] and this includes heterogeneous networks and display environments. Hence the need for adaptation of these digital items in quality (bit-rate, spatial resolution, and frame rate, within the network. The digital item adaptation (DIA) engine, shown in Figure 12.22, is proposed to accomplish this.

Digital items such as video packets enter on the left and are operated upon by the resource adaptation engine for transcoding to the output-modified digital

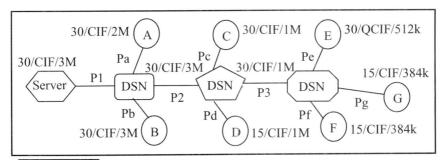

FIGURE 12.23   *Example overlay network*

item. At the same time, the descriptor field of this digital item needs to be adapted also, to reflect the modified or remaining content properly, e.g., if 4CIF is converted to CIF, descriptors of the 4CIF content would be removed, with only the CIF content descriptors remaining. There would also be backward flows or signaling messages from the receivers, indicating the needs of the downstream users as well as information about network congestion, available bitrates, and bit-error rates, that the DIA engine must consider also in the required adaptation. A general overview of video adaptation can be found in Chang and Vetro [9].

12.4.2 FINE–GRAIN ADAPTIVE FEC

We have seen that JSNC uses an overlay infrastructure to assist video streaming to multiple users simultaneously by providing lightweight support at intermediate overlay nodes. For example, in Figure 12.23, overlay *data service nodes* (DSNs) construct an overlay network to serve heterogeneous users. Users A to G have different video requirements (frame rate/resolution/available bandwidth), and Pa–Pg are the packet-loss rates of different overlay virtual links. We do not consider the design of the overlay network here. While streaming, the server sends out a single *fine-grain-adaptive-FEC* (FGA-FEC) coded bitstream based on the highest user requirement (in this case 30/CIF/3M) and actual network conditions. FGA-FEC divides each network packet into small fine-grain blocks and packetizes the FEC coded bitstream in such a way that if any original data packets are actively dropped (adapted by the DIA engine), the corresponding information in parity bits is also completely removed. The intermediate DSNs can adapt the FEC coded bitstream by simply dropping a packet or shortening a packet by removing some of its blocks. Since there is no FEC decoding/re-encoding, JSNC is very efficient in terms of computation. Furthermore, the data manipulation is at block level, which is precise in terms of adaptation.

In this section, we use the scalable MC-EZBC video coder [18] to show the FGA-FEC capabilities. As seen in Chapter 11, MC-EZBC produces embedded bitstreams supporting a full range of scalabilities—temporal, spatial, and SNR. Here

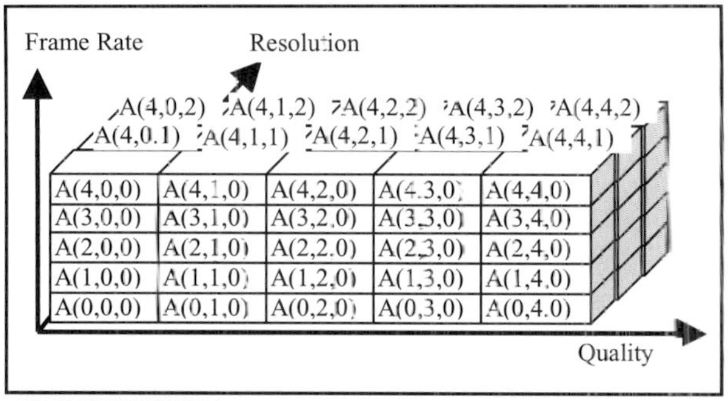

FIGURE 12.24   *Atom diagram of video scalability dimensions (from Hewlett-Packard Labs web site [17])*

we use the same notation as [18]. Each GOP coding unit consists of independently decodable bitstreams $\{Q^{MV}, Q^{YUV}\}$. Let $l_t \in \{1, 2, \ldots, L_t\}$ denote the temporal scale, then the MV bitstream Q^{MV} can be divided into temporal scales and consists of Q_{lt}^{MV} for $2 \leqslant l_t \leqslant L_t$. Let $l_s \in \{1, 2, \ldots, L_s\}$ denote the spatial scales. The subband coefficient bitstream Q^{YUV} is also divided into temporal scales and further divided into spatial scales as $Q_{lt,ls}^{YUV}$, for $2 \leqslant l_t \leqslant L_t$ and $1 \leqslant l_s \leqslant L_s$. For example, the video at one-quarter spatial resolution and one-half frame rate is obtained from the bitstream as $Q = \{Q_{lt}^{MV}: 1 \leqslant l_t \leqslant L_t - 1\} \cup \{Q_{lt,ls}^{YUV}: 1 \leqslant l_s \leqslant L_s - 1\}$. In every sub-bitstream $Q_{lt,ls}^{YUV}$, the luma and chroma subbands Y, U, and V are progressively encoded from the most significant bitplane (MSB) to the least significant bitplane (LSB) (cf., Section 8.6.6).

Scaling in terms of quality is obtained by stopping the decoding process at any point in bitstream Q. The MC-EZBC encoded bitstream can be further illustrated as digital items as in Figure 12.24 [17], which shows the video bitstream in view of three forms of scalability. The video bitstream is represented in terms of *atoms*, which are usually fractional bitplanes [18]. The notation $A(F, Q, R)$ represents an atom of (frame rate, quality, resolution). Choosing a particular (causal) subset of atoms corresponds to scaling the resulting video to the desired resolution, frame rate, and quality. These small pieces of bitstream are interlaced in the embedded bitstream. Intermediate DSNs adapt the digital items according to user preferences and network conditions. Since the adaptation can be implemented as simple dropping of corresponding atoms, DSNs do not need to decode and re-encode the bitstream, which is very efficient. On the other hand, the adaptation is done based on atoms in a bitstream, which can approximate the quality of pure source coding.

EXAMPLE 12.4-1 *(adaptation example)*

Using the MC-EZBC scalable video coder we have adapted the bitrate in the network, by including limited overhead information in the bitstream, allowing bitrate adaptation across a range of bitrates for the *Coastguard* and

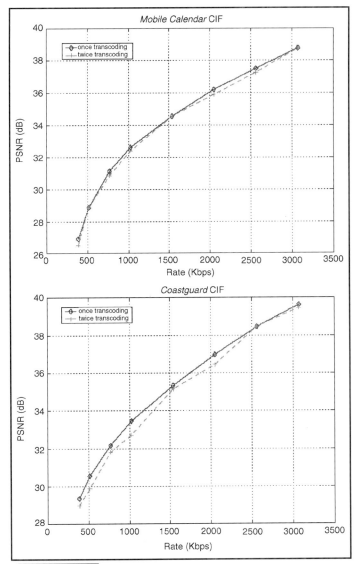

FIGURE 12.25   *PSNR performance comparison (Mobil Calendar and Coastguard in CIF format) versus bitrate for DIA operation in network*

Mobile Calendar CIF test clips. Figure 12.25 shows the resulting PSNR plots versus bitrate. Two plots are given, the first corresponding to precise extraction at the source or server, and the second corresponding to DIA within the network using six quality layers, whose starting location is carried in a small

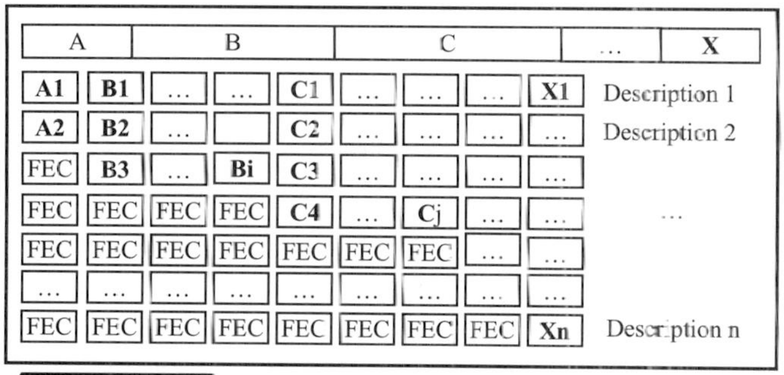

A	B			C				...	X	
A1	B1	...	...	C1	...	...	...		X1	Description 1
A2	B2	...		C2	...	...	...			Description 2
FEC	B3	...	Bi	C3	...	...	...			
FEC	FEC	FEC	FEC	C4	...	Cj	...	...		...
FEC	FEC	FEC	FEC	FEC	FEC	FEC	...	...		
...	...	...	...	...	...	...	...			
FEC	FEC	FEC	FEC	FEC	FEC	FEC	FEC		Xn	Description n

FIGURE 12.26 *FGA-FEC coding scheme*

header. We can see that the PSNR performance is only slightly reduced from that at the video server, where full scaling information is available.

Digital item adaptation can be done for nonscalable formats too, but only via the much more difficult method of transcoding. Next we show a method of combining DIA with FEC to increase robustness.

DSNs adapt the video bitstream based on user requirements and available bandwidth. When parts of the video bitstream are actively cropped by the DIA engine, FEC codes need to be updated accordingly. This update of FEC codes has the same basic requirements as does the video coding—efficiency (low computation cost) and precision (if a part of the video data is actively dropped, parity bits protecting that piece of data should also be removed). Based on these considerations, we have proposed a precise and efficient FGA-FEC scheme [42] based on RS codes. FGA-FEC solves the problem by fine granular adapting of the FEC to suit multiple users simultaneously, and it works as follows.

Given a segment of the video bitstream, shown in Figure 12.26 (top line), divided into chunks as $A, B, C, \ldots, X$, the FGA-FEC further divides each chunk of bitstream into small equal-sized blocks. The precision or fine granularity of the FGA-FEC scheme is determined by this blocksize. Smaller blocksize means finer granularity and better adaptation precision. In the lower part of Figure 12.26, the bitstream is divided into blocks as $(A1, A2; B1, \ldots, Bi; C1, \ldots, Cj; \ldots; X1, \ldots, Xn)$. The RS coding computation is applied vertically across these blocks to generate the parity blocks, denoted "FEC" in the figure. Each vertical column consists of data blocks, followed by their generated parity blocks. More protection is added to the important part of the bitstream (A, B) and less FEC is allocated to data with lower priority $(C, D, \ldots)$. The optimal allocation of FEC to different chunks of data has been described in Section 12.2 and in [5,38]. After FEC encoding, each horizontal row of blocks is packetized as one description, i.e., one description is equivalent to one network packet.

Similar to MD-FEC of Section 12.2, FGA-FEC transforms a priority-embedded bitstream into nonprioritized descriptions to match a best-efforts network. In addition, the FGA-FEC scheme has the ability of fine granular adaptation

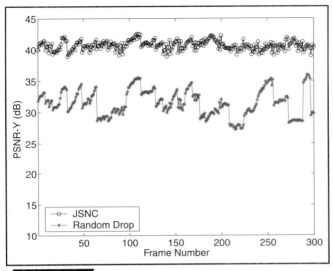

FIGURE 12.27   *JSNC versus random drop*

at block level. Based on our experiments, with a 1 Mbps *Foreman* CIF bitstream and GOP size of 16 frames, MD-FEC can generate blocksizes varying from 1 to 475 bits, which is difficult to perform as in-network adaptation. So in FGA-FEC, we use blocksizes that are a multiple of bytes, the RS symbol size. To facilitate intermediate overlay node adaptation, an information packet is sent ahead of one GOP to tell the intermediate nodes about the blocksize, FEC codes, and bitstream information at the block level.

EXAMPLE 12.4-2 (*overlay network adaptation*)
Traditionally, when network congestion occurs, data packets are randomly dropped due to congestion. To avoid this, the JSNC scheme adapts the packets in the intermediate network nodes to reduce the bandwidth requirement. Given a 1.5 Mbps bitstream and an available bandwidth of 1455 Kbps, in Figure 12.27 we compare PSNR-Y of JSNC vs. a random packet-drop scheme with a 3% packet-loss rate. Observe that the proposed scheme significantly outperforms random drop by about 10 dB. In Figure 12.28 we show objective video quality (PSNR) when the available bandwidth changes.

Originally, the user is receiving a 2 Mbps CIF format, 30 fps bitstream, but, starting with frame 100, the user has only 512 Kbps available. There are three possible choices for the user: (1) SNR adaptation to 512 Kbps, (2) temporal adaptation to one-quarter of the original frame rate, and (3) spatial adaptation down to QCIF resolution. Both (2) and (3) need additional SNR adaptation to 512 Kbps. With FGA-FEC, the users can choose their preference based on their application needs. We note from Figure 12.28, though, that there is a significant PSNR penalty to pay (about 7 dB) for staying with

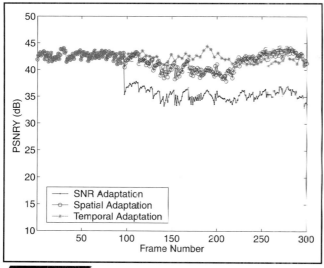

FIGURE 12.28 *3-D adaptation in frame rate, resolution, and SNR or bitrate*

the full frame rate and spatial resolution, and just cutting down on the bitrate. Please see video results on enclosed CD-ROM in the Network Video folder.

12.5 CONCLUSIONS

Compressed video data must be protected and adapted to the channel in the case of lossy transmission. This chapter has introduced the concept of error resilience of video coders and their use in transport-layer error protection for wired and wireless channels, the chief difference being the significant probability of packet bit errors in the latter. We have shown examples of scalable video coders together with the concept of multiple description coding and introduced the concept of extension of MD-FEC to problems of video adaptation necessary for heterogeneous networks.

A nice overview of video communication networks is found in Schonfeld [43]. An overview of JSCC for video communications is contained in Zhai *et al.* [56].

12.6 PROBLEMS

1   This problem concerns the synch word concept, the problem being how to segment a VLC bitstream into the proper characters (messages). One way to do this, together with Huffman coding, is to add an additional low-probability (say $0 < p \ll 1$) message to the message source called *synch*. If we already had a set of independent messages $m_1, m_2, \ldots, m_N$ with probabilities $p_1, p_2, \ldots, p_M$, comprising a message source with entropy H, show

that for sufficiently small p, the total entropy, including the synch word, is negligibly more than H, for p sufficiently small. Hint: Assume that the synch word constitutes an independent message, so that the *compound source* uses an original message with probability $(1 - p)p_i$ or a synch word with probability p. Note that

$$\lim_{p \searrow 0} p \log \frac{1}{p} = \lim_{n \nearrow \infty} 2^{-n} \log 2^n = \lim_{n \nearrow \infty} n 2^{-n} = 0.$$

2 In Example 12.1-1, the outputs of the even and odd scalar quantizers are the two descriptions to be coded. Call the even quantizer output description 1, then the odd output description 2.

(a) For description 1, what is the entropy (rate) and distortion, assuming an input random variable uniformly distributed on $[1, 10]$?

(b) Compare the total rate for the two descriptions to the entropy for the original single-description quantizer, assuming the same random-variable input?

3 Assume that bit errors on a certain link are independent and occur at the rate (i.e., with probability) 1×10^{-3}. Assume that symbols (messages) are composed of 8-bit bytes. What then is the corresponding byte-error rate? Assume that packets are 100 bytes long. What then is the corresponding packet-error rate? Repeat for 300-byte packets.

4 This problem is about calculating the total number of bits needed for MD-FEC.

(a) Refer to Figure 12.15. Assume that R_1 through R_4 are expressed in bits. Calculate the total number of bits (both information bits and parity bits) needed for the four packets (descriptions), in terms of R_1 through R_4.

(b) Now assume the goal is to create N packets. Calculate the total number of bits needed for N packets, in terms of $R_1, R_2, \ldots, R_N$.

5 In channel coding theory, *code rate* is defined as the ratio of information bits to the total (information + parity) number of bits.

(a) Using the results from the previous problem, what is the code rate of MD-FEC for N packets, expressed in terms of $R_1, R_2, \ldots, R_N$?

(b) Based on the definition of code rate, any nontrivial code must have a rate in the interval $(0, 1]$. Let N be given. Assuming that $R_k, k = 1, 2, \ldots, N$ can be real numbers, show that MD-FEC can achieve any rate $r \in [\frac{1}{N}, 1]$. In other words, given a number $r \in [\frac{1}{N}, 1]$, find $R_1, R_2, \ldots, R_N$, such that the rate of the corresponding MD-FEC code is r.

6 In Example 12.4-1 on digital item adaptation, two coded videos were bitrate adapted (reduced) as shown in Figure 12.25. These figures show a slight decrease in PSNR performance at the second transcoding (adaptation), which

Table 12.1. Sub-bitstream sizes

$l_t \backslash l_s$	1	2	3	4	5
1	100	100	200	100	200
2	200	200	400	400	500
3	800	800	1200	1000	1000
4	2200	2400	3000	2200	2000
5	4400	5300	5800	3200	2400
6	4600	6400	5000	2400	2000

uses only six quality layers. Would there be any further losses in quality at subsequent bitrate reductions (adaptations) that may happen further on in the network?

7 One GOP consists of MC-EZBC encoded bitstreams $\{Q_{lt}^{MV}, Q_{lt,ls}^{YUV}\}$, where $\{1 \leqslant l_t \leqslant 5\}$ and $\{1 \leqslant l_s \leqslant 6\}$ (please refer to Section 12.4.2 for notation). For simplicity, we set the sizes $|Q_1^{MV}| = 4000$, $Q_2^{MV}| = 1000$, $|Q_3^{MV}| = 700$, $|Q_4^{MV}| = 300$, and $|Q_5^{MV}| = 0$, all in bytes. The sizes of the sub-bitstream partitions $\{Q_{lt}^{MV}, Q_{lt,ls}^{YUV}\}$ are given in Table 12.1.

(a) What is the size of the adapted GOP if it is adapted to half resolution and one-quarter frame rate?

(b) Suppose the original bitstream is encoded with FGA-FEC as in Figure 12.26. Each motion vector sub-bitstream is encoded with a one-quarter code rate FEC code, and each YUV sub-bitstream is encoded with a one-half code rate FEC code. What is the size of the adapted GOP with FEC if the GOP is again adapted to half resolution and one-quarter frame rate?

REFERENCES

[1] A. Albanese, J. Blömer, J. Edmonds, M. Luby, and M. Sudan. "Priority Encoding Transmission," *IEEE Trans. Inform. Theory*, 42, 1737–1744, November 1996.

[2] J. B. Anderson, *Digital Transmission Engineering*, IEEE Press, Piscataway, NJ, 1999.

[3] I. V. Bajić, "Adaptive MAP Error Concealment for Dispersively Packetized Wavelet-Coded Images," *IEEE Trans. Image Process.* In press.

[4] I. V. Bajić and J. W. Woods, "Maximum Minimal Distance Partitioning of the $\mathbb{Z}^2$ Lattice," *IEEE Trans. Inform. Theory*, 49, 981–992, April 2003.

[5] I. V. Bajić and J. W. Woods, "EZBC Video Streaming with Channel Coding and Error Concealment," *Proc. SPIE VCIP*, 5150, 512–522, July 2003.

[6] I. V. Bajić and J. W. Woods, "Domain-Based Multiple Description Coding of Images and Video," *IEEE Trans. Image Process.*, 12, 1211–1225, October 2003.

[7] H. Balakrishnan, V. N. Padmanabhan, S. Seshan, and R. H. Katz, "A Comparison of Mechanisms for Improving TCP Performance over Wireless Links," *IEEE/ACM Trans. Network.*, 5, 756–769, Dec. 1997.

[8] M. Bystrom and J. W. Modestino, "Combined Source-Channel Coding Schemes for Video Transmission over an Additive White Gaussian Noise Channel," *IEEE J. Select. Areas Commun.*, 18, 880–890, June 2000.

[9] S.-F. Chang and A. Vetro, "Video Adaptation: Concepts, Technologies, and Open Issues," *Proc. IEEE*, 93, 148–158, Jan. 2005.

[10] J. H. Conway and N. J. A. Sloane, *Sphere Packings, Lattices and Groups*, Springer-Verlag, New York, NY, 1988.

[11] C. D. Creusere, "A New Method of Robust Image Compression Based on the Embedded Zerotree Wavelet Algorithm," *IEEE Trans. Image Process.*, 6, 1436–1442, October 1997.

[12] D. E. Dudgeon and R. M. Mersereau, *Multidimensional Digital Signal Processing*, Prentice-Hall, Englewood Cliffs, NJ, 1983.

[13] N. Farber, T. Wiegand, and B. Girod, *Error Correcting RVLC*, ITU/VCEG Q15-E-32, British Columbia, CA, July 1998.

[14] R. G. Gallager, *Information Theory and Reliable Communication*, John Wiley and Sons, New York, NY, 1968.

[15] M. Ghanbari, *Video Coding: An Introduction to Standard Codecs*, IEE Telecommunications Series, London, UK, 1999.

[16] B. Girod, "Bidirectionally Decodable Streams of Prefix Code-Words," *IEEE Commun. Lett.*, 3, 245–247, Aug. 1999.

[17] Hewlett-Packard Labs. Web site: http://www.hpl.hp.com. See links "Research (Printing and Imaging)" for Structure Scalable Meta-format (SSM).

[18] S.-T. Hsiang and J. W. Woods, "Embedded Video Coding Using Invertible Motion Compensated 3-D Subband Filter Bank," *Signal Process. Image Commun.*, 16, 705–724, May, 2001.

[19] International Organization for Standardization. *MPEG-21 Overview V.4*, ISO/IEC JTC1/SC29/WG11/N4801, Shanghai, Oct. 2002.

[20] International Organization for Standardization. *MPEG-21 Digital Item Adaptation WD (v3.0)*, ISO/IEC JTC1/SC29/WG11/N5178, Shanghai, Oct. 2002.

[21] Internet Engineering Taskforce (IETF). Website: www.ietf.org (see links 'concluded working groups,' then 'diffserv').

[22] N. S. Jayant, "Subsampling of a DPCM Speech Channel to Provide Two 'Self-Contained' Half-Rate Channels," *Bell Syst. Tech. J.*, 60, 501–509, April 1981.

[23] W. Jiang and A. Ortega, "Multiple Description Coding via Polyphase Transform and Selective Quantization," *Proc. SPIE VCIP*, 3653, 998–1008, January 1999.

[24] M. Karczewicz and R. Kurceren, "The SP and SI Frames Design for H.264/AVC," *IEEE Trans. Circuits Syst. Video Technol.* 13, 637–644, July 2003.

[25] A. K. Katsaggelos and N. P. Galatsanos, eds., *Signal Recovery Techniques for Image and Video Compression*, Kluwer Academic Publishers, 1998.

[26] B. G. Kim and W. A. Pearlman, "An Embedded Wavelet Video Coder Using Three-Dimensional Set Partioning in Hierarchical Trees,' *Proc. Data Compress. Conf.*, pp. 251–260, March 1997, Snowbird, Utah

[27] S. Kumar, L. Xu, M. K. Mandal, and S. Panchanathan, "Overview of Error Resiliency Schemes in H.264/AVC Standard," preprint for *Signal Processing: Image Communication*, 2005.

[28] J. F. Kurose and K. W. Ross, *Computer Networking*, 3rc edn., Pearson–Addison Wesley, Boston, MA, 2005.

[29] W. M. Lam, A. R. Reibman, and B. Liu, "Recovery of Lost or Erroneously Received Motion Vectors," *Proc. ICASSP.* 5, 417–420, March 1993.

[30] A. Luthra *et al.*, eds., "Special Issue on the H.264/AVC Video Coding Standards," *IEEE Trans. Circ. Syst. Video Technol.*, 13, July 2003.

[31] M. Mathis, J. Semke, J. Mahdavi, and T. Ott, "The Macroscopic Behavior of the TCP Congestive Avoidance Algorithm," *Computer Commun. Rev.*, 27, July 1997.

[32] J. W. Modestino and D. G. Daut, "Combined Source-Channel Coding of Images," *IEEE Trans. Commun.*, COM-27. 1644–1659, Nov. 1979.

[33] A. E. Mohr, E. A. Riskin, and R. E. Laedner, "Unequal Loss Protection: Graceful Degradation Over Packet Erasure Channels Through Forward Error Correction," *IEEE J. Select. Areas Commun.*, 18, 819–328, June 2000.

[34] J. Padhye, J. Kurose, and D. Towsley, *A TCP-friendly Rate Adjustment Protocol for Continuous Media Flows Over Best Effort Networks*, University of Massachusetts. CMPSCI Tech. Report 98-04, Oct. 1998.

[35] V. Paxson, "End-to-End Internet Packet Dynamics," *IEEE/ACM Trans. Network.*, 7, 277–292, June 1999.

[36] J. G. Proakis and M. Salehi, *Communication Systems Engineering*, 2nd edn., Prentice-Hall, Upper Saddle River, NJ, 2002.

[37] R. Puri, K.-W. Lee, K. Ramchandran, and V. Bharghavan, "An Integrated Source Transcoding and Congestion Control Paradigm for Video Streaming in the Internet," *IEEE Trans. Multimedia* 3. 18–32, March 2001.

[38] R. Puri and K. Ramchandran, "Multiple Description Source Coding Using Forward Error Correction Codes," *Proc. 33rd ACSSC*, Pacific Grove, CA, Oct. 1999.

[39] T. Rappaport, A. Annamalai, R. M. Buehrer, and W. H. Tranter, "Wireless Communications: Past Events and a Future Perspective," *IEEE Commun. Magazine*, pp. 148–161, May 2002 (50th anniversary commemorative issue).

[40] J. K. Rogers and P. C. Cosman, "Wavelet Zerotree Image Compression with Packetization," *IEEE Signal Process. Lett.*, 5, 105–107, May 1998.

[41] A. Said and W. A. Pearlman, "A New, Fast, and Efficient Image Codec Based on Set Partitioning in Hierarchical Trees," *IEEE Trans. Circuits Syst. Video Technol.*, 6, 243–250, June 1996.

[42] Y. Shan, I. V. Bajić, S. Kalyanaraman, and J. W. Woods, "Joint Source–Network Error Control for Scalable Overlay Video Streaming," *Proc. IEEE ICIP 2005*, Genoa, Italy, Sept. 2005.

[43] D. Schonfeld, "Video Communications Networks," in *Handbook of Image and Video Processing*, 2nd edn. (A. C. Bovik, ed.), Elsevier Academic Press, Burlington, MA, 2005. Chapter 9.3.

[44] T. Stockhammer, M. M. Hannuksela, and T. Wiegand, "H.264/AVC in Wireless Environments," *IEEE Trans. Circuits Syst. Video Technol.*, 13, 657–673, July 2003.

[45] Y. Takashima, M. Wada, and H. Murakami, "Reversible Variable Length Codes," *IEEE Trans. Commun.*, 43, 158–162, Feb. 1995.

[46] W. Tan and A. Zakhor, "Real-time Internet Video Using Error Resilient Scalable Compression and TCP-friendly Transport Protocol," *IEEE Trans. Multimedia*, 1, 172–186, June 1999.

[47] A. S. Tenenbaum, *Computer Networks*, 3rd edn., Prentice-Hall, Upper Saddle River, NJ, 1996.

[48] V. A. Vaishampayam, "Design of Multiple Description Scalar Quantizers," *IEEE Trans. Inform. Theory*, 39, 821–834, May 1993.

[49] Y. Wang and D.-M. Chung, "Robust Image Coding and Transport in Wireless Networks Using Non-Hierarchical Decomposition," in *Mobile Multimedia Communications* (D. J. Goodman and D. Raychaudhury, eds.), Plenum Press, New York, NY, 1997.

[50] Y.-K. Wang, M. M. Hannuksela, V. Varsa, A. Hourunranta, and M. Gabbouj, "The Error Concealment Feature of the H.26L Test Model," *Proc. ICIP*, 2, 729–732, Sept. 2002.

[51] Y. Wang, M. Orchard, V. Vaishampayam, and A. R. Reibman, "Multiple Description Coding Using Pairwise Correlating Transform," *IEEE Trans. Image Process.*, 10, 351–366, March 2001.

[52] Y. Wang, J. Ostermann, and Y.-Q. Zhang, In *Video Processing and Communications*, Prentice-Hall, Upper Saddle River, NJ, 2002. Chapters 14 and 15.

[53] J. Wen and J. Villasenor, "A Class of Reversible Variable Length Codes for Robust Image and Video Coding," *Proc. IEEE ICIP 1997*, pp. 65–68, Santa Barbara, CA, Oct. 1997.

[54] S. Wenger, M. M. Hanuksela, T. Stockhammer, M. Westerlund, and D. Singer, *RTP Payload Format for H.264 Video*, IETF RFC 3984, Feb. 2005.

[55] W. Zeng and B. Liu, "Geometric Structure Based Error Concealment with Novel Applications in Block-Based Low-Bit-Rate Coding," *IEEE Trans. Circuits Syst. Video Technol.*, 9, 648–665, June 1999.

[56] F. Zhai, Y. Eisenberg, and A. G. Katsaggelos, "Joint Source-Channel Coding for Video Communications," in *Handbook of Image and Video Processing*, 2nd edn. (A. C. Bovik, ed.), Elsevier Academic Press, Burlington, MA, 2005. Chapter 9.4

[57] W. Zhu *et al.*, eds., "Special Issue on Advances in Video Coding and Delivery," *Proc. IEEE*, **93**, Jan. 2005.

Index